Autonomous Underwater Vehicles

Autonomous Underwater Vehicles

Modeling, Control Design and Simulation

Martin Azese
Mohammed Bouchahdane

Kruger Brentt
Publishers

2025

Kruger Brentt Publishers UK. LTD.
Company Number 9728962

Regd. Office: 68 St Margarets Road, Edgware, Middlesex HA8 9UU

ISBN: 978-1-78715-409-4

PREFACE

"Autonomous Underwater Vehicles: Modeling, Control Design, and Simulation" delves into the fascinating realm of underwater robotics, offering readers a comprehensive exploration of the principles, techniques, and applications that define the field of autonomous underwater vehicles (AUVs). This book serves as an essential resource for students, researchers, engineers, and practitioners seeking to understand the challenges and opportunities in modeling, controlling, and simulating AUVs.

Autonomous underwater vehicles play a crucial role in ocean exploration, environmental monitoring, underwater mapping, underwater archaeology, offshore energy exploration, and marine research. These robotic platforms enable scientists and engineers to access remote and hazardous underwater environments, collect data, and perform tasks autonomously with minimal human intervention.

The primary objective of this book is to provide readers with a comprehensive overview of AUVs, covering key topics such as vehicle dynamics, hydrodynamics, propulsion systems, sensor technologies, navigation and localization methods, control algorithms, and simulation techniques. Each chapter offers insights, mathematical formulations, simulation examples, and practical guidance to help readers understand the fundamental principles and challenges of AUV modeling, control design, and simulation.

Contributions from leading experts in the field explore various aspects of AUV technology, including vehicle design considerations, sensor fusion techniques, path planning strategies, fault detection and recovery methods, and emerging trends such as swarm robotics and underwater autonomy. Special emphasis is placed on real-world applications and case studies showcasing the capabilities and limitations of AUVs in challenging underwater environments.

As AUV technology continues to advance and find new applications, it is essential for researchers and practitioners to develop robust and efficient modeling, control, and simulation methods. By leveraging interdisciplinary approaches, innovative technologies, and collaborative efforts, we can address the complex challenges of underwater exploration and contribute to our understanding of the world's oceans and marine ecosystems.

We extend our sincere gratitude to the contributors whose expertise and dedication have enriched this book. It is our hope that "Autonomous Underwater Vehicles: Modeling, Control Design, and Simulation" will serve as a valuable resource and inspiration for researchers, engineers, and students engaged in advancing the frontier of underwater robotics and marine technology.

Martin Azese

Mohammed Bouchahdane

CONTENTS

1 Chapter

CREATION OF A VECTORED WATER-JET-PROPELLED SPHERICAL UNDERWATER VEHICLE

1. INTRODUCTION: AN OVERVIEW

The utilization of underwater vehicles has significantly surged in recent years, including applications in mine clearance operations, feature tracking, cable or pipeline monitoring, and deep ocean exploration. The mechanical and electrical setup and design of an undersea vehicle vary according to its applications. For example, manipulators are essential in mine-clearing operations or other jobs that need interaction with the environment. For underwater environment sensing or observation, it is advantageous to design the vehicle to be compact and agile, enabling it to navigate confined spaces with ease. A streamlined body is necessary for the vehicle to achieve high speeds in water. Various structures of underwater vehicles of differing sizes are being developed. Many of these underwater vehicles possess torpedo-like, streamlined bodies, as noted by Sangekar et al. (2009). Additionally, there exist compact AUVs, as noted by Allen et al. (2002) and Madhan et al. (2006). Additionally, several AUVs employ alternative body shapes, as noted by Antonelli and Chiaverini (2002). Simultaneously, the propulsion system is a crucial factor in the performance of underwater vehicles, since it underpins the control layers of the entire system. Propulsion devices exhibit several configurations, such as paddle wheels, poles, magnetohydrodynamic drives, sails, and oars. Paddle wheel thrusters represent the most prevalent and conventional propulsion systems for underwater vehicles. Typically, an underwater vehicle is equipped with a minimum of two thrusters: one for horizontal propulsion and another for vertical propulsion. The drawbacks of paddle wheel thrusters are evident; for instance, they readily disrupt the surrounding water of underwater vehicles. Concurrently, the increased utilization of paddle wheel thrusters results in heightened weight, noise, and energy consumption. The steering mechanisms of conventional underwater vehicles involve altering the angle of rudders or use differential propulsion forces from two or more thrusters. Indeed, vectored propellers are utilized in undersea vehicles. Cavallo et al. (2004) and Le Page & Holappa (2002a) describe underwater vehicles equipped with vectored thrusters. Duchemin et al. (2007) propose multi-channel Hall-effect thrusters that incorporate vector propulsion and vector composition. Le Page and Holappa (2002b) propose an autonomous underwater vehicle including a vectored thruster. The design of vectoring thrusters employed in airplanes serves as an illustration of a vectored propulsion system (Kowal, 2002; Beal, 2004; Lazic & Ristanovic, 2007). The objective of this research is to create an underwater vehicle capable of freely adjusting its orientation by altering the direction of its propulsive forces. Simultaneously, we want to enhance the vehicle's adaptability while navigating in aquatic environments. Motivated by jet aircraft, we utilize vectored water jet propellers as the propulsion mechanism. A symmetrical structure is preferable for our underwater vehicle, as per the design objective (Guo et al., 2009). This spherical underwater vehicle has numerous applications. Due to its adaptability, our vehicle is suitable for observing underwater organisms. For instance, we can equip the vehicle with underwater cameras. It can monitor and snap images of fishes. Another instance is that, because to its diminutive dimensions, it can be employed to ascertain the internal conditions of submerged oil pipelines.

2. MECHANICAL AND ELECTRICAL ENGINEERING DESIGN

2.1 Mechanical system design

Prior to the practical fabrication, we want to provide a conceptual design of the entire framework for this spherical underwater vehicle. At this stage, we must evaluate the dimensions, weight distribution, materials, and component installation, among other factors. We must also evaluate the configuration of the propulsion system, namely the optimal number of water-jet propellers to enhance power efficiency without compromising propulsion capability. Consequently, based on the aforementioned details, we provide the conceptual design of our spherical underwater vehicle as illustrated in Fig. 1. It assumes a spherical configuration, with all components housed within the structure. The radius is 20 cm, which is less than that reported in Antonelli et al. (2002). The total weight is around 6.5 kg. The operational depth is calibrated to 0-10 meters, with a maximum velocity of approximately 1.5 meters per second. The vehicle will be equipped with three water-jet propellers serving as the propulsion system, sufficient for surge, yaw, and heave maneuvers. A single watertight enclosure is utilized for all electronic components, including sensors, batteries, and control boards. All of these are affixed to a triangular support that is secured to the spherical hull. The entire structure exhibits symmetry along the z-axis. Consequently, it may rotate about the z-axis, thereby enabling the vehicle to alter its orientation with ease.

2.1.1 The spherical hullof this underwater vehicle

As illustrated in Fig. 2, the spherical hull of this underwater vehicle is constructed from acrylic, which is lightweight and easily machinable. The thickness is around 3mm, and the diameter is 40cm. This spherical hull consists of two transparent hemispherical shells. Three apertures exist that accommodate the rotation of water-jet propellers for various movements.

2.1.2 The waterproof box

Waterproofing is crucial for submersible vehicles. Figure 3 illustrates the design of the waterproof enclosure. The dimensions of this box are approximately 22 cm (height) × 14 cm (inside diameter). An O-ring is utilized as a seal, capable of ensuring watertight integrity in this context. Within the waterproof enclosure, there will be two control boards and one or two lithium batteries, contingent upon the tasks. Additionally, a digital rate gyro sensor for orientation feedback will be positioned at the upper section within the box. The waterproof box's body is transparent, allowing for easy observation of the internal operational state.

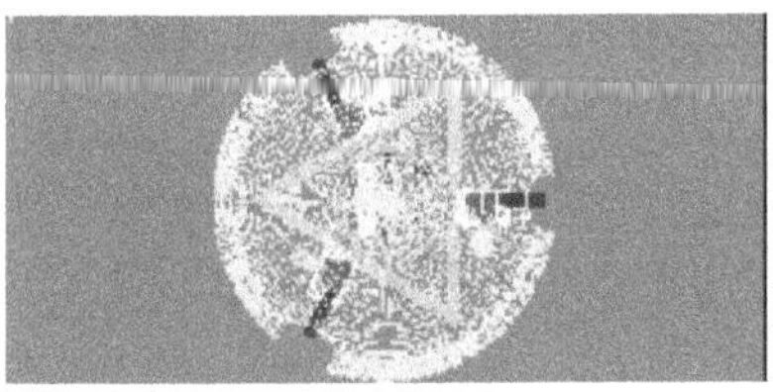

(a) Top View

(b) Front View

(c) Side View

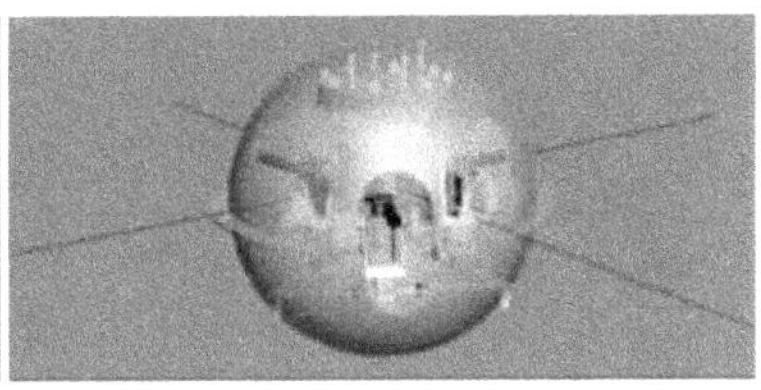

(d) Computer Rendering

Fig. 1. Mechanical System Schematics of the Spherical Underwater Vehicle

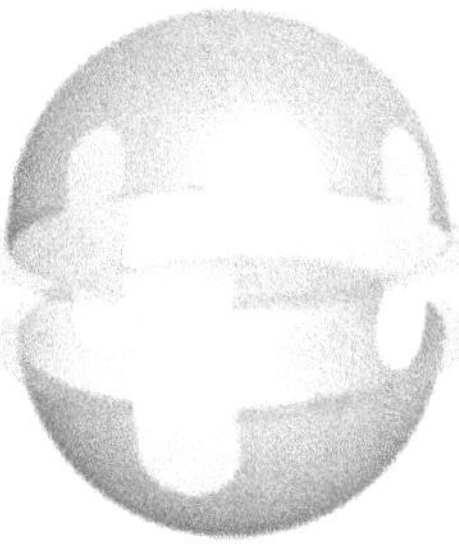

(a) Design

(b) Prototype

Fig. 2. Spherical Hull

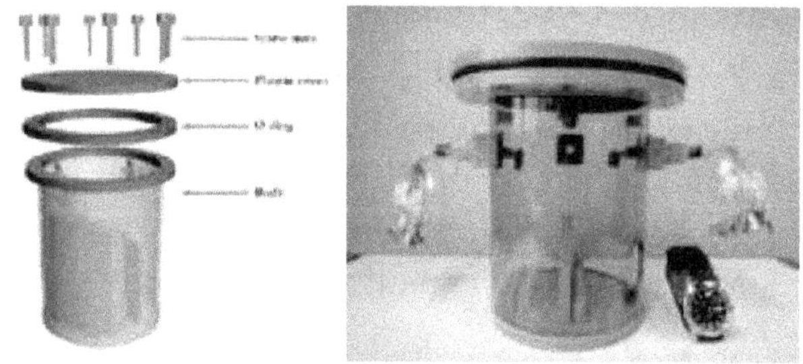

(a) Design (b) Prototype

Fig. 3. Design of Waterproof Box

2.1.3 The water-jet propulsion system's mechanism

The structure of a single water-jet propeller is shown in Fig. 4. It is made up of two servo motors (side and above) and one water-jet thruster. To ensure waterproofness, the water-jet thruster is enclosed within a plastic box. Additionally, to make servo motors waterproof, we utilize waterproof glue. These two servo motors have the ability to rotate the thruster, which allows for the direction of the water jet to be adjusted in the X-Y and X-Z planes, respectively.

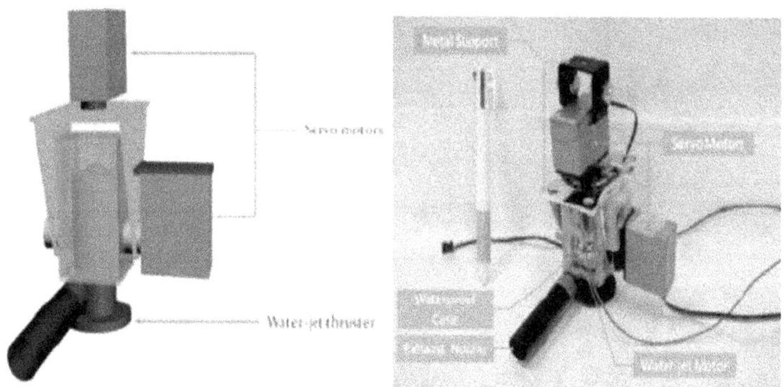

Fig. 4. Structure of a Water-jet Propeller

Three of the water-jet propellers are mounted on the metal support frame, as shown in Fig.5. Three of them are circumferentially $2\pi/3$ apart from each other.

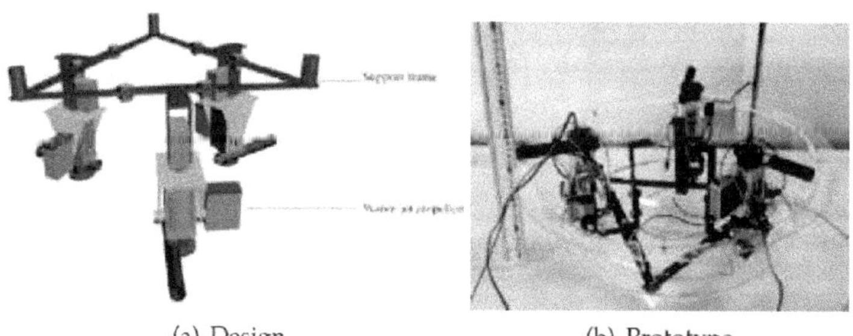

(a) Design (b) Prototype

Fig. 5. Water-jet Propellers mounted on Support Frame

2.2 Design of electrical systems

We implement a simple hardware configuration for the experimental prototype vehicle. For a solitary spherical underwater vehicle, there exist three principal electrical categories: the sensor group, the control group, and the actuator group. Figure 6 presents the electrical schematics. Currently, we utilize a single pressure sensor for depth regulation and one gyroscopic sensor for surge management. A control board based on ARM7 serves as the central unit for control, data collecting, algorithm implementation, and strategic decision-making. An AVR-based board serves as the coprocessor unit for motor control. It receives commands from the ARM and converts them into driving signals for the water-jet propellers.

Fig.7 gives the main hardware for this vehicle. Fig.7(a) is the ARM7 based board with S3C44B0X on it, which can fulfill our requirement at present. Fig.7(b) is the AVR based board with ATmega2560 on it. RS232 bus is used for the communication between ARM7 and AVR. In Fig.7(c) is the set of pressure sensor with the sensor body(right) and its coder (left). It use RS422 bus for data transmission. Digital gyro sensor CRS10 is shown in Fig.7(d), we use the build in AD converter of S3C44B0X for data acquisition.

Fig. 6. Electrical Schematics for Prototype System

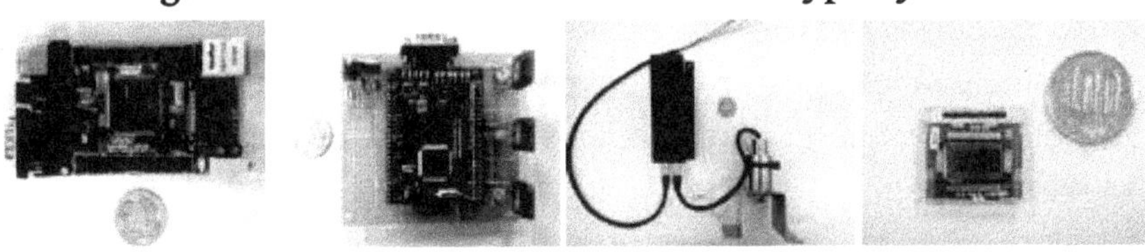

(a) S3C44B0X Board (b) ATmega2560 Board (c) Pressure Sensor (d) Gyro Sensor

Fig. 7. Electrical Components for the Experimental Prototype Underwater Vehicle

2.3 Power supply for the spherical underwater vehicle

We adopt two power supply for the spherical underwater vehicle. The highest power consumption components in our vehicle are propellers. For each of them, the thruster has a working voltage of $7.2V$ and $3.5A$ current drain, servo motors can work under 5V with relatively small current. Therefore, we use two 2-cells LiPo batteries as the power supply for the propellers. The capacity of each battery is 5000mAh with parameter of $50c - 7.4V$. Besides, we use 4 AA rechargeable batteries for the control boards. We carried out the power consumption test for one LiPo battery, and Fig. 8 gives the battery discharge graph of the power system.

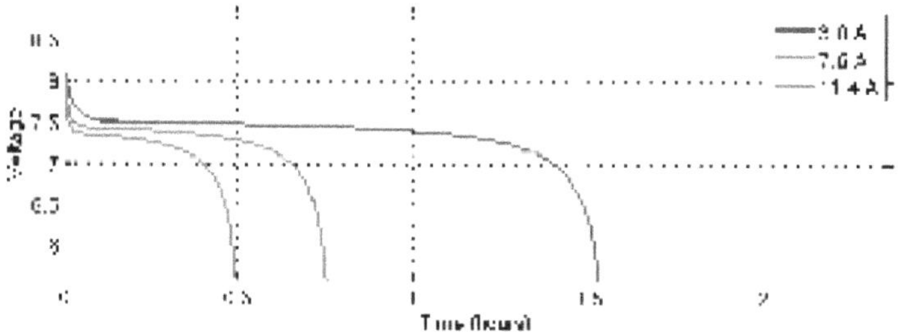

Fig. 8. Power Consumption of the Whole System

3. PRINCIPLES AND MODELING OF THE PROPULSION SYSTEM

This section will address the operational concepts, modeling methodology, and identification experiment for the water-jet propeller. Numerous studies have delineated the computational formula for the torque and thrust produced by a thruster. Most of them are base on the lift theory, and mainly focus on blades type propellers (Newman, 1977), (Fossen, 1995) and (Blanke et al., 2000). Our propellers are different with blades type propellers, therefore, we try to find another method for the modeling of water-jet propellers. In (Kim & Chung, 2006), the author presented a dynamic modeling method in which the flow velocity and incoming angle are taken into account. This modeling method will be employed for our water-jet propellers.

3.1 Working principles

Prior to the modeling of the propulsion system, we want to elucidate certain fundamental ideas about water-jet propellers. Figure 9(a) illustrates the above perspective of the arrangement of three propellers. They can collaborate to achieve various movements, including surge and yaw.

If we let θ be the interval angle of each water-jet propeller, as shown in Fig.9(b), then, for the purpose of kinematics transform, three propeller-fixed coordinates are introduced for propellers, which are fixed in the rotation center of the propellers. So we can see, these three propeller-fixed coordinates are actually transform results of vehicle-fixed coordinate reference frame. Meanwhile, it should be noted that, this transform only happens in *X*-*Y* plane. Let the matrix form of the coordinates transform be given as:

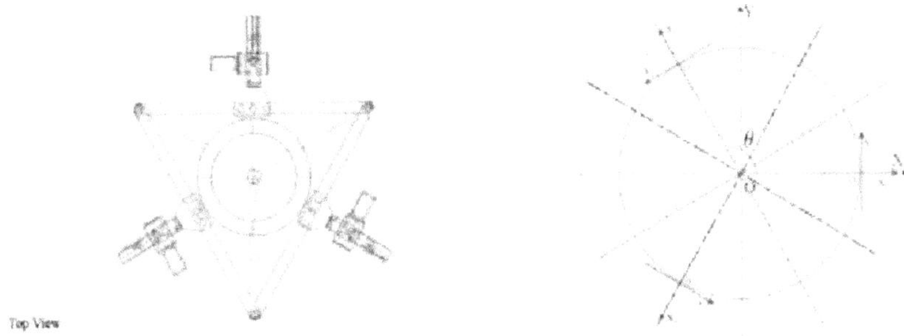

(a) Propeller Distribution (b) propeller-fixed Coordinates

Fig. 9. Distribution and Coordination of Multiple Propellers

$$\begin{pmatrix} X_1 \\ Y_1 \\ Z_1 \end{pmatrix} = \begin{pmatrix} X \\ Y \\ Z \end{pmatrix} + \begin{pmatrix} -R \\ 0 \\ 0 \end{pmatrix} \tag{1}$$

$$\begin{pmatrix} X_2 \\ Y_2 \\ Z_2 \end{pmatrix} = \begin{pmatrix} c\theta & s\theta & 0 \\ -s\theta & c\theta & 0 \\ 0 & 0 & 1 \end{pmatrix} \begin{pmatrix} X \\ Y \\ Z \end{pmatrix} + \begin{pmatrix} \frac{1}{2}Rc\theta - Rs\theta c\frac{\pi}{6} \\ -\frac{1}{2}Rc\theta - Rc\theta c\frac{\pi}{6} \\ 0 \end{pmatrix} \tag{2}$$

$$\begin{pmatrix} X_3 \\ Y_3 \\ Z_3 \end{pmatrix} = \begin{pmatrix} c2\theta & s2\theta & 0 \\ -s2\theta & c2\theta & 0 \\ 0 & 0 & 1 \end{pmatrix} \begin{pmatrix} X \\ Y \\ Z \end{pmatrix} + \begin{pmatrix} \frac{1}{2}Rc2\theta + Rs2\theta c\frac{\pi}{6} \\ -\frac{1}{2}Rc2\theta + Rc2\theta c\frac{\pi}{6} \\ 0 \end{pmatrix} \tag{3}$$

where *R* is the radius of the vehicle, $s(\cdot) \equiv \sin(\cdot)$ and $c(\cdot) \equiv \cos(\cdot)$.

So, a general transform matrix can be obtained:

$$^{p}\boldsymbol{P}_b = \boldsymbol{\Phi}_p^b \cdot {}^{p}\boldsymbol{P}_p + \boldsymbol{C} \tag{4}$$

where $^{p}P_b$ is the position vector of propeller-fixed coordinate expressed in vehicle-fixedcoordinate $\boldsymbol{\Phi}_p^b = (\boldsymbol{\Phi}_{p1}^b, \boldsymbol{\Phi}_{p2}^b, \boldsymbol{\Phi}_{p3}^b)^T$ is the transform matrix from propeller-fixed coordinateto vehicle-fixed coordinate, $^{p}P_p$ is the position vector in propeller-fixed coordinate and the **C**is a constant vector.

Now, let us take a look at three motions, surge, heave and yaw. The definition of these three motions can be found in (Fossen, 1995). Before that, we define two angles which will be used for orientation of propellers. Fig. 10 gives the definition of α and β. Fig. 11 gives a demonstration of surge, heave and yaw.

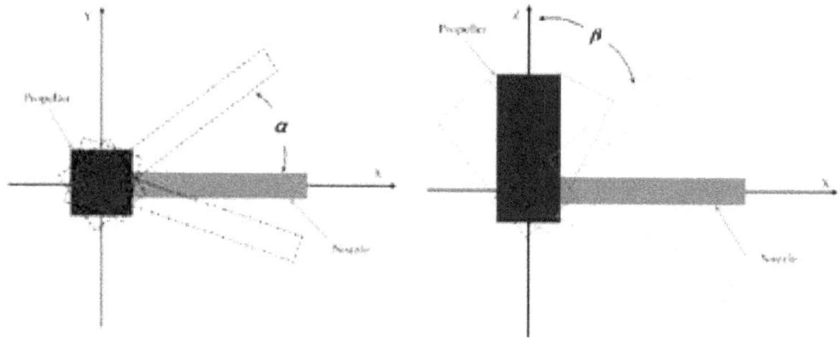

(a) Rotation in X-Y Plane (b) Rotation in X-Z Plane

Fig. 10. Orientation of Propellers

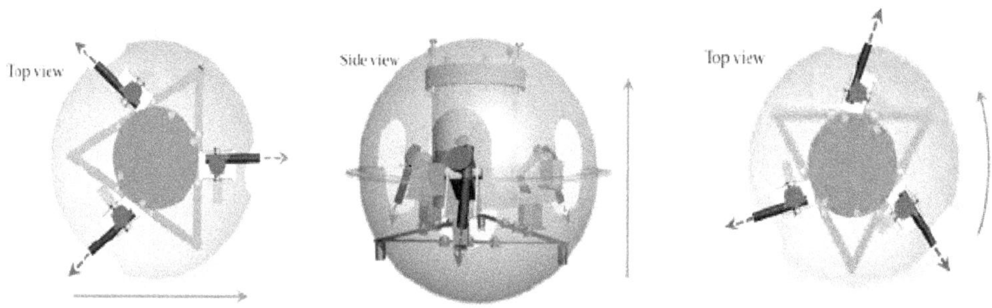

Fig. 11. Propulsion Forces for Surge, Heave and Yaw

The first case is surge. In this case, two of the water-jet propellers will work together, and the other one could be used for brake. So, from Fig.11(a), two water-jet propellers in the left will be used for propulsion, and if we want to stop the vehicle from moving, the third propeller can act as a braking propeller. From Equation 4, the resultant force for surge can be expressed in vehicle-fixed coordinate as:

$$\begin{cases} {}^{p}F_{xb} = \mathbf{\Phi}_{p1}^{bT} \sum\limits_{i=1}^{3} ({}^{p}\mathbf{F}_{ip} + \mathbf{e}_1 C_i) \neq 0 \\ {}^{p}F_{yb} = 0 \\ {}^{p}F_{zb} = 0 \end{cases} \quad (5)$$

where, $\boldsymbol{e}_1 = (1, 0, 0)^T$.

Then, for the heave case, all the three water-jet propellers will work and the side servo motor will rotate to an angle that $\beta > \pi/2$. Therefore, in this case, the resultant force for heave canbe expressed in vehicle-fixed coordinate as:

$$\begin{cases} {}^{P}F_{xb} = 0 \\ {}^{P}F_{yb} = 0 \\ {}^{P}F_{zb} = \mathbf{\Phi}_{p3}^{bT} \sum_{i=1}^{3} ({}^{P}\mathbf{F}_{ip} + \mathbf{e}_3 C_i) \neq 0 \end{cases} \tag{6}$$

where, $e_3 = (0, 0, 1)^T$.

The third case is yaw which is rotating on z-axis. By denoting in propeller-fixed coordinates, α should have the same orientation, clockwise or counterclockwise, that means, $\alpha_i > 0$ or $\alpha_i < 0$. So in yaw, rotation moment will take effect. We can write the equation for yaw in vehicle-fixed coordinate as:

$$\begin{cases} {}^{P}F_{xb} = \mathbf{\Phi}_{p1}^{bT} \sum_{i=1}^{3} ({}^{P}\mathbf{F}_{ip} + \mathbf{e}_1 C_i) \neq 0 \\ {}^{P}F_{yb} = \mathbf{\Phi}_{p2}^{bT} \sum_{i=1}^{3} ({}^{P}\mathbf{F}_{ip} + \mathbf{e}_2 C_i) \neq 0 \\ {}^{P}F_{zb} = 0 \\ {}^{P}M_{xb} + {}^{P}M_{yb} + {}^{P}M_{zb} \neq 0 \end{cases} \tag{7}$$

where, $e_2 = (0, 1, 0)^T$, $e_3 = (0, 0, 1)^T$.

3.2 Modeling of single water-jet propeller

In the author's previous research (Guo et al., 2009), the modeling for orientation of water-jet propeller is presented. Therefore, in this part, we will only discuss about the hydrodynamics modeling of the water-jet thruster. The method we refer to is presented in (Kim & Chung, 2006). For the purpose of dynamic modeling of water-jet propeller, we give the flow model of the water-jet thruster, which is shown in Fig. 12. The shaft is perpendicular to the nozzle, and there are two blades.

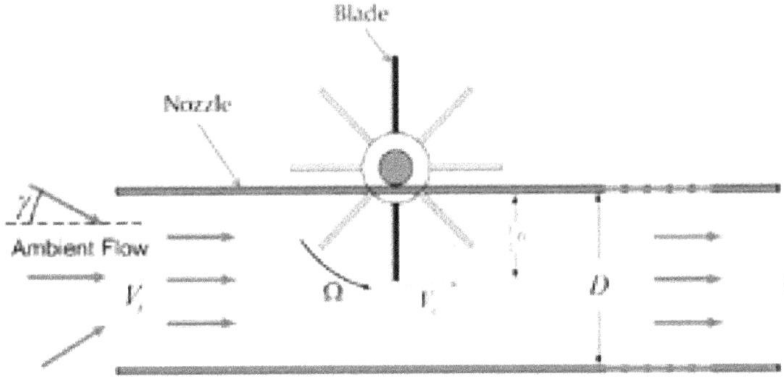

Fig. 12. Flow Model of the Water-jet Thruster (top view)

where,

Ω is angular velocity of the thruster

V_i is velocity of incoming flow

V_c is central flow velocity in the nozzle

D is diameter of the nozzle

V_o is velocity of outlet flow

γ is incoming angle of ambient flow

Because the diameter of the nozzle is small, the velocity difference in the nozzle can be ignored, so we consider the axis flow velocity V_a as a linear combine of incoming flow velocity and the central flow velocity,

$$\begin{aligned} V_a &= k_1 V_i + k_2 V_c \\ V_c &= \frac{1}{2} D\Omega \\ V_i &= V_f cos\gamma \end{aligned} \tag{8}$$

By assuming that the flow is incompressible, therefore, from equation of continuity, we know that the volume of incoming flow must equal to the outlet flow, then we get:

$$\rho_a V_a A_a = \rho_o V_o A_o \tag{9}$$

where, $\rho_a = \rho_o$ is density of flow, $A_a = A_o$ is cross-section of the nozzle.

Therefore, we can also get:

$$V_a = V_o \tag{10}$$

Meanwhile, we know that, the propulsive force of the water-jet thruster is:

$$F_t = \rho A V_a^2 \tag{11}$$

By substituting Equation 8 in Equation 11, we can get:

$$F_t = \frac{\pi}{4} \rho D^2 (k_1^2 V_i^2 + 2k_1 k_2 V_i D\Omega + k_2^2 D^2 \Omega^2) \tag{12}$$

By rewriting Equation 12, we get:

$$\frac{F_t}{\rho D^4 \Omega^2} = \frac{\pi}{4}(k_1^2(\frac{V_i}{D\Omega})^2 + 2k_1k_2\frac{V_i}{D\Omega} + k_2^2) \quad (13)$$

Then, we can let the non-dimensional parameter be:

$$K_T(J_0) = \frac{\pi}{4}(k_1^2(\frac{V_i}{D\Omega})^2 + 2k_1k_2\frac{V_i}{D\Omega} + k_2^2) \quad (14)$$

where

$$J_0 = \frac{V_i}{D\Omega} = \frac{V_f cos\gamma}{D\Omega} \quad (15)$$

J_0 is the advance ratio.

Now, the modeling becomes measuring of three parameters, flow velocity, incoming angle and angular velocity of thruster. For this purpose, we designed an experiment to measure these parameters and find out their relationship.

Table 1. Experiment Condition

Flow velocities	Depth	Incoming angles	Control voltages
$0.1m/s$	$80cm$	$0-?$	$3V-7V$ (DC)
$0.2m/s$			

3.3 Dynamic modeling experiments

This section aims to experimentally identify the dynamics model of the water-jet propeller. We are focused on the relationship between incoming flow angles, flow velocities, and propulsive forces. The experimental conditions are detailed in Table 1.

3.3.1 Experiment design

Fig.13 gives the experiment principle. We use one NEC 2301 stain gage as the force sensor and use NEC AC AMPLIFIER AS 1302 to amplify the output signal from strain gage, which are shown in Fig.14.

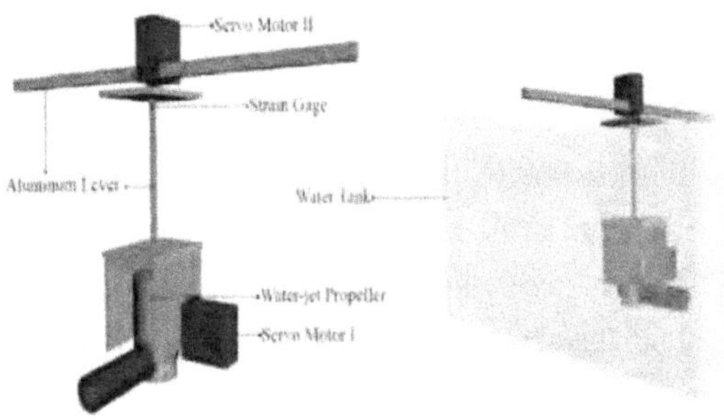

Fig. 13. Experiment Design for Identification

Fig. 14. Strain Gage and Amplifier

Firstly, we give a brief illustration for the strain gage measurement. Let F_d be deformation force, and ε be the deformation of aluminium lever used in the experiment. Therefore, from the theory of mechanics of materials, we can get the relation of F_d and ε as:

$$F_d = \frac{ZE}{X}\varepsilon \tag{16}$$

where, Z is second moment of area, E is the Young's modulus, X is the distance from acting point of force to stain gage.

3.3.2 Experiments and analysis

In the experiment, we must consider four variables: the equivalent cross-section of the propeller, flow velocity, entering angle, and control voltage. We are focused on the fluctuation of propulsive force at varied incoming angles and control voltages.

3.3.2.1 Variation of equivalent cross-section of propellers

The vectored water-jet propulsion technology allows for alterations in both the propulsive force and its direction. Consequently, when the propeller alters its orientation, the angle of incoming flow concurrently changes, resulting in a variation of the propeller's effective cross-section. Equation 11 indicates that the propulsive force will vary with alterations in cross-section A. Figure 15 illustrates this scenario. As the propeller rotates from position I to II, the corresponding cross-section will transition from cross-section I to cross-section II. We endeavor to formulate an equation that characterizes this fluctuation.

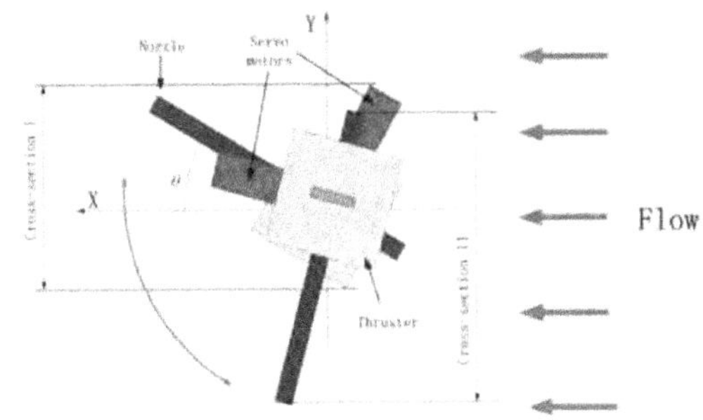

Fig. 15. Variation of Equivalent Cross-section of Propellers

Considering that the measured force from stain gage is actually a resultant force of propulsive force and fluid force. And we also know that the fluid force acted on the propeller depends on the equivalent cross-section.

So the first experiment is measurement of the equivalent cross-section variation. The propeller is submerged in the flow which has a speed of $0.2m/s$, propeller is powered off. And we only change its orientation in $X-Y$ plane. Because of experiment limits, we can not change the flow direction, so in the experiment, the incoming angle equals to the orientation angle of thepropeller. Fig.15 gives a demonstration of the equivalent cross-section. We give some special angles, 0, $\pi/6$, $\pi/3$, $2\pi/3$, $5\pi/6$, π, for this experiment. Fig.16 gives the experiment data of equivalent cross-section. You may notice that, we did not adopt the orientation angle of $\pi/2$. Because, when the propeller rotate to $\pi/2$, which means that the measure surface of the strain gage is parallel to the flow direction, the strain gage can not measure the flow force.

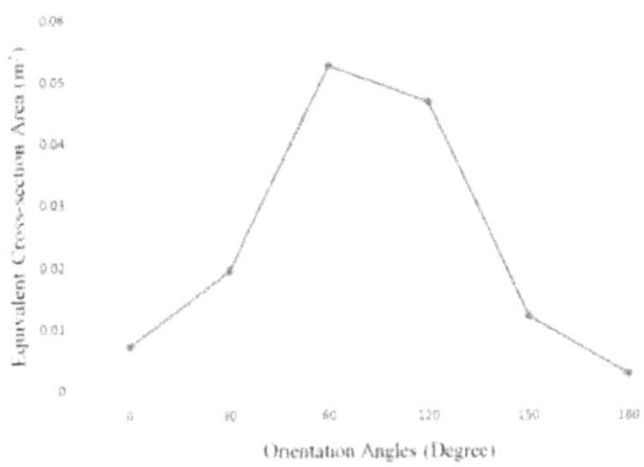

Fig. 16. Variation of Equivalent Cross-section

From Fig.16 we can see, the data curve is similar with a sinusoid, so we use a sine function to fit this experiment data:

$$A_e(\phi) = \lambda_1 sin(\lambda_2 \phi + \lambda_3) \tag{17}$$

where ϕ is incoming angle, $\lambda_1, \lambda_2, \lambda_3$ are coefficients.

Incoming angle and deformation force

In this case, the flow velocity is seen as constant. Two groups of experiment are carried out at flow velocity of $0.1m/s$ and $0.2m/s$. The control voltage to thruster is from $3V$ to $7V$ every $1V$. From the data shown in Fig.17, we can see that the deformation force does not simply increase with the increasing of incoming angle, the maximum deformation of the lever happens at about 60 degree of the incoming angles. They are not a linear relation. Then, how about the real propulsive force?

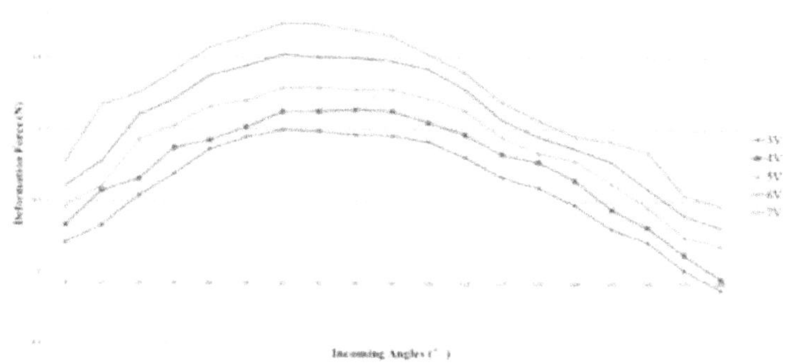

(a) $V_f = 0.1m/s$

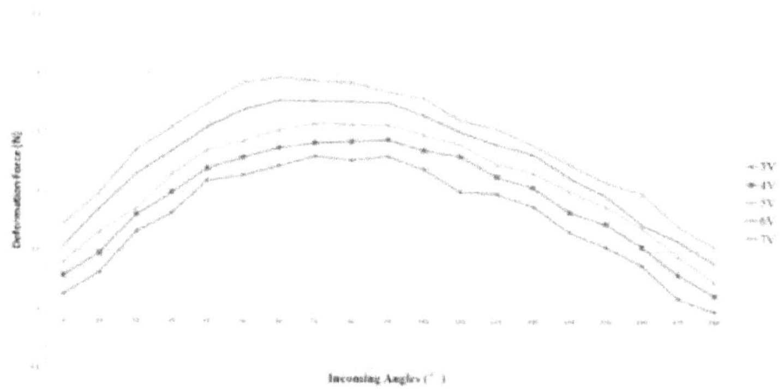

(b) $V_f = 0.2m/s$

Fig. 17. Deformation Forces with Different Incoming Angles

3.3.2.3 Incoming angle and propulsive force

As we mentioned, what we measured by stain gage is actually a resultant force of propulsive force and fluid force. Therefore, the real propulsive force F_t should be calculated using deformation force F_d and the equivalent cross-section A_e.

$$F_d = F_t - F_f cos\gamma \tag{18}$$

If we consider about the fluid force of the flow, we can refer to Equation 11 and 17, then write the fluid force as:

$$F_f = \rho V_c^2 A_e(\phi) \tag{19}$$

We substitute Equation 19 into 18 we get:

$$F_t = F_d + \rho V_c^2 A_e(\phi) cos\gamma \tag{20}$$

So now, we can calculate the real propulsive force by using equivalent cross-section, deformation force and incoming angles. The results is shown in Fig.18. Fig. 18(a) and Fig. 18(b) are results at the flow velocity of V_f = 0.1m/s and V_f = 0.2m/s, respectively.

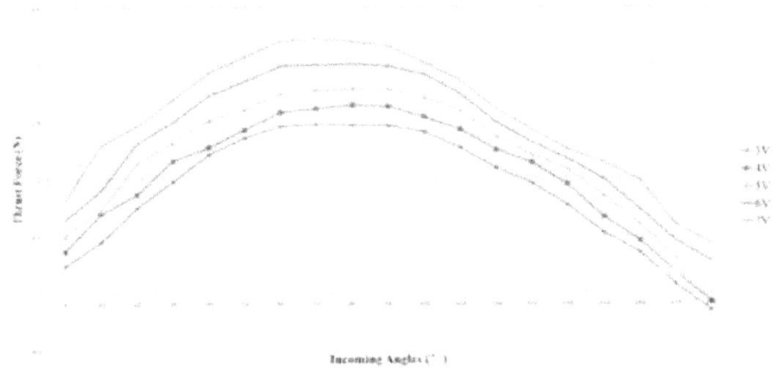

(a) $V_f = 0.1m/s$

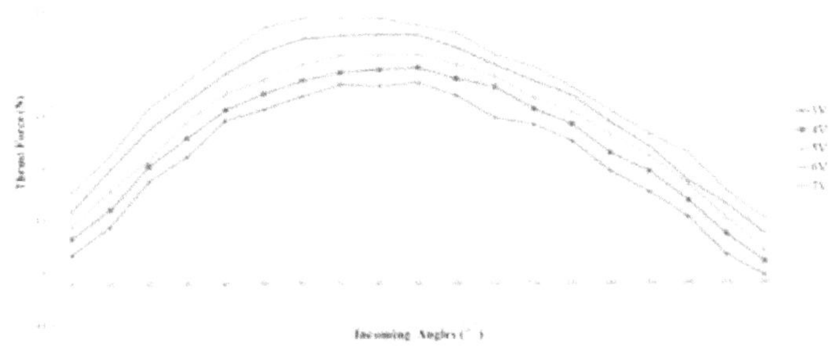

(b) $V_f = 0.2m/s$

Fig. 18. Propulsive forces with Different Incoming Angles

3.3.2.4 Control voltage and propulsive force

The results of 3.3.2.3 yield the relationship between control voltage and propulsive force. Initially, we present the experimental findings in Fig. 19. The diagram illustrates that the relationship between control voltage and propulsive force can be represented by a linear equation.

4. UNDERWATER EXPERIMENTS FOR BASIC MOTIONS

Because of the symmetrical shape of the hull, it is obvious that motion characteristics of surge, sway and heave should be similar. However, from another point of view, surge and sway are motions in $X-Y$ plane while heave is a motion that its motion surface perpendicular to $X-Y$ plane. Therefore, we carry out experiments for horizontal motion surface and vertical motion surface respectively. Besides, for the experimental prototype vehicle, we only consider one rotational DOF in Z axis, so the third experiment is yaw motion.

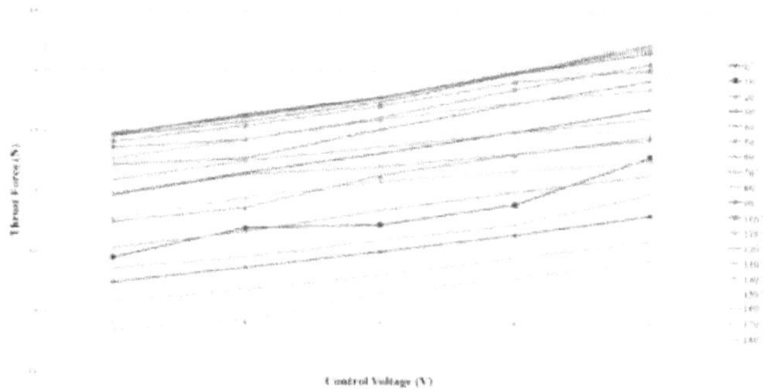

(a) $V_f = 0.1m/s$

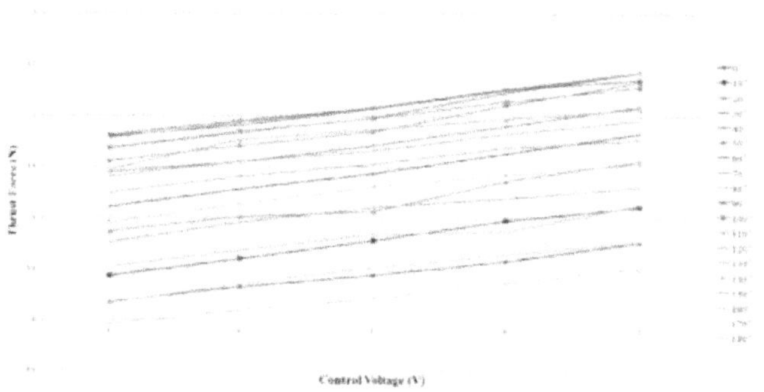

(b) $V_f = 0.2m/s$

Fig. 19. Propulsive forces with Different Control Voltages

4.1 Experiment of horizontal motion

This experiment integrates surge and sway to validate the vehicle's motion characteristics in the horizontal plane. We conducted three experiments:

Case 1:

step 1. Surge (Move forward in X axis);

step 2. Right steering (Turn right about 90°);

step 3. Sway (Move forward along Y axis.)

Case 2:

step 1. Surge (Move forward in X axis);

step 2. Left steering (Turn left about 90°);

step 3. Sway (Move forward along Y axis.)

Case 3:

step 1. Surge (Propeller I and II work together, propeller III powered off);

step 2. Brake (Propeller I and II powered off, Propeller III works to produce brake force).

In scenario one, the duration of step one is around 10 seconds, while step two requires roughly 12 seconds. The timing of instance 2 is comparable to that of example 1 due to the analogous hydrodynamic properties of right and left turns. In scenario three, achieving a stable velocity requires 15 seconds, but the braking effect occurs within approximately 3 seconds, which is advantageous for low-speed underwater vehicles. Figure 20 illustrates that the experimental results align closely with the simulated results during the surge phase; however, significant discrepancies arise when the vehicle is in rotation. This is due to the simulation experiment's exclusive consideration of linear and quadratic damping forces, although, in reality, additional hydrodynamic forces also influence the vehicle.

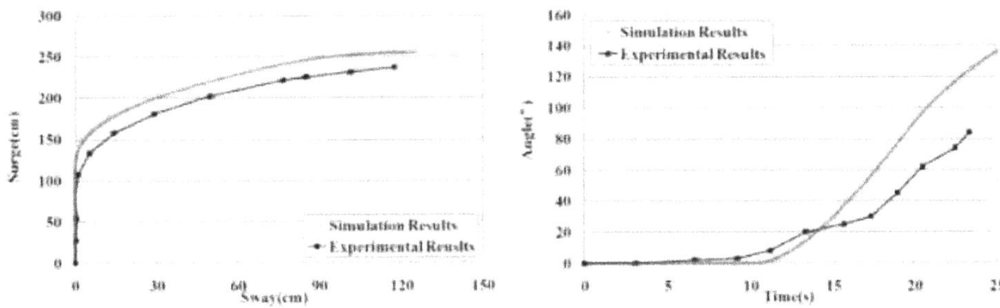

(a) The trajectory of case 1 (b) Steering angle of case 1

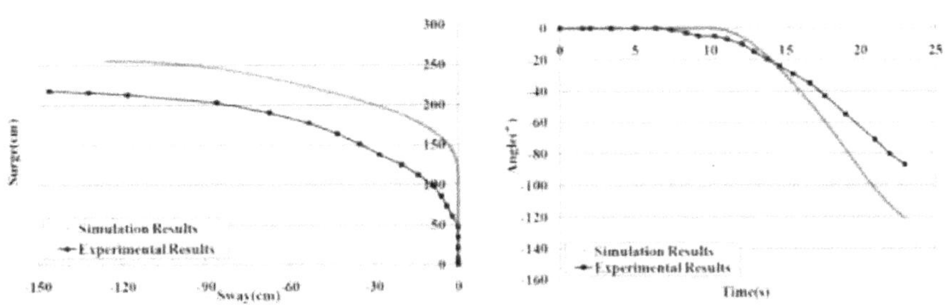

(c) The trajectory of case 2 (d) Steering angle of case 2

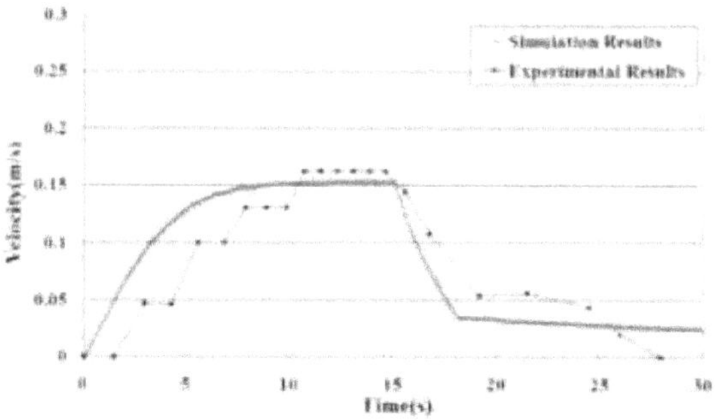

(e) Surge and Brake of case 3

Fig. 20. Experimental Results of Horizontal Motion

4.2 Experiment of vertical motion

Although we have designed the vehicle's operational depth to 10 meters, the testing pool's depth of only 1.2 meters restricts our trials to shallow water. We established the duration of vertical motion within a certain range. We conducted two experiments:

Case 1:

step 1. Set the top point of the spherical hull as the start point;

step 2. Move downward in Z axis for about 7s;

step 3. Float up to the surface.

Case 2:

step 1. Set the top point of the spherical hull as the start point;

step 2. Move downward in Z axis for about 7s;

step 3. Stop the vehicle.

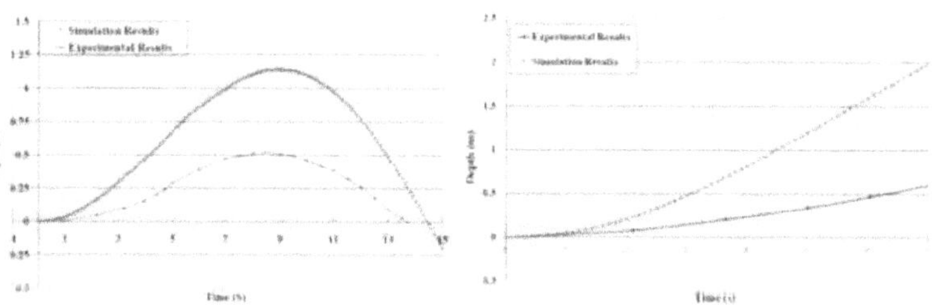

(a) Submerge and Float up (b) Submerge Only

Fig. 21. Experimental Results of Vertical Motion

From Fig. 21(a) and Fig. 21(b), we can see, the experimental results does not fit well with simulation results very well, errors exceed 100%. When we analyze the reasons, we find that, the simulation experiment does not consider the variation of water pressure. The control voltage to the thrusters is 7V as a constant. That means, the propulsive force will not change. But with the increasing of depth, water pressure increases. As a result, the effective propulsive force are weaken by water pressure.

4.3 Experiment of yaw

We permitted the vehicle to rotate approximately 90 degrees and then cease movement. According to Fig. 22(a) and Fig. 22(b), the peak discrepancy between simulation and experimental results occurs at around 2.8 seconds, coinciding with the near-maximum angular velocity. The cause of this outcome is the simplification of our vehicle model, particularly for the hydrodynamic damping forces. In our scenario, only linear and quadratic damping forces are considered. However, in the actual experiment, numerous other velocity-related hydrodynamic damping forces exist; hence, as the angular velocity increases, the damping effects of the neglected forces become pronounced.

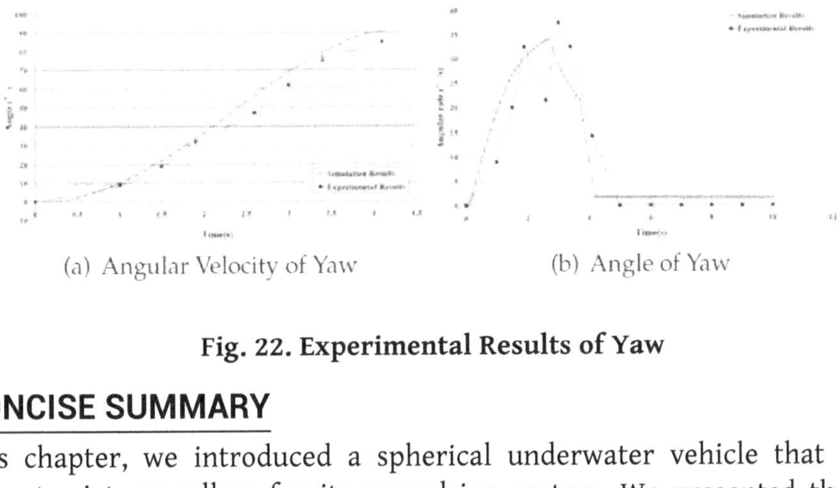

(a) Angular Velocity of Yaw (b) Angle of Yaw

Fig. 22. Experimental Results of Yaw

5. CONCISE SUMMARY

In this chapter, we introduced a spherical underwater vehicle that employs three water-jet propellers for its propulsion system. We presented the design specifications of the mechanical and electrical systems. We included the ideas of the water-jet propulsion system into the vehicle's design, including the force distribution among three water-jet propellers and the operational principles of various motions. Subsequently, we deliberated on the modeling of an individual propeller using identification studies. The modeling incorporates the flow velocity and equivalent cross-section of the propeller for the dynamics model. An experimental prototype of this spherical underwater vehicle has been constructed for evaluation purposes. Underwater studies are conducted to assess the motion properties of this spherical underwater vehicle. Experimental results and analyses are provided for each experiment. The underwater testing of the prototype vehicle demonstrate the viability of the concept, and the water-jet propulsion system functions effectively for various movements. However, there are other issues that require resolution. Initially, the propulsive force of the water-jet propellers must be augmented; subsequently, the fluctuation of water pressure on the propulsive force should be incorporated into the dynamics model of the propellers; thirdly, the distribution of gravity should be recalibrated to enhance stability; ultimately, empirical data necessitates the refinement of the dynamics model of the vehicle for precise control.

REFERENCES

Allen, B., Stokey, R., Austin, T., Forrester, N., Goldsborough, R., Purcell, M. & von Alt,C. (2002). REMUS: a small, low cost AUV; system description, field trials and performance results, OCEANS'97. *MTS/IEEEConference Proceedings,* Vol. 2, IEEE, pp. 994–1000.

Antonelli, G. & Chiaverini, S. (2002). Adaptive tracking control of underwater vehicle-manipulator systems, *Proceedings of the 1998 IEEE International Conference on Control Applications, 1998,* Vol. 2, IEEE, pp. 1089–1093.

Antonelli, G., Chiaverini, S., Sarkar, N. & West, M. (2002). Adaptive control of an autonomous underwater vehicle: experimental results on ODIN, *IEEE Transactions on Control Systems Technology* 9(5): 756–765.

Beal, B. (2004). *Clustering of Hall effect thrusters for high-power electric propulsion applications,* PhD thesis, The University of Michigan.

Blanke, M., Lindegaard, K. & Fossen, T. (2000). Dynamic model for thrust generation of marine propellers, *Proceedings of the IFAC Conference on Maneuvering and Control of Marine Craft (MCMC 2000),*Citeseer.

Cavallo, E., Michelini, R. &Filaretov, V. (2004). Conceptual design of an AUV Equipped with a three degrees of freedom vectored thruster, *Journal of Intelligent & Robotic Systems* 39(4): 365–391.

Duchemin, O., Lorand, A., Notarianni, M., Valentian, D. & Chesta, E. (2007). Multi-Channel Hall-Effect Thrusters: Mission Applications and Architecture Trade-Offs, *30th International Electric Propulsion Conference,* Florence, Italy.

Fossen, T. I. (1995). *Guidance and Control of Ocean Vehicles,* John Wiley & Sons Ltd., USA.

Guo, S., Lin, X. & Hata, S. (2009). A conceptual design of vectored water-jet propulsion system, *International Conferenceon Mechatronics and Automation, 2009. ICMA 2009,* IEEE, pp. 1190–1195.

Kim, J. & Chung, W. (2006). Accurate and practical thruster modeling for underwater vehicles,*Ocean Engineering* 33(5-6): 566–586.

Kowal, H. (2002). Advances in thrust vectoring and the application of flow-control technology,*Canadian aeronautics and space journal* 48(2): 145–151.

Lazic, D. &Ristanovic, M. (2007). Electrohydraulic thrust vector control of twin rocket engines with position feedback via angular transducers, *Control Engineering Practice* 15(5): 583–594.

Le Page, Y. & Holappa, K. (2002a). Hydrodynamics of an autonomous underwater vehicle equipped with a vectored thruster, *OCEANS 2000 MTS/IEEE Conference and Exhibition,* Vol. 3, IEEE, pp. 2135–2140.

Le Page, Y. & Holappa, K. (2002b). Simulation and control of an autonomous underwater vehicle equipped with a vectored thruster, OCEANS 2000 *MTS/IEEE Conference and Exhibition,* Vol. 3, IEEE, pp. 2129–2134.

Madhan, R., Desa, E., Prabhudesai, S., Sebastião, L., Pascoal, A., Desa, E., Mascarenhas, A., Maurya, P., Navelkar, G., Afzulpurkar, S. et al. (2006). Mechanical design and development aspects of a small AUV–Maya, *7th IFAC Conference MCMC2006.*

Newman, J. (1977). *Marine hydrodynamics,* The MIT press.

Sangekar, M., Chitre, M. & Koay, T. (2009). Hardware architecture for a modular autonomous underwater vehicle STARFISH, *OCEANS 2008,* IEEE, pp. 1–8.

Chapter 2 CREATION OF THE P-SURO, AN INTELLIGENT AUTONOMOUS UNDERWATER VEHICLE WITH A HOVERING-TYPE

1. INTRODUCTION: AN OVERVIEW

P-SURO (PIRO-Smart Underwater Robot) is a hovering-type autonomous underwater vehicle (AUV) designed for the development of diverse underwater core technologies (Li et al., 2010). In contrast to the comparatively developed torpedo-type AUV technology (Prestero, 2001; Marthiniussen et al., 2004), there have been limited presentations of commercial hovering-type AUVs to date. This is partly due to the fact that certain underwater missions of hovering-type AUVs can be executed using ROV (Remotely Operated Vehicle) systems. The primary cause is the underdeveloped basic technology for hovering-type AUVs. To perform its underwater functions, a hovering-type AUV may require precise underwater localization, obstacle avoidance, and agile maneuverability, among other capabilities. Conversely, due to the constraints of current underwater communication bandwidth, great autonomy has become a fundamental requirement for hovering Autonomous Underwater Vehicles (AUVs) (Li et al., 2010).As a test-bed AUV, P-SURO has been constructed to develop various underwater coretechnologies, such as underwater vision, SLAM, and vehicle guidance & control. There are four thrusters mounted to steer the vehicle's underwater motion: two vertical thrusters for up/down in the vertical plane, and 3DOF horizontal motion is controlled by two horizontal ones, see Fig. 1. Three communication channels are designed between the vehicle and the surface control unit. Ethernet cables are utilized in the initial phases of development and for the uploading and downloading of programs and files. The RF channel facilitates the interchange of information and user commands, whilst the acoustic channel (ATM: Acoustic Telemetry Modem) operates underwater. A color camera is affixed to the front of the car. Three range sonars, oriented forward, backward, and below, are engineered to facilitate vehicle navigation, obstacle avoidance, and simultaneous localization and mapping (SLAM). An AHRS integrated with a one-axis gyro, one-axis accelerometer, and depth sensor constitutes the vehicle's navigation system. This chapter presents the current advancements in the vehicle's development, encompassing SLAM, obstacle detection/path planning, and various vehicle control algorithms. The subsequent sections of this chapter are structured as follows. Section II presents the vehicle's overall specifications and several of its features. The underwater vision for the P-SURO AUV is addressed in Section III, while the SLAM algorithm in the basin environment is detailed in Section IV. Section V addresses several control concerns pertaining to the P-SURO AUV. In Section VI, we provide a concise summary of the findings and address some future research topics.

Fig. 1. P-SURO AUV and its open frame.

2. P-SURO AUV

As aforementioned, P-SURO AUV is a test-bed for developing underwater technologies. And most of its experimental tests will be carried out in an engineering basin in the PIRO with dimension of 12(L)?8(W)?6(D)m. Under these considerations, the vehicle is designed to be compact size with easiness of various algorithm tests (Li et al., 2010). The general specification of the vehicle is as Table. 1.

Table 1. General specification of P-SURO AUV.

Item	Specifications
Depth rating	100m
Weight	53kg
Dimension	1.05(L)?0.5(W)?0.3(H)m
Max. speed	FW: 2.5knot; BW, UP/DW: 1.5knot
Battery system	400W·hr, Lithium Ion, Endurance: 2.5hrs
Payload	?4kg

2.1 Mechanical system

For the convenience of maintenance and also under the security consideration, we separate the battery system from other electronics systems, see Fig. 2. Main frame is made of AL- 6061, fixing parts for camera and range sonar are made of POM. To increase the hydrodynamic mobility in the underwater horizontal plane, the open frame of vehicle is wrapped in a two-piece of FRP (Fibre-Reinforced Plastic) shell (Li et al., 2010).

Throughout its underwater missions, P-SURO is always keeping zero pitch angle using two vertical thrusters. With this kind of stability in its pitch dynamics, the vehicle's horizontal 3DOF motion is steered by two horizontal thrusters. From control point of view, this is a typical underactuated system. And how to design path tracking or following scheme for this kind of underactuated system has become one of most intense research area in the nonlinear control community (Jiang, 2002; Do et al., 2004; Li et al., 2008b).

Fig. 2. Mechanical arrangement of P-SURO AUV.

Table 2. Sensor & thrust system of P-SURO AUV.

Sensor Model (Maker)	Specifications
Super SeaSpy **(Tritech)**	- >480 TV lines - 1/3" Interline Transfer CCD - Composite output
Micron Echo Sounder **(Tritech)**	- Operating frequency: 500KHz - Beamwidth: 6° conical - Operating range: 50m
AHRS **(Innalabs)**	- Update rate: 1-100Hz - Heading accuracy: 1.0 deg - Attitude accuracy: <0.4 deg
CVG25 **(Innalabs)**	- Measurement range: ?200deg/sec - Bandwidth: 50Hz - Bias stability: 1.5deg/hr

Sensor Model (Maker)	Specifications
AL-15M2.5 **(Innalabs)**	- Input range: ?2.5g - Bias: <4mg - Bandwidth: 300Hz
PR-36XW **(Keller)**	- Depth rating: 100m - Accuracy: 0.1% FS - Output rate: 100Hz
UWM2000H **(LinkQuest)**	- Payload data rate: 300-1200bits/sec - Working range: 1200m - Beam width: 210deg (omni-directional)
LinkWiser™-HP400 **(Cellution)**	- Half-duplex - Operating frequency: 400-470MHz - Air data rate: 4.8kbps
BTD-156 **(SeaBotix)**	- Continual thrust: 2.2kgf - Input voltage: 28V - Interface: RS485, 115200bps

2.2 Sensor, thrust, and power system

For underwater vision, there is one colour camera mounted at the vehicle nose. And three range sonar (forward, backward and downward) are mounted on the vehicle. There sonar are designed for obstacle detection and also for assisting vehicle's underwater localization. For P-SURO AUV, we design a relatively simple but low grade of inertial navigation system which consists of AHRS, 1-axis Gyro, 1-axis accelerometer, one depth sensor.

SeaBotix BTD-156 thrusters are selected to steer the vehicle's underwater motion. This is a small size underwater thruster with 90W of average power consumption. For power system, the calculated total power consumption of vehicle system is about 450W. And correspondingly, we design the 1.2kW Lithium Ion battery system, which can support more than two hours of the vehicle's underwater continuous operation. The overall sensor & thrust system for P-SURO AUV is listed in Table. 2.

2.3 Embedded system

Three of PC104 type PCM3353 SBCs (Single Board Computers) are chosen as core modules, each of vision, navigation, and control. PCM3353 SBC provides 4 RS232 channels plus 4 USB channels. And using these USB channels, we can easily extend the necessary serial channels (RS232/422/485) using proper USB to serial converters. PCM3718HG analogue and digital I/O board is used for various peripheral interface. In addition, two peripheral boards, including DC/DC converter system, magnetic switch circuit, leakage detection circuit, are also designed. Fig. 3 shows the inner view of electronic pressure hull.

Fig. 3. Electronics system of P-SURO AUV.

3. SOFTWARE ARCHITECTURE

As aforementioned, we choose Windows Embedded CE 6.0 as the near real-time OS for three of core modules; vision module, navigation module, and control module. For this, we design three different WinCE 6.0 BSPs (Board Support Package) for each of three core modules. Furthermore, these three core modules are connected to each other through Ethernet channel, and constructing a star topology of network structure.

Software frame for each core module consists of thread-based multi tasking structure. For each module, there are various sensors connected through serial and analogue channels. And these serial sensors, according to their accessing mechanism, can be classified into two types: active sensor (frequently output measurement) and passive sensor (trigger mode). For these passive sensors as well as analogue sensors, we read the measurements through *Timer( )* routine. And for each of active sensors, we design a corresponding thread. In most of time, this thread is in *Blocking* mode until there is measurement output. And this kind of real-time sensor interface also can be used to trigger other algorithm threads. For example, in the navigation module, there is a thread designed for interfacing with AHRS sensor (100kHz of output rate). After accessing each of attitudes, gyro, and accelerometer output measurement, the thread will trigger *Navigation( )* thread. Moreover, some of these threads are cautiously set with different priority values.

As with the most of other AUVs so far, the P-SURO AUV has the similar overall software frame, which can be divided into two parts: surface remote control system and the vehicle software system. For surface system, the main functions of it are to monitor the vehicle and deliver the user command. According to the user command (mission command in this case), the vehicle will plan a series of tasks to accomplish the mission. For P-SURO AUV, its most experimental field is in a small cuboid. In this kind of environment, it is well known that underwater acoustic channel is vulnerable. For this reason, the vehicle is required to possess relatively high level of autonomy, such as autonomous navigation, obstacle avoidance, path planning and so on.

From the control architecture point of view, the software architecture of P-SURO AUV can be classified into hybrid architecture (Simon et al., 1993; Healey et al., 1996; Quek & Wahab, 2000), which is a certain combinaiton of hierarchical architecture (Wang et al., 1993; Peuch et al., 1994; Li et al., 2005) and behavioral architecture (Brooks, 1986; Zheng, 1992; Bennett, 2000). As aforementioned, because of the limitation of underwater acoustic communication in the engineering basin in PIRO, it is strongly recommended for the vehicle to self- accomplish its mission without any of user interface in the water. For this consideration, the control architecture of P-SURO AUV is featured as a behavioral architecture based hybrid system (see Fig. 4).

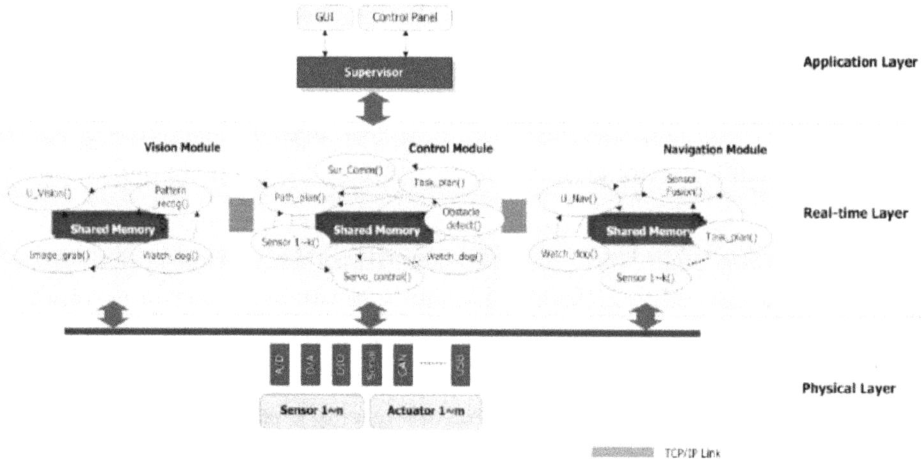

Fig. 4. Hybrid control architecture for P-SURO AUV.

If a pattern emerges in a specific area in front of the vehicle, the vision module will identify the pattern and frequently relay the vehicle's pose information to the control module to assist in course planning. In accordance with the received mission directive, the control module organizes a sequence of actions to fulfill the mission. This module executes several thruster and actuator controls. The primary function of the navigation module is to execute the real-time navigation algorithm utilizing obtained attitude, gyroscope, and accelerometer data. Additional data such as range sonar, depth sensors, and underwater vision serve as supplementary information for this inertial navigation system.

4. VISION-BASED UNDERWATER LOCALIZATION

Visual localization approaches are typically categorized into two types: one utilizes natural feature points in the surroundings for robot posture identification, while the other employs artificial landmarks, which are predefined patterns put in the environment (Oh et al., 2010).The PIRO engineering basin is enclosed by flat concrete walls, making the extraction of individual feature points challenging. Consequently, we employ artificial landmarks for the optical localization of P-SURO.

4.1 Artificial landmarks

The primary function of underwater vision for the P-SURO AUV is to furnish localization data for the vehicle's path planning operations. In recent decades, vision has emerged as a prominent study domain within the robotics community, leading to the development of numerous artificial landmarks and associated recognition methodologies (Hartley & Zisserman, 2000). Figure 5 illustrates the pattern created for the P-SURO AUV, comprising two rectangles and two groups of four dots exhibiting identical cross-ratios. The prominent figure at the center is utilized to differentiate the pattern (Li et al., 2010).The eight dots in the pattern contain 3D pose information. For extracting these dots, we use the cross-ratio invariant, which is a basic projective invariant of perspective space (Hartley & Zisserman, 2000). The cross-ratio is defined as following

$$Cross(x_1, x_2, x_3, x_4) = \frac{|x_1x_2||x_3x_4|}{|x_1x_3||x_2x_4|}, \qquad (1)$$

where

$$|x_ix_j| = det\begin{bmatrix} x_{i1} & x_{i2} \\ x_{j1} & x_{j2} \end{bmatrix}.$$

Fig. 5. Designed underwater pattern for P-SURO AUV.

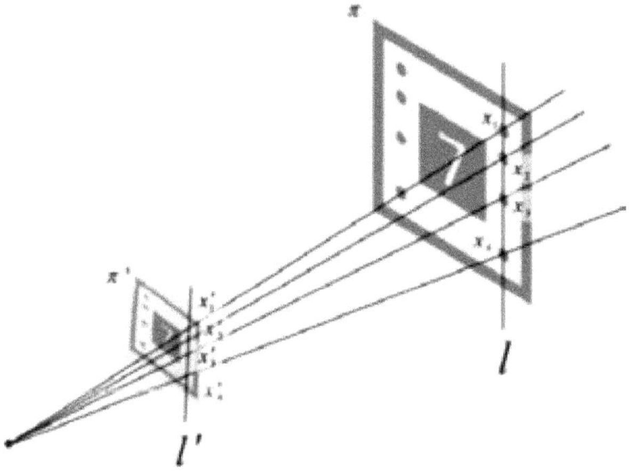

Fig. 6. Cross-ratio in the projective space.

The eight dots in the pattern contain 3D pose information. For extracting these dots, we use the cross-ratio invariant, which is a basic projective invariant of perspective space (Hartley & Zisserman, 2000). The cross-ratio is defined as following

$$Cross(x_1, x_2, x_3, x_4) = \frac{|x_1x_2||x_3x_4|}{|x_1x_3||x_2x_4|}, \tag{2}$$

where

$$|x_ix_j| = det\begin{bmatrix} x_{i1} & x_{i2} \\ x_{j1} & x_{j2} \end{bmatrix}.$$

The cross-ratio value defined in (2) is invariant under any projective transformation of the line. If $x' = H_{2\times2}x$, then $Cross(x'_1, x'_2, x'_3, x'_4) = Cross(x_1, x_2, x_3, x_4)$. As shown in Fig. 6, suppose the plane π is the pattern and the plane π' is the projected pattern image, then four dots on the line l have the same cross-ratio with the four dots on the line l'. Using this invariant, we can find the match between pattern dots and its image projection.

4.2 Auto-calibration of underwater camera

It is difficult to directly calibrate the camera in the underwater environment. For this reason, we apply a camera auto-calibration method using cross-ratio invariant (Zhang, 2000).

If we denote the pattern point as $x = [u, v, 1]^T$ and the image point as $x' = [u', v', 1]^T$, then the relationship between them is given by

$$sx' = Hx, \tag{3}$$

where s is an arbitrary scale factor. In (3), homography H is defined as

$$H = [h_1\ h_2\ h_3]^T = \lambda A[r_1\ r_2\ t], \tag{4}$$

where ? is a scale factor, r_1, r_2 are part of rotation matrix $R = [r_1\ r_2\ r_3]$, t is a translation matrix and A is a camera intrinsic matrix. If camera moves, each matching point from the scene makes a homography. We can get the camera intrinsic matrix$A$ from homography$H$(Zhang, 2000).

Given the camera intrinsic matrix$A$and the homography H in the image, we can get the three-dimensional relationship between the pattern and the camera (robot) using following equations

$$\begin{aligned} r_1 &= \lambda A^{-1} h_1 \\ r_2 &= \lambda A^{-1} h_2, \\ r_3 &= r_1 \times r_2, \\ t &= \lambda A^{-1} h_3. \end{aligned} \tag{5}$$

4.3 Lab-based evaluation

We conduct a series of ground tests to assess the created vision algorithm. Figure 7 presents the test findings. Initially, eight dots in the configuration are recovered from the image (Fig. 7-a). Figure 7-b illustrates the chosen photos for camera auto-calibration, while the extracted camera posture is depicted in Figures 7-c and 7-d. In assessing the efficacy of the proposed visual localization method, we primarily focus on two aspects: the pattern recognition rate and the precision of pose estimation. To assess the pattern recognition rate, we randomly reposition the camera and capture 306 photos. 150 were accurately recognized, 85 were not recognized, 61 were blurred due to camera movement, and 10 did not contain the complete pattern. Excluding 61 blurred photos and 10 images with missed patterns, the recognition rate is approximately 64%. Nevertheless, it is important to note that approximately fifty percent of the unrecognized images are rotated beyond 40 degrees from the pattern, resulting in a recognition rate of approximately 81%. To assess the precision of pose estimation, we stabilize the camera and position the pattern at 12 predetermined locations ranging from 400mm to 700mm in distance. The computed average posture estimation error is 0.155 mm, with a standard deviation of 1.798 mm.

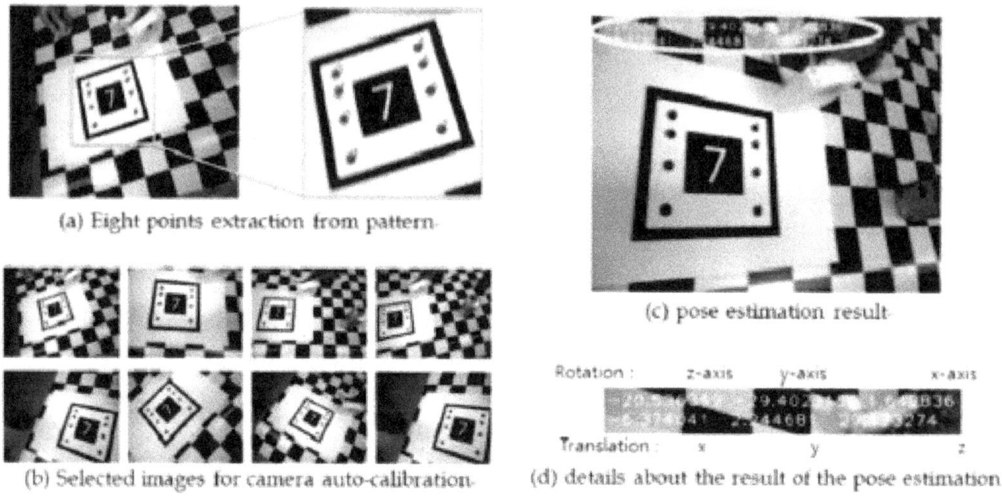

(a) Eight points extraction from pattern

(b) Selected images for camera auto-calibration

(c) pose estimation result

(d) details about the result of the pose estimation

Fig. 7. Process of pose estimation.

4.4 Camera underwater test

With a specified pattern (landmark) and a camera, the minimum and maximum recognition range of the pattern can be established in advance. This information will be utilized for the underwater path planning of vehicles. In light of this, we conduct a series of tests to measure the minimum and maximum recognition ranges in both air and water. The test findings depicted in Figures 8 and 9 indicate that the maximum recognition range in water is roughly fifty percent of that in air. For safety reasons, we require the vehicle to maintain a distance of at least 1.5 meters from the basin wall during the various basin tests.

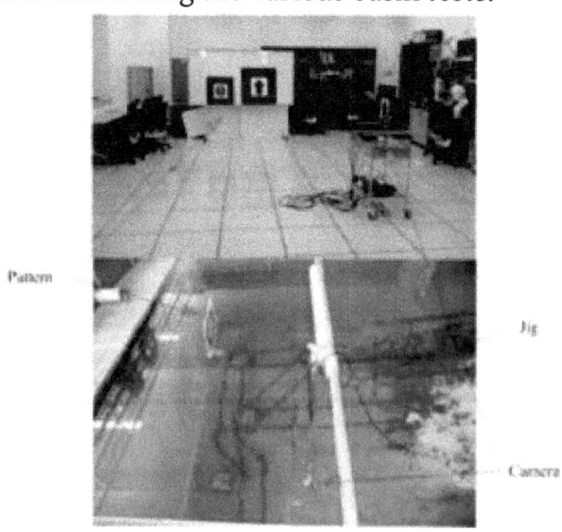

Fig. 8. Test environments.

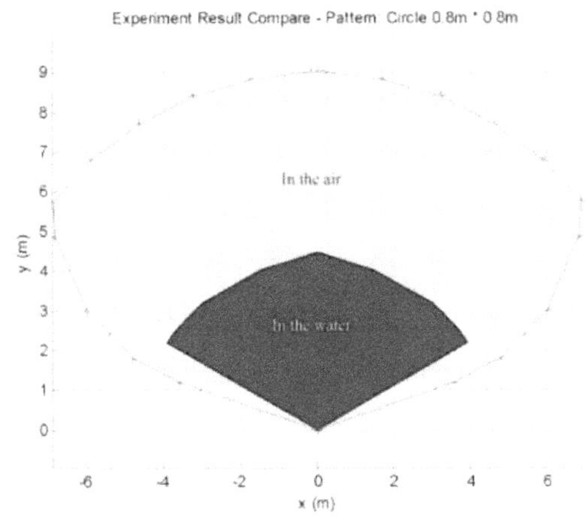

Fig. 9. Test results

5. SIMULTANEOUS LOCALIZATION AND MAPPING (SLAM) UTILIZING RANGE DATA SONAR ARRAY

In the past decades, SLAM (Simultaneous Localization and Mapping) has been one of most hot issues in the robotics community (Leonard &Burrant-Whyte, 1992; Castellanos & Tardos, 1999; Thrun et al., 2005). SLAM problems arise when the robot does not have access to a map of the environment; nor does it have access to its own poses (Thrun et al., 2005).

For P-SURO AUV, as aforementioned, most of its underwater operations are carried out in the engineering basin in PIRO. For this reason, we have designed a relatively simple SLAM method with partially known environment.

5.1 Range sonar model

There are three Micron Echosounder@Tritech (6o conical beamwidth with 500kHz operating frequency) mounted on the vehicle; each of forward, backward, and downward. Because of their narrow beam-angle, we apply simple centreline model (Thrun et al., 2005) for all of these sonar behaviour.

Throughout its underwater operation, the vehicle is forced to keep away from the basin wall at least 2m. In this case, the maximum range error is about 2.2m (dM in Fig. 10). Another significant error source is misalignment of range sonar with AHRS. For vehicle's dynamics, we observed that the yaw angular velocity of P-SURO was less than 10o/s. Consider the 1500m/s of acoustic velocity in the water, 10o/s of yaw motion may cause less than 0.1o of azimuth error. Therefore, the effect of vehicle dynamics on the range measurement can be neglected.

To investigate the range sonar behavior in a basin which is of a small cuboid, we performed a simple test where the vehicle is rotated (about 7.6o/s) several cycles with the center at the same point. Throughout the test, we force the vehicle to keep zero pitch angle. The resulted basin profile image is shown in Fig. 11, from which we can see that closer to the basin corner, more singular measurements are occurred.

5.2 Obstacle detection

At any point (x_t, y_t, ψ_t), through rotating the vehicle on the horizontal plane, we can easily get a 2D profile image of basin environment, see Fig. 12. And for 3D case, we simply extend the horizontal rotating motion with constant velocity of descent/ascent motion and get rough 3D profile image, see Fig. 13. According to this profile image, we detect the obstacle and further design corresponding vehicle path.

The obstacle block A in Fig. 12 is modelled as (a,b) with a obstacle start point and b the end point. Here $R_i = |P_c i|$ and ψ_i is yaw angle of $\overrightarrow{P_c i}$ with $i = a,b,c$. In the case of the vehicle facing point a and c, if $|ac| = |P_c c| - |P_c a| > d_c$ where $d_c = f(\psi c)$ is a design parameter, then point a is taken as the start point of an obstacle. And in the case of b and d, if $|bd| = |P_c d| - |P_c b| > d_d$ with $d_d = f(\psi a)$, then b is taken as the end point of the obstacle.

5.3 Path planning

Path planning is an important issue in the robotics as it allows a robot to get from point A to point B. Path planning can be defined as "determination of a path that a robot must take in order to pass over each point in an environment and path is a plan of geometric locus of the points in a given space where the robot has to pass through" (Buniyamin et al., 2011). If the robot environment is known, then the global path can be planned off line. And local path planning is usually constructed online when the robot face the obstacles. There are lots of path planning methodologies such as roadmap, probability roadmap, cell decomposition method, potential field have been presented so far (Choset et al., 2005; Khatib, 1986; Valavanis et al., 2000; Elfes, 1989; Amato & Wu, 1996; Li et al., 2008a).

For P-SURO AUV, there is only one range sonar mounted in front of vehicle. To get a profile image of environment, the vehicle has to take some specific motions such as rotating around it, and this usually takes quite an amount of time. In other word, it is not suitable for the vehicle to frequently take obstacle detecting process. For this reason, we design a relativelysimple path planning method for autonomous navigation of P-SURO AUV. Consider the Fig. 12, to get to the point $P_{END} = (x_e, y_e, depth_e)$ the vehicle will turn around at start pointP_c. According to detected obstacle $A(a, b)$, we calculate d_a, d_b and d_{max}=max$\{d_a, d_b\}$. If$d_{max} > d_{threshold}$ with $d_{threshold}$design parameter, then we design the target point as$P_{TARGET} = (x_t, y_t, depth_e)$ with $y_t = R_a \sin\psi_a$ (see Fig. 12, in this case, we assumed$d_a > d_b$and $d_a > d_{max}$). In the case of $d_{max} \leq d_{threshold}$, the vehicle will take descent motion as shown in Fig. 13 until $depth = h_d$. Here d_R (see

Fig. 13) is a design parameter. And in this case, the target point is set to P_{TARGET} =(x_e, y_e, h_d).

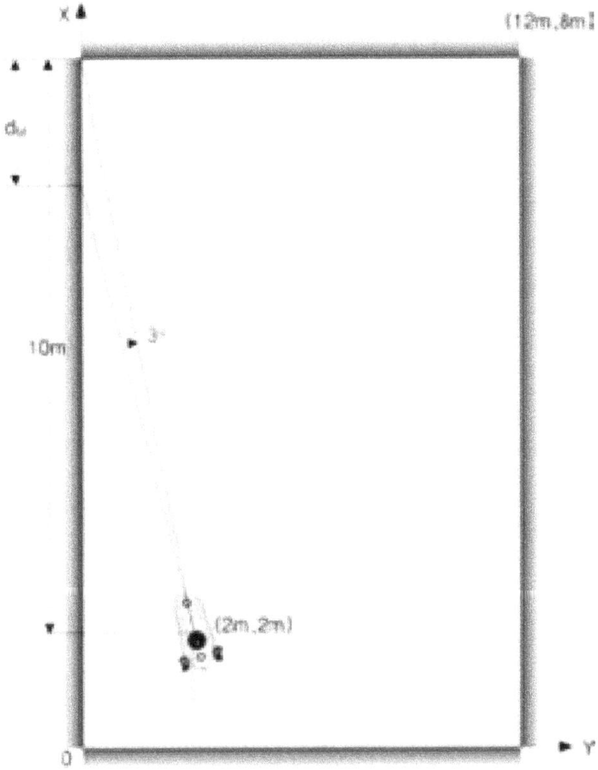

Fig. 10. Maximum range error.

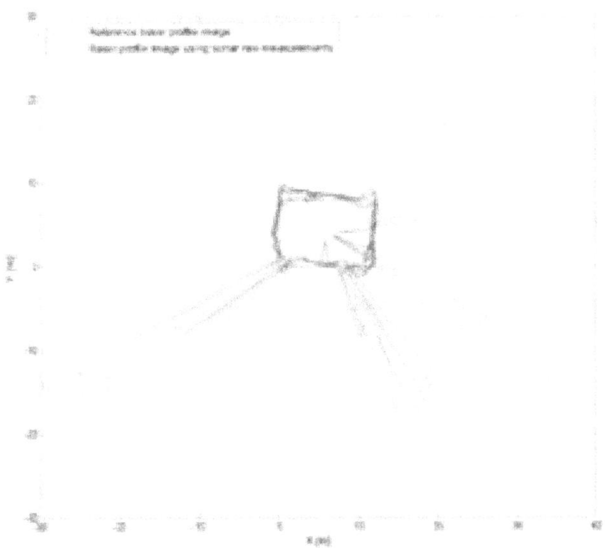

Fig. 11. Basin profile image using range sonar.

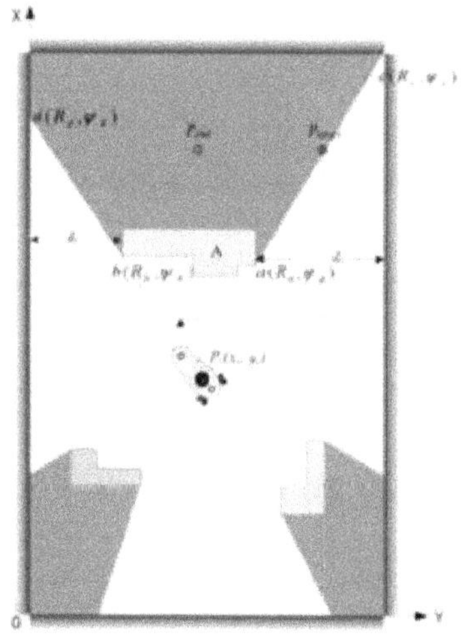

Fig. 12. Acquisition of 2D profile image.

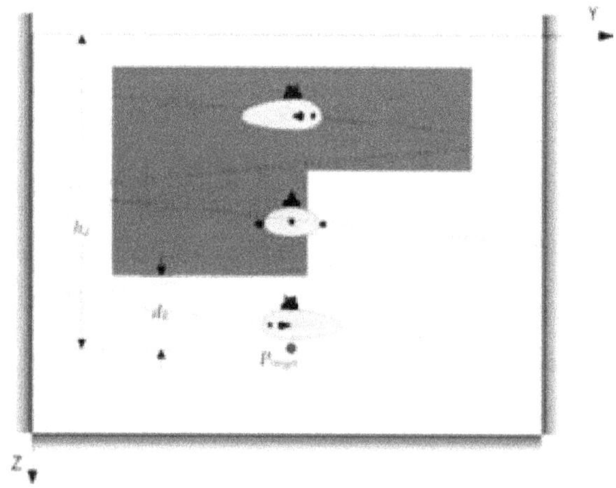

Fig. 13. Acquisition of 3D profile image.

6. BASIN TEST

To illustrate the suggested vision-based underwater localization and SLAM techniques, we conducted a series of field tests in the engineering basin at PIRO.

6.1 Preliminary of basin test

During its underwater operation, the vehicle is consistently required to maintain a zero pitch angle. In the horizontal plane, we configure the vehicle's reference route to consistently align parallel to the X or Y axis, as illustrated in Fig. 12. In this instance, the vehicle's position can be readily obtained by a straightforward rotational mode. However, given that the vehicle does not maintain a fixed position during its rotation, there is a positional drift in the rotational mode. Although the precision of range sonar measurements is at the centimeter level, the overall positional inaccuracy in this rotation mode is considerable. Numerous basin testing reveal that this type of positional mistake can reach 0.5m.Consider this kind of forward/backward motion; the vehicle's forward/backward velocity can be calculated using range sonar measurements. For this purpose, the following filter is designed for acquisition of range sonar raw measurements

$$d_{FR}(k) = (1 - 2^{-n})d_{FR}(k-1) + 2^{-n}d_R(k), \tag{6}$$

where d_{FR} and d_R denote each of filtered and raw measurements of range sonar, and $n$is filtering order. The filtering results can be seen in Fig. 14.

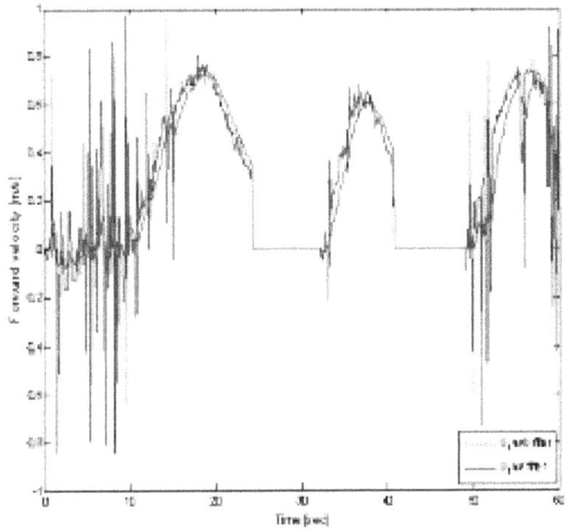

Fig. 14. Calculated forward speeds.

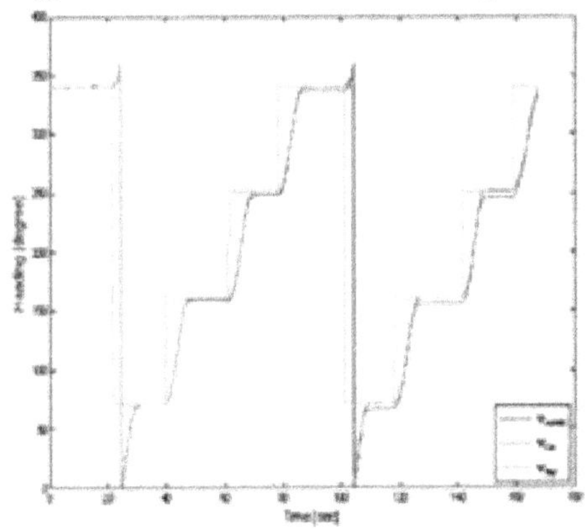

Fig. 15. Comparison of heading measurements

An additional significant concern regarding the basin test is to the vehicle's AHRS sensor. The engineering basin at the PIRO is situated in the basement of a building mostly composed of steel elements. In such a scenario, due to significant distortion of the Earth's magnetic field, the AHRS is unable to achieve appropriate initialization and the Kalman filter compensation process fails. Consequently, there is considerable deviation in the AHRS heading output. Fortunately, a highly accurate 1-axis gyro sensor is horizontally mounted on the vehicle for motion control purposes. We estimate the vehicle's heading value utilizing the Gyro output, with its bias value assessed through numerous basin tests. Figure 15 illustrates the comparability of these measurements.

6.2 P-SURO SLAM

To illustrate the SLAM methodology proposed for the P-SURO AUV, we conduct three autonomous navigation experiments: a) without impediment, b) with one impediment, c) with two impediments, refer to Fig. 16. The autonomous navigation mission can be categorized into the following four phases.

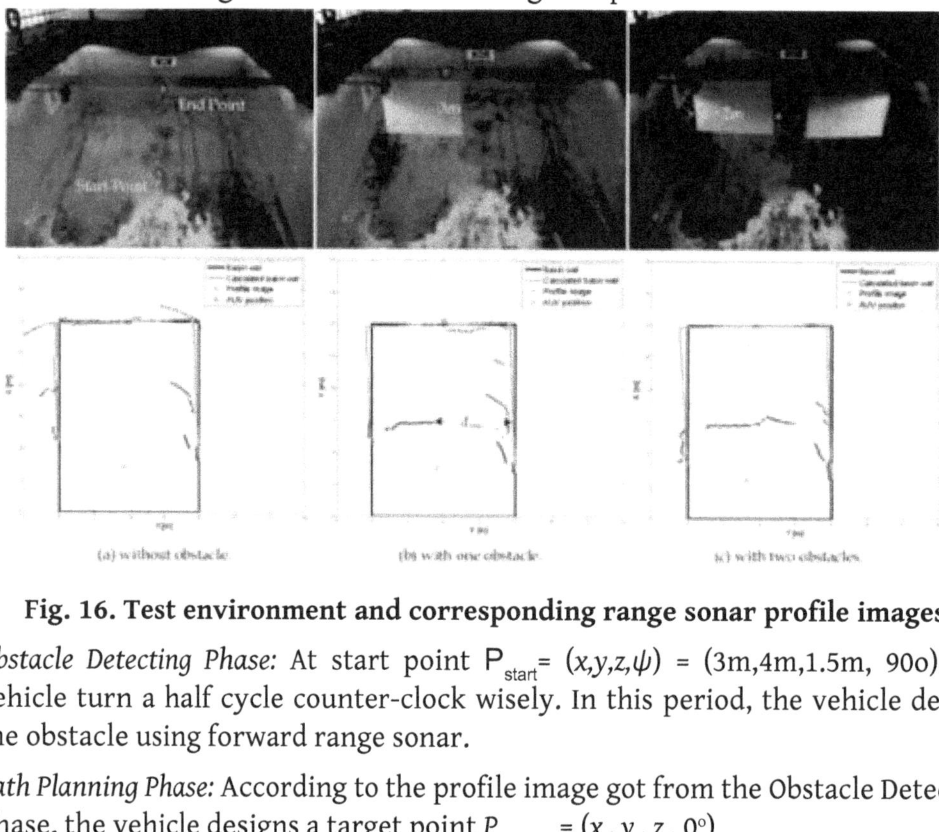

Fig. 16. Test environment and corresponding range sonar profile images.

Obstacle Detecting Phase: At start point $P_{start} = (x,y,z,\psi) = (3m,4m,1.5m, 90o)$, the vehicle turn a half cycle counter-clock wisely. In this period, the vehicle detects the obstacle using forward range sonar.

Path Planning Phase: According to the profile image got from the Obstacle Detecting Phase, the vehicle designs a target point $P_{TARGET} = (x_t, y_t, z_t, 0^\circ)$

Vision-based Underwater Localizing Phase: While approaching to P_{TARGET}, the vehicle recognizes the underwater pattern, from which defines the end point P_{END} $(9, y_c \ y_v, z_c - z_v, 0^\circ)$.Here (x_c, y_c, z_c) denotes the vehicle's current position, and (x_v, y_v, z_v) is the vehicle's pose information acquired from pattern recognition.

Homing Phase: After approaching P_{END}, or failed to recognize the pattern, the vehicle returns to P_{START} along with its previous tracking trajectory.

In Fig. 16, the blue line (calculated basin wall) is got through $f(x_c, y_c, \psi_c, R_{FSonar}, t)$ where R_{FSonar} is the forward range sonar measurement.

In the Path Planning Phase, the target points are set to different values according to three different cases. In the case of without obstacle, we set P_{TARGET} = (8m, 4m, 1.5m, 0°) in the case of one obstacle, set P_{TARGET}=(8m,y_c+0.5d_{max},1.5m,0°) with d_{max} is shown in Fig.16(b); and in the case of two obstacles, P_{TARGET}= (8m,4m,h_d,0°) where$h_d$is defined in Fig. 13 .

Autonomous navigation with obstacle avoidance and underwater pattern recognition testresults are shown in Fig. 17. Through these field tests, we found that the proposed SLAM method for P-SURO AUV shown a satisfactory performance. Also, we found that the aforementioned drift in the vehicle's rotating motion is the main inaccuracy source of both the navigation and the path planning (specially, in the calculation of d_i with i=a,b,c in Fig. 12).

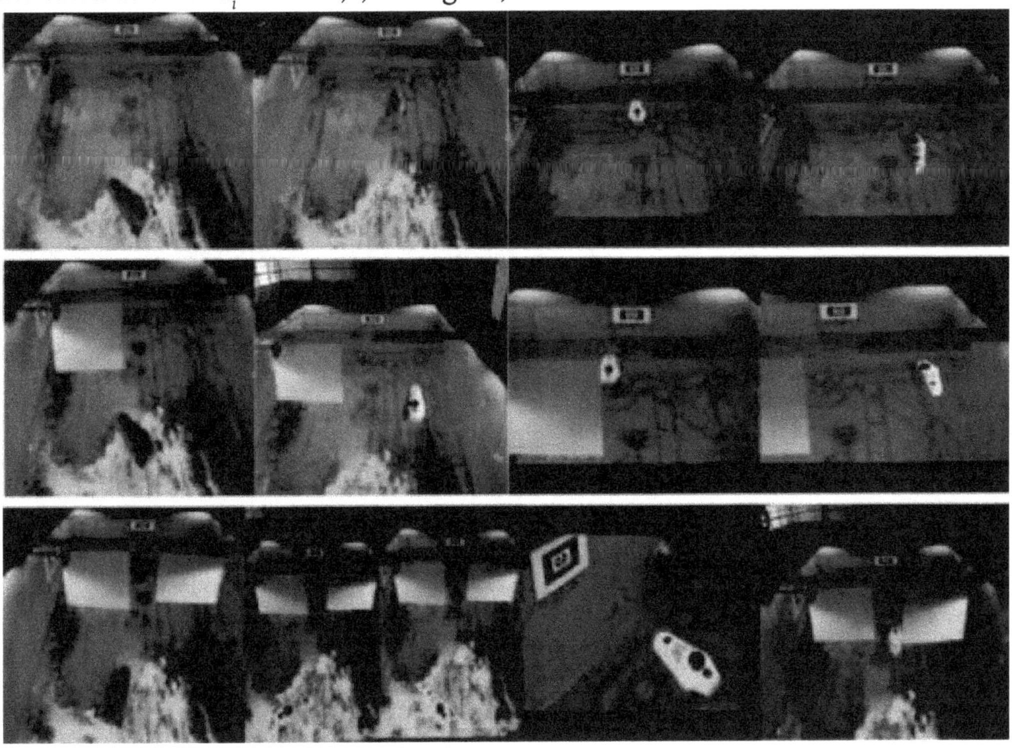

Fig. 17. Autonomous navigation test results.

7. CONCISE SUMMARY

Recently, enhancing vehicle autonomy has emerged as a prominent topic within the underwater robotics field. This chapter addresses various underwater intelligent technologies, including vision-based underwater localization and the SLAM approach, which utilize simply video cameras and range sonar—both of which are relatively inexpensive underwater equipment. Field experiments conducted in the engineering basin at PIRO utilizing the P-SURO AUV demonstrated that the suggested technologies delivered adequate accuracy for the autonomous navigation of hovering-type AUVs in the basin. Nevertheless, the basin testing revealed that the suggested vision algorithm exhibited excessive sensitivity to ambient factors. Enhancing the robustness of underwater vision is a significant focus of our future endeavors. Furthermore, the advancement of specific low-cost underwater navigation technology under imperfectly understood environmental circumstances is among our future priorities.

REFERENCES

Amato, N. M. & Wu, Y. (1996). A randomized roadmap method for path and manipulation planning, *Proceedings of IEEE International Conference on Robotics and Automation,* pp. 113-120, Osaka, Japan, November 4-8, 1996

Bennet, A. A & Leonard, J. J. (2000). A behavior-based approach to adaptive feature detection and following with autonomous underwater vehicles. *IEEE Journal of Oceanic Engineering,* Vol.25, pp. 213-226, 2000

Brooks, R. A. (1986). A robust layered control system for a mobile robot. *IEEE Journal of Robotics and Automation,*Vol.2, pp. 14-23, 1986

Buniyamin, N., Sariff, N., Wan Ngah, W. A. J., Mohamad, Z. (2011). Robot global path planning overview and a variation of ant colony system algorithm. *International Journal of Mathematics and Computers in Simulation,* Vol.5, pp. 9-16, 2011

Castellanos, J. A. & Tardos, J. D. (1999). *Mobile Robot Localization and Map Building: A Multisensor Fusion Approach.* Boston, Mass.: Kluwer Academic Publishers, 1999

Choset, H., Lynch, K. M., Hutchinso, S., Kantor, G., Burgard, W., Kavraki, L. E., Thrun, S. (2005). *Principles of Robot Motion.* MIT Press, 2005

Do, K. D., Jiang, J. P., Pan, J. (2004). Robust adaptive path following of underactuated ships.*Automatica,* Vol.40, pp. 929-944, 2004

Elfes, A. (1989). Using occupancy grids for mobile robot perception and navigation. IEEE Transactions on Computer, Vol.22, pp. 46-57, 1989

Khatib, O. (1986). Real-time obstacle avoidance for manipulators and mobile robots. International Journal of Robotics Research, Vol.5, pp. 90-98, 1986

Hartley, R & Zisserman, A. (2000). *Multiple View Geometry in Computer Vision.* Cambridge University Press, June 2000

Healey, A. J., Marco, D. B., McGhee, R. B. (1996). Autonomous underwater vehicle control coordination using a tri-level hybrid software architecture, *Proceedings of the IEEE International Conferences on Robotics and Automation,* Minneapolis, Minnesota, pp. 2149-2159, 1996

Jiang, J. P. (2002). Global tracking control of underactuated ships by Lyapunov's direct method. *Automatica,* Vol.40, pp. 2249-2254, 2002

Leonard, J. & Durrant-Whyte, H. (1992). *Directed Sonar Sensing for Mobile Robot Navigation.*London: Kluwer Academic Publishers, 1992

Li, J. H., Jun, B. H., Lee, P. M., Hong, S. K. (2005). A hierarchical real-time control architecture for a semi-autonomous underwater vehicle. *Ocean Engineering,* Vol.32, pp. 1631-1641, 2005

Li, J. H., Lee, P. M., Jun, B. H., Lim, Y. K. (2008a). *Underwater Vehicle,*InTech, ISBN 978-953- 7619-49-7, Vienna, Austria

Li, J. H., Lee, P. M., Jun, B. H., Lim, Y. K. (2008b). Point-to-point navigation of underactuated ships. *Automatica,* Vol. 44, pp. 3201-3205, 2008

Li, J. H., Yoon, B. H., Oh, S. S., Cho, J. S., Kim, J. G., Lee, M. J., Lee, J. W. (2010). Developmentof an Intelligent Autonomous Underwater Vehicle, P-SURO, *Proceedings of Oceans'10 IEEE Sydney, Sydney,* Australia, May 24-27, 2010

Oh, S. S., Yoon, B. H., Li, J. H. (2010). Vision-based localization for an intelligent AUV, P-SURO, Proceedings of KAOSTS *Annual conference,* pp. 2602-2605, Jeju, Korea, 2010

Marthiniussen, R., Vestgard, K., Klepaker, R., Storkersen, N. (2004). HUGIN-AUV concept and operational experience to date, *Proceedings of IEEE/MTS Oceans'04,* pp. 846-850, Kobe, Japan, November 9-12, 2004

Peuch, A., Coste, M. E., Baticle, D., Perrier, M., Rigaud, V., Simon, D. (1994). And advanced control architecture for underwater vehicles, *Proceedings of Oceans*'94, Brest, France, pp. I-590-595, 1994

Prestero, T. (2001). *Verification of a six-degree of freedom simulation model for the REMUS autonomous underwater vehicles,*Masters Thesis, MIT, USA

Quek, C & Wahab, A. (2000). Real-time integrated process supervision. *Engineering Applications of Artificial Intelligence,* Vol.13, pp. 645-658, 2000

Simon, D., Espiau, B., Castillo, E., Kapellos, K. (1993). Computer-aided design of a generic robot controller handling reactivity and real-time control issues. IEEE *Transactions on Control Systems Technology,* Vol.1, pp. 213-229, 1993

Thrun, S., Burgard, W., Fox, D. (2005). *Probabilistic Robotics.* The MIT Press, 2005

Valavanis, K. P., Hebert, T., Kolluru, R., Tsourveloudis, N. C. (2000). Mobile robot navigation in 2-D dynamic environments using electrostatic potential fields.

IEEE Transactions on Systems, Man and Cybernetics-part A, Vol.30, pp. 187-197, 2000

Wang, H. H., Marks, R. L., Rock, S. M., Lee, M. J. (1993). Task-based control architecture for an untethered, unmanned submersible, *Proceedings of the 8th Symposium on Unmanned Untethered Submersible Technology,* Durham, New Hampshire, pp. 137- 148, 1993

Zhang, Z. (2000). A flexible new technique for camera calibration. *IEEE Transactions on Pattern Analysis and Machine Intelligence,* Vol. 22, No. 11, pp. 1330-1334, 2000

Zheng, X. (1992). Layered control of a practical AUV, *Proceedings of the Symposium on Autonomous Underwater Vehicle Technology,* Washington DC, pp. 142-147, 1992

3 Chapter

HYDRODYNAMIC PROPERTIES OF THE PRINCIPAL COMPONENTS OF THE HYBRID-DRIVEN UNDERWATER GLIDER PETREL

1. INTRODUCTION: AN OVERVIEW

The Autonomous Underwater Vehicle (AUV), Remotely Operated Vehicle (ROV), and Autonomous Underwater Glider (AUG) are the primary autonomous underwater platforms presently available, which significantly contribute to marine environmental monitoring. Figure 1 illustrates the links among the three categories of vehicles.

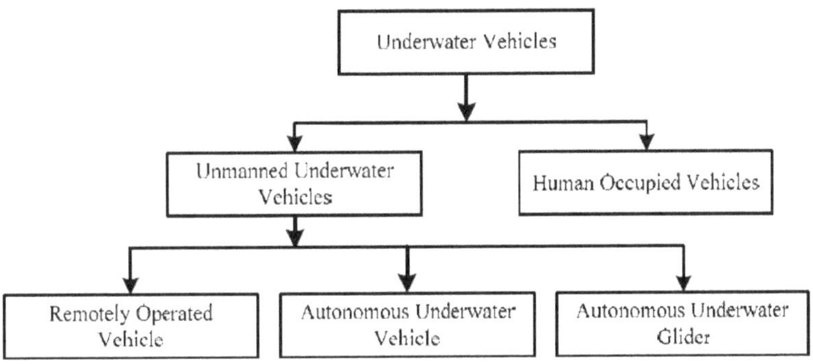

Fig. 1. Underwater Vehicles

As a special type of AUV, underwater gliders have many advantages, such as long endurance, low noise and low energy cost. A glider can periodically change its net buoyancy by a hydraulic pump, and utilize the lift from its wings to generate forward motion. The inherent characteristics of a glider can be summarized as buoyancy-driven propulsion, sawtooth pathway, high endurance and slow speed. There exist three legacy gliders named respectively Seaglider, Spray and Slocum [1~6]. In spite that underwater gliders features low level of self noise and high endurance, they also have weaknesses like the lack of maneuverability and the inability to perform a fixed depth or level flight [7].

Driven by a propeller with carried energy source, autonomous underwater vehicles is preprogrammed to carry out an underwater mission without assistance from an operator on the surface. However, they can only cover a relatively short range after each recharge due to the high power consumed for propulsion and generate much more noise than the AUGs because of its propeller and motors [8~10]. The range of AUV's is restricted by the amount of energy carried on board, can was not more than several hundreds kilometers in general [11]. The performances of the underwater vehicle are compared in Figure 2.

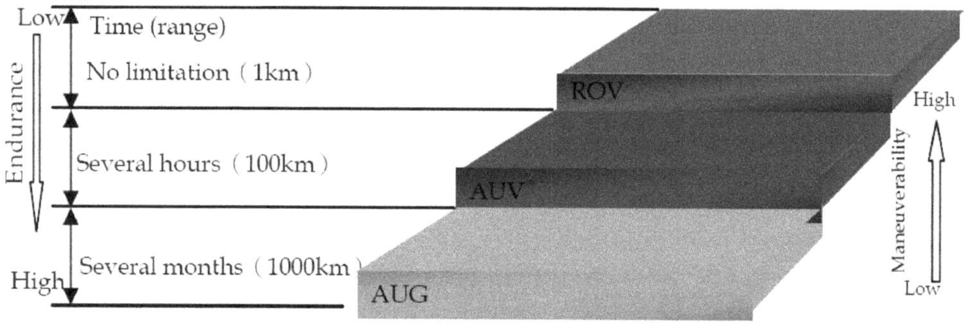

Fig. 2. Performances of three Underwater Vehicles

By combining the advantages of the glider and the propeller-driven AUVs, A hybrid-driven underwater glider PETREL with both buoyancy-driven and propeller-driven systems is developed. Operated in buoyancy-driven mode, the PETREL carries out its mission to collect data in a wide area like a legacy glider. When more exact measurements of a smaller area or level flight are needed, the PETREL will be operated by using the propeller-driven system [5, 7]. This flexible driven glider contributes to have a long range while operated in the buoyancy driven mode like a glider, as well as improve the robust performance to deal with some wicked circumstances by the propeller driven system [7].

Proper hydrodynamic design is important for the improvement of the performance of an underwater vehicle. A bad shape can cause excessive drag, noise, and instability even at low speed. At the initial stage of design, there are two ways to obtain the hydrodynamic data of the underwater vehicle, one is to make model experiment and the other is to use the computational fluid dynamics (CFD). With the development of the computer technology, some accurate simulation analysis of hydrodynamic coefficients have been implemented by using the computational fluid dynamic (CFD) software, instead of by experiments at a much higher cost over the past few years [12-13] . In consideration of the reduced time, lower cost, more flexible and easier optimum design, the CFD method was used in this article. The fluent Inc.'s (Lebanon,New Hampshire) CFD software FLUENT 6.2 was adopted by this article.

This chapter examines the hydrodynamic impacts of the primary components of a hybrid-driven underwater glider, particularly during the glide phase. Through the examination of the outcomes pertaining to the three primary hydrodynamic components—the wings, rudders, and propeller—the attributes of drag, glide efficiency, and stability will be evaluated, accompanied by recommendations for modifying the HUG's design to enhance its hydrodynamic performance.

2. COMPUTATIONAL DETAILS

2.1 Mathematical model

A criterion for determining of the flow regime of the water when the vehicles moving in it is proposed by Reynolds number [14-15]:

$$R_e = \rho v L / \mu \tag{1}$$

Here ρ is the density of water, *v* is the velocity of vehicle, *L* is the characteristic length, μ is the dynamic coefficient of viscosity. The transition point occurred when the Reynoldsnumber is near 10^6 for the external flow field, which is called critical Reynolds numbers. It was laminar boundary layer when the Re < 5 x 10^5, it was seem as turbulent flow while Re > 2 x 10^6. The Reynolds number of the hybrid underwater glider PETREL at two different steering modes is shown in table 1.

Table 1. The Reynolds number at different steering modes

steering mode	velocity v / (*m/s*)	Reynolds number
Glider	0.5	1.25×10^6
AUV	2	5×10^6

The turbulence model will be adopted because the Reynolds numbers of the PETREL in two steering modes are all above the critical Reynolds numbers. Computations of drag, lift and moment and flow field are performed for both the model over a range of angles of attack by using the commercially available CFD solver FLUENT6.2. The Reynolds averaged Navier– Stokes equation based on SIMPLAC algorithm and the finite volume method were used by our study. In our study RNG k-ε model was adopted and the second-order modified scheme was applied to discrete the control equations to algebra equations. Assuming that the fluids were continuous and incompressible Newtonian fluids. For the incompressible fluid, the *RNG* $k-\varepsilon$ transport equations are [12, 16]:

$$\rho\frac{\partial}{\partial t}(k)+\rho\frac{\partial}{\partial x_i}(ku_i)=\frac{\partial}{\partial x_j}\left(\alpha_k\mu_{eff}\frac{\partial k}{\partial x_j}\right)+G_k-\rho\varepsilon+S_k \tag{2}$$

$$\rho\frac{\partial}{\partial t}(\varepsilon)+\rho\frac{\partial}{\partial x_i}(\varepsilon u_i)=\frac{\partial}{\partial x_j}\left(\alpha_\varepsilon\mu_{eff}\frac{\partial \varepsilon}{\partial x_j}\right)+C_{1\varepsilon}\frac{\varepsilon}{k}G_k-C_{2\varepsilon}\rho\frac{\varepsilon^2}{k}-R_\varepsilon+S_\varepsilon \tag{3}$$

Here S_k and S are source items, μ_{eff} is effective viscosity, G_k is turbulence kinetic energy induced by mean velocity gradient.

$$G_k = \rho\overline{u_i'u_j'}\frac{\partial u_j}{\partial x_i} \tag{4}$$

σ_k and σ_ε is respectively the reversible effect Prandtl number for k and ε. $C_{1\varepsilon}=1.42$, $C_{2\varepsilon}=1.68$

In the *RNG* model,a turbulence viscosity differential equation was generated in the non- dimensional treatment.

$$\mathrm{d}\left(\frac{\rho^2 k}{\sqrt{\varepsilon\mu}}\right)=1.72\frac{\hat{v}}{\sqrt{\hat{v}^3-1+C_v}}\mathrm{d}\hat{v} \tag{5}$$

Here, $\hat{v}=\mu_{eff}/\mu$, $C_v\approx 100$ .Taking the integral of the(5),the exact description of active turbulence transport variation with the effective Reynolds number can be acquired, which makes the mode having a better ability to deal with low Reynola number and flow near the wall. For the large Reynola number, the equation(5) can be changed into (3-6).

$$\mu_t=\rho C_\mu\frac{k^2}{\varepsilon} \tag{6}$$

Here , C_μ = 0.0845 The RNG $k-\varepsilon$ model was adopted due to the initial smaller Reynola number of boundary layer, and the more exact results can be gained by substituting the differential model into the RNG $k-\varepsilon$ model.

2.2 Meshing and boundary conditions

The size function and unstructured meshes were utilized to provide a reasonable distribution of the meshes and facilitate efficient mesh generation. Figures 3, 4, and 5 illustrate the examples of meshing.

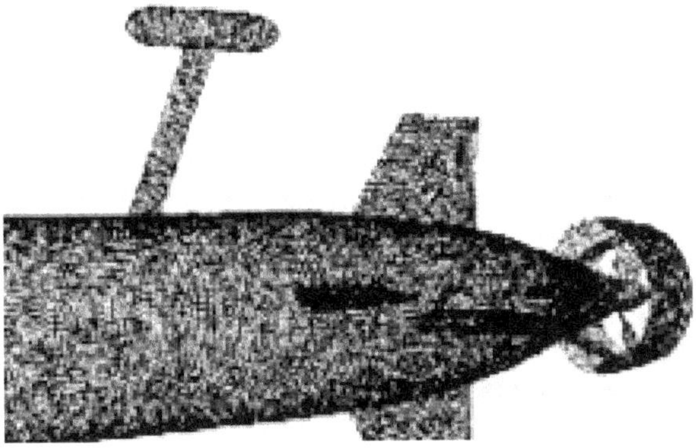

Fig. 3. Stern meshes

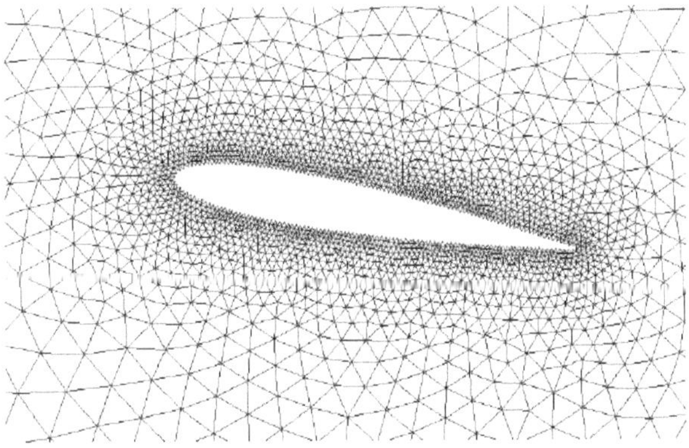

Fig. 4. Two-dimension rudder meshes

Fig. 5. Whole meshes of the vehicle

Boundary conditions:

- Inlet boundary condition: setting the velocity inlet in front of the head section with a distance of one and a half times of the length .
- Outlet boundary condition: setting the free outflet behind the foot section with a distance of double length of the vehicle.
- Wall boundary condition: setting the vehicle surface as static non-slip wall.
- Pool wall boundary condition: non-slip wall.

2.3 Results verification

To verify the precision of the calculation, we computed the drag coefficients of Slocum underwater glider [17] at different angle of attack as shown in table 2. The table 3 shows the verification of numerical simulation results of drag of AUV shell of Tianjin University. The error percentage of our calculation is less than 9.35%.

Table 2. Verification of numerical simulation results of CD

Angle of attack α (degree)	Reynolds number R_e	C_D (experiment)	C_D (CFD)	Error percentage
-2.9	7.5×105	0.31	0.281	9.35%
2.3	6.3×105	0.25	0.268	7.20%
2.7	5.8×105	0.27	0.274	1.46%

Table 3. Verification of numerical simulation results of drag of AUV shell

Velocity / (m/s)	Reynolds number	Drag Experiment (N)	Drag CFD(N)	Error percentage
0.81	2.5×106	7.4	6.903	6.72%
1.4	4.4×106	20.3	19.92	1.87%
2.0	6.2×106	37.5	37.34	0.427%

3. WING HYDRODYNAMIC DESIGN [18]

3.1 Orthogonal experimental design and results analysis

3.1.1 Orthogonal experimental design

An orthogonal experimental with four factors and three levels was conducted by keeping the main body size of the vehicle as constant. The four factors are wing chord, aspect ratio, backswept and distance between the center of wing root and the center of body. The simulation experiments were done at the situation of angle of attack is $\alpha = 6°$ and the velocity is $v = 0.5m/s$. The airfoil of the wings was NACA0010. The orthogonal experimental table was shown as table 4.

Table 4. Orthogonal experimental table

level	chord(mm)	aspect ratio	backswept (°)	distance(mm)
1	100	6	20	100
2	150	8	40	0
3	200	10	60	-100

3.1.2 Analysis indexes

The design of the wing will generate important impacts on glide efficiency and glide stability of the vehicle. The lift to drag ratio L D is chosen for measurement of the glide efficiency, the bigger values correspond to the more efficient gliding. The inverse of *L/D* expresses the glide slope [7, 19]. Existing oceanographic gliders are designed for static stability in steady glides, and the static stability can be measured by the non-dimensional hydrodynamic lever l'_α, the equations are [20~21]:

$$l'_\alpha = l_\alpha / l \tag{7}$$

$$l_\alpha = -M_\alpha / L_\alpha \tag{8}$$

Here, l is the vehicle length, Mα is the hydrodynamic moment induced by angle of attack α , Lα and is the Lift induced by the angle of attack α . It is static instability while $l'_\alpha > 0$, the moment induced by incremental angle of attack makes the angle of attack become bigger; It is neutral stability while $l'_\alpha = 0$; It is called static stability while $l'_\alpha < 0$, the moment induced by incremental angle of attack makes the vehicle to turn to the original state.

3.1.3 Influencing factors analysis

The orthogonal experimental table L18(37) and the simulation results are shown in table 5. The trend charts were shown as Figure 6 and Figure 7. The *L / D* increase with the growth of chord and aspect ratio, and decrease with the growth of backswept, it has little relationship with the location of the wings. The l'_α increase as chord and backswept increase when the wings is located after the hydrodynamic center, which means the stability increase as chord and backswept raise. The stability gets higher as the wing location becomes father away from behind the center of the body.

The ranges of chord, aspect ratio, backswept and distance of the wings is separately 2.448, 1.077, 1.303 and 0.312 for the *L/D,* which was gained by the range method. It is shows that the effects significance series for glide efficiency is chord, backswept, aspect ratio and the location of wings. The chord was dramatic for the index *L/D* at the significance level 0.10 and 0.05 adopted by the range method.

In like manner, the range of chord, aspect ratio, backswept and distance of the wings is separately 0.051, 0.037, 0.095 and 0.031 for the l'_α , which was gained by the range method. It is shows that the effects significance series for glide stability is backswept, chord, aspect ratio and the location of wings. The backswept was dramatic for the index l'_α at the significance level 0.10 and 0.05 adopted by the range method.

Table 5. Orthogonal experimental table and the results

Simulation	Factors				Results	
Times	Chord(mm)	Aspect ratio	Backswept (°)	Distance(mm)	L/D	I'_{α}
1	100	6	20	100	2.86	0.0274
2	100	8	40	0	3.11	-0.0323
3	100	10	60	-100	2.74	-0.109
4	150	6	20	0	4.07	-0.00059
5	150	8	40	-100	4.37	-0.0854
6	150	10	60	100	3.82	-0.127
7	200	6	40	100	4.88	-0.0158
8	200	8	60	0	4.47	-0.172
9	200	10	20	-100	6.81	-0.0713
10	100	6	60	-100	2.33	-0.0630
11	100	8	20	100	3.31	0.0212
12	100	10	40	0	3.44	-0.0444
13	150	6	40	-100	3.78	-0.0594
14	150	8	60	100	3.40	-0.0858
15	150	10	20	0	5.30	-0.0237
16	200	6	60	0	3.94	-0.115
17	200	8	20	-100	6.17	-0.0562
18	200	10	40	100	6.21	-0.0756

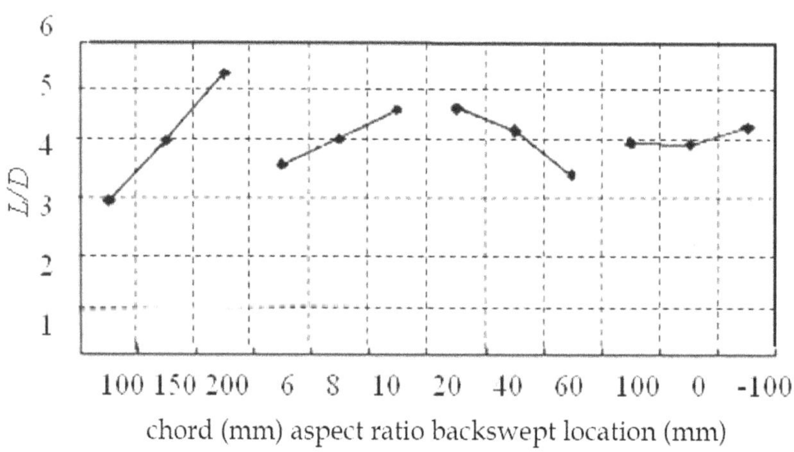

Fig. 6. L/D tendency chart

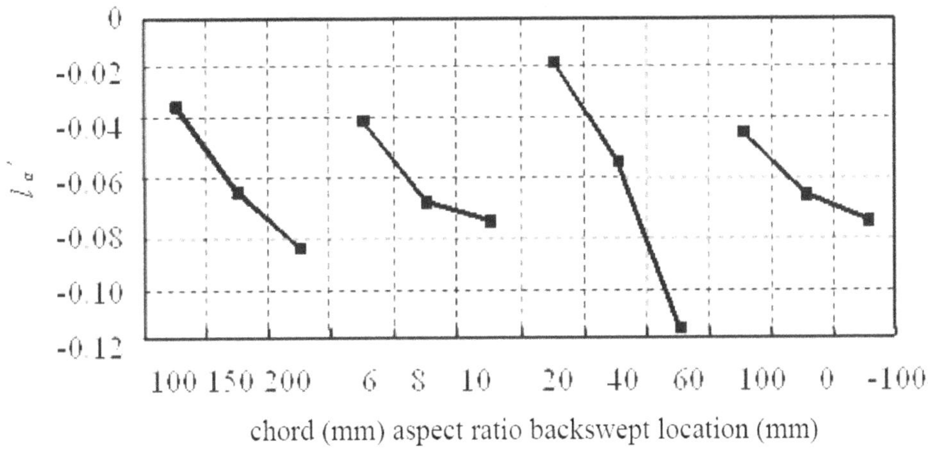

Fig. 7. I'$_\alpha$ Tendency chart

It is well known that the chord and aspect ratio of the wings should be increased, and the backswept decreased for the higher glide efficiency when PETREL is operated in the gilde mode. Simutaneously, the backswept of the wings should be increased and the wings should be moved backward father behind the center of the vehicle for the higher stability. It indicates that the effects from the increment of the backswept of the wings are inversed in increasing the glide efficiency and the stability. The backswept of the wings should be determined in terms of other capability indexes of the underwater vehicle.

3.2 Concrete models analysis

Four concrete models with varied wing parameters listed in table 6 were choosen for some further investigation. We carried out this new series of experiments in the hope of providing the effects of wings, rudders and propeller on the L / D and l'$_\alpha$ at different glide angle of attack when the velocity is 0.5m/s and the angle of attack on the rang of 0°~20°.The model one has the highest glide efficiency and glide stability in the table 6; The models 2~4 were proposed in order to evaluate the affects of location, aspect ratio and chord of wings on the analysis index as shown in the table 6. The Figure 8 gives the pressure distribution chart of model 3. The calculation results of different models are shown as Fig. 9~ Fig. 12.

Table 6. The parameter of the concrete model

models	chord(mm)	aspect ratio	backswept(°)	location(mm)
1	200	10	40	0
2	200	10	40	100
3	200	8	40	0
4	150	10	40	0

The location of the wings has little influence on the *L/D*, which means it has little influence on the glide efficiency illustrated in Figure 9, but it has dramatic effects on the glide stability which can be seen in the Figure 10. From the figure 9 and 10, it can be seen that the *L/D* decreased and the $l\alpha'$ increased obviously when the aspect ratio and chord reduced, but the effects is more dramatically to decrease the chord of the wings for the *L/D*. It has the biggest lift to drag ratio when the angle of attack at 6 degree shown in Figure 9, that is means the maximum glide efficiency can be gain when the angle of attack at 6 degree.

Fig. 8. The pressure distribution chart of model 3

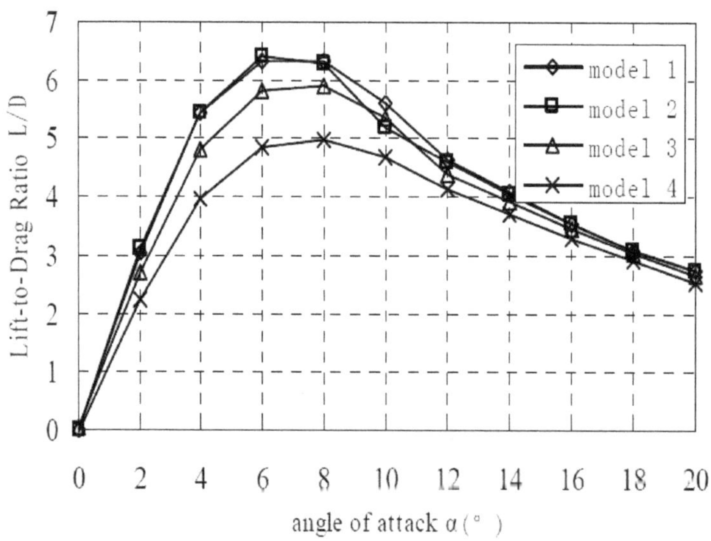

Fig. 9. The relationship between *L/D* and angle of attack

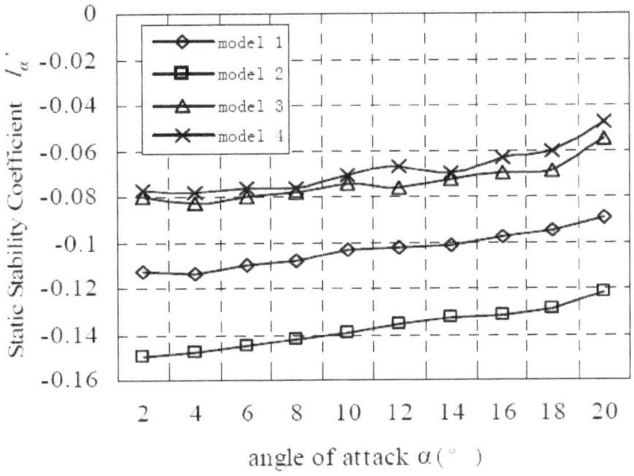

Fig. 10. The relationship between lα′ and angle of attack

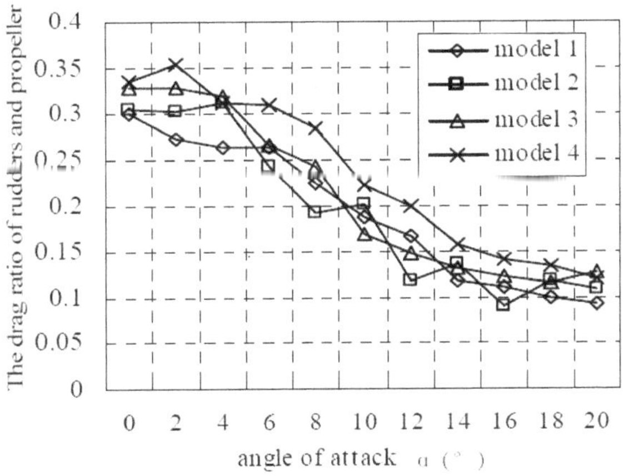

Fig. 11. The drag ratio of rudders and propeller

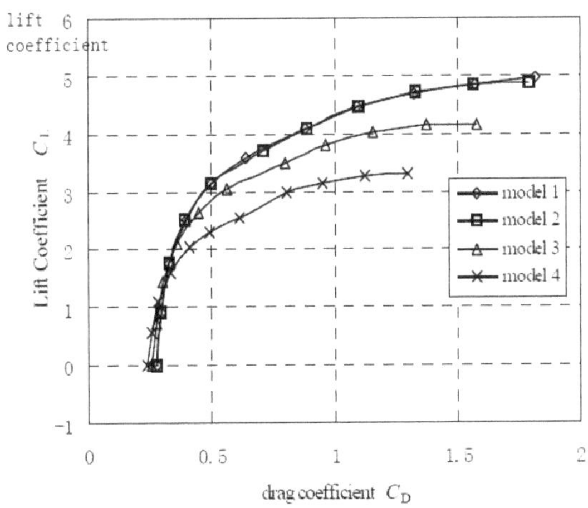

Fig. 12. The lift-drag polar curve of four concrete models

The drag of the hybrid glider will be increased because of the drag generated by the rudders and propeller compared with the legacy gliders in the glide mode. The range in the glide mode will be decreased because of the drag of these parts. The ratio of drag on the propeller and rudders to whole drag is illustrated in Fig. 11, where we find that the ratio changed as the angle of attack increases, and the values is within the range of 10%~35%. Compared with the legacy gliders, the range of the vehicles with the same configuration as PETREL will be decrease 10 %~35%. The Lift to Drag polar curves of the four concrete models are shown as figure 12. The model 3 and model 4 have the bigger lift than the model 1 and model 2 when the drag coefficients from the figure 3-9 is less than 0.5, but the lift of model 1 and model 2 increases greatly when drag coefficients gets bigger than 0.5. Due to the drag of the vehicle need overcome by the variable buoyancy B in the end and there is equation (9), so the net buoyancy supplied by the buoyancy driven system and glide angle should be taken into consideration.

$$B\sin\theta = D \qquad (9)$$

Here B is the net buoyancy, θ is the glide angle, *D* is the drag of the glider.

3.3 Results and discussion

The orthogonal experiment indicates that glide efficiency is predominantly driven by chord length, whereas vehicle stability is largely affected by sweep angle. Further numerical calculations based on four specific models with the attack angle in the range of 0°-20° indicate that location of the wings mainly affects glide stability but has little influence on glide efficiency.

When the vehicle glides at about 6° attack angle it has the maximum ratio of lift to drag. The range of the hybrid glider with the same configuration as PETREL will be decrease 10%~35% compared with the legacy gliders.

4. RUDDER HYDRODYNAMIC DESIGN[22]

4.1 Rudder parameters

The rudders parameters include root chord, half span, aerofoil and backswept, which are shown in Figure 13. As defined in the [23], the chord is denoted by C, the distance from the leading edge to trailing edge in a given two-dimensional section. The chord is measured in parallel with the section at the root of the rudder. In general, the chord can vary along the span, in which case the geometric mean chord, $\bar{C}$, is used in computations unless noted[21]. The $\bar{C}$ is defined based on Figure 14 as

$$\overline{C} = \frac{C_t + C_r}{2} \qquad (10)$$

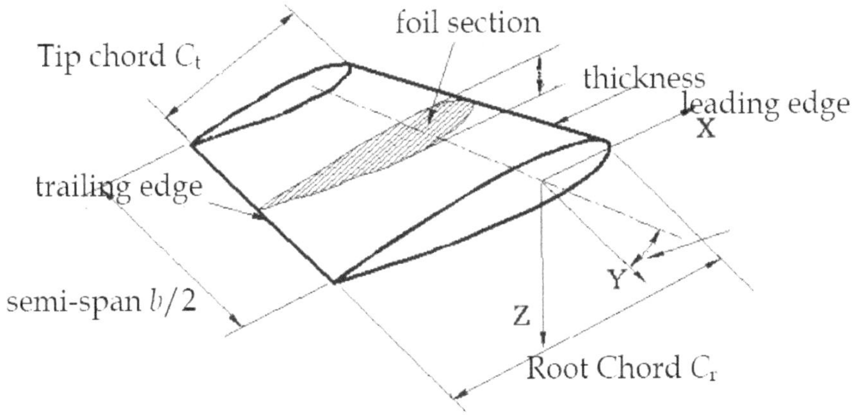

Fig. 13. Rudders parameters

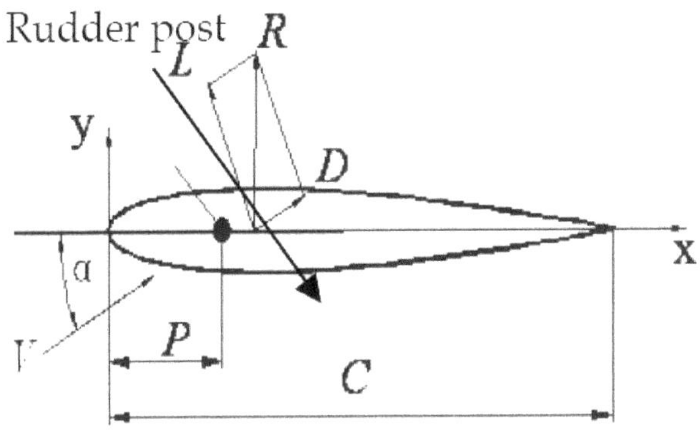

Fig. 14. Foil section and hydrodynamic force

The semi-span, denoted by $b / 2$, measures the distance from the rudder root to tip along the line perpendicular to the root section. The span, in this work, is twice as long as the root-to- tip distance for an isolated plan. The hydrodynamic forces including lift and drag acted on the aerofoil is shown in Figure 14 and can be expressed as

$$L = 1/2\,\rho C_L A V^2 \qquad (11)$$

$$D = 1/2\,\rho C_D A V^2 \qquad (12)$$

Here, ρ is the density of the water; C_L is the lift coefficient; C_D is the drag coefficient; *A* is the area of rudder; *V* is the velocity of water; α is the angle of attack. The rudderpost location is expressed by *P*, which is shown in Figure 14.

4.2 Foil section

The geometry of a rudder is mainly defined by the two-dimensional foil section. The symmetrical foil sections are generally used by the underwater vehicles. Many types of the foil sections are proposed by many countries to improve the hydrodynamic performance. The famous foil sections series include NACA series, НЕЖ series, ЦАГИ series, and JFS series [21], among which the four-digit NACA sections are most widely used for underwater vehicle rudders in that it provides the higher lift and the lower drag. The four-digit NACA section series is a low velocity foil sections series, and have a bigger radius of leading edge and a plumpy head section, which is suitable for the rudder of underwater vehicles at low velocity. In this work, the four digit NACA00×× section was used, where the ×× denote the thickness-to-chord ratio. The lift coefficient and drag coefficient of the foil sections can be calculated as

$$C_L = \frac{L}{1/2\,\rho V^2 C} \tag{13}$$

$$C_D = \frac{D}{1/2\,\rho V^2 C} \tag{14}$$

Here, *L* is the profile lift, *D* is the profile drag, C is the chord. The NACA0008, NACA0012, NACA0016, NACA0020 and NACA0025 are usually used for the rudders of miniature underwater vehicles, their hydrodynamic characteristics were calculated by using computational fluid dynamics. According to the most often adopted velocity of the autonomous underwater vehicles and the velocity of PETREL in AUV mode, the calculation velocity was determined as 2m/s. An example of CFD meshing result is shown in figure 15, where the unstructured mesh was adopted and the wall of section was made dense. The calculating results were shown in the Figure 16~ Figure 18

The relationship of lift coefficient and angle of attack is illustrated in Figure 16, where we can see that there was a bigger angle of stalling and bigger maximal lift coefficient when the section becomes much thicker. From the figure 17 we can see that the thinner wing section has a lower drag cofficient when the angle of attack is small, but the thicker wing section has a lower drag cofficient when the angle of attack is bigger than a certain critical angle of attack. The NACA0008 section has the maximal L/D and NACA0025 has the minimal L/D than other sections which is shown in the Figure 18. The NACA0012 section with angle of stall about 20° and a higher L/D was adopted by the Hybrid glider PETREL.

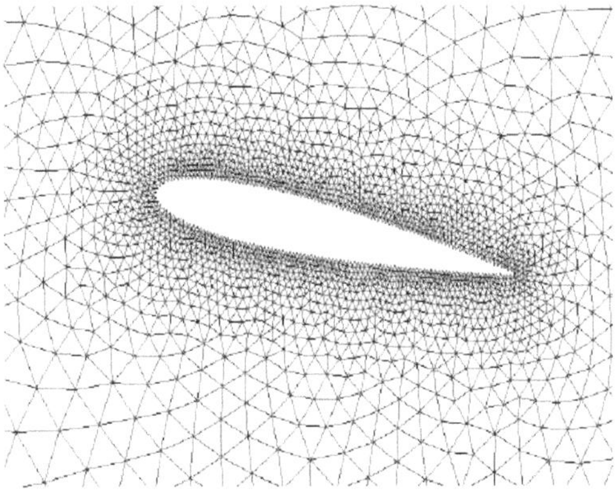

Fig. 15. CFD meshing results

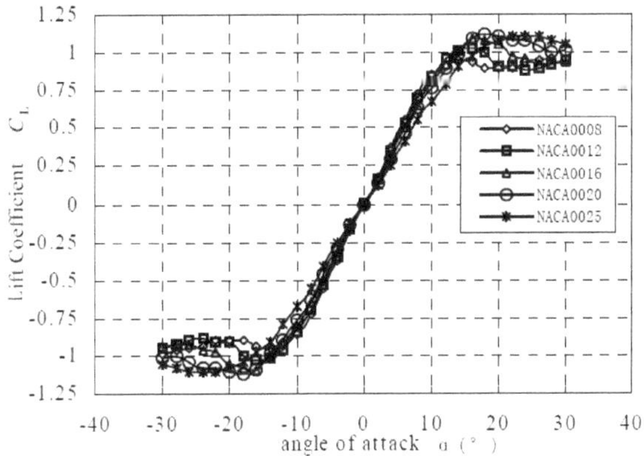

Fig. 16. The relationship of profile lift coefficient and angle of attack

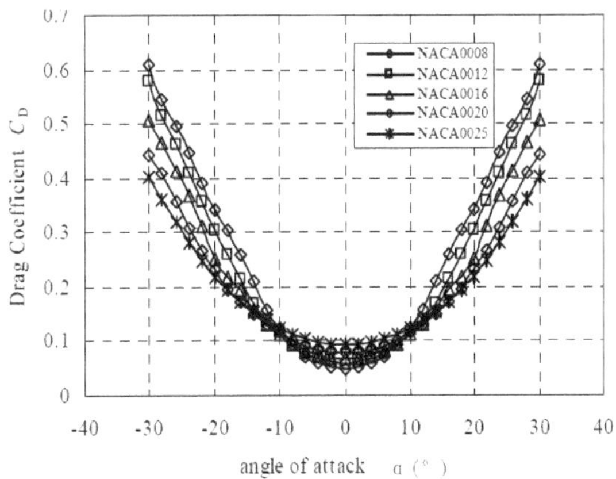

Fig. 17. The relationship of profile drag coefficient and angle of attack

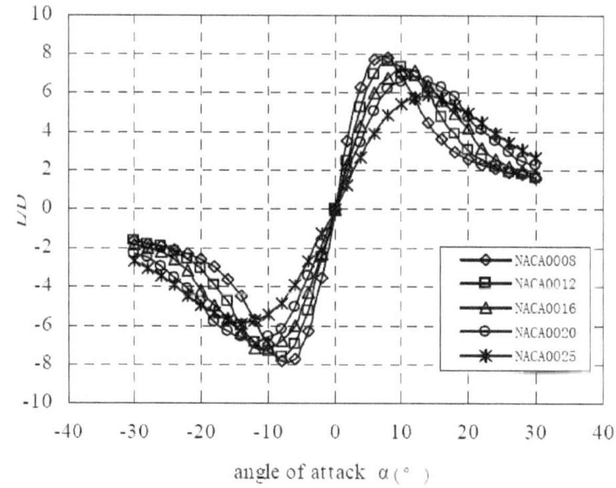

Fig. 18. The relationship of L/D and angle of attack

4.3 Area of rudder calculation

The rudder's area, a crucial factor for the mobility of the underwater vehicle, is associated with the dimensions and configuration of the body. The rudder area can be designed using the trial-and-error approach, master model method, and empirical formula design method. The control surfaces for the highly maneuverable vessel might be constructed in accordance with the rudder sizing regulations established by Det Norske Veritas (DNV) [24].

$$Area = \frac{DL}{100}[1 + 25(\frac{B}{L})^2] \tag{15}$$

Here, *D* is the diameter of the vehicle, *L* is the length of the vehicle, B is the width of the vehicle, and $B = D$ for revolution body. It suggested 30% increase in area if rudders in front of the propeller, and then increased by an additional 50% to match empirical data from other underwater vehicles by the DNV rules. The turn diameter induced by single rudder is about triple-length of the vehicle in terms of the design by DNV rules. The rudder design for the hybrid glider PETREL is shown in Figure 19 and the parameters of the rudder shown in table 7.

Fig. 19. The photo of the rudder of PETREL

Table 7. The parameters of the rudder

parameters	Tip chord C_t	Root chord C_r	Semi-span $b/2$	section
Value	125mm	200mm	120mm	NACA0012

4.4 Hinge moment analysis

The hinge moment is generated by a hydrodynamic force acting around the hinge line of a control surface.The hinge moment must be surmounted when steering, affecting the maneuverability of the underwater vehicle. The increased hinge moment will reduce the turning velocity of the rudders, resulting in sluggish control responses.

The hydrodynamic performance of three dimension rudders at different angles of attack was simulated by using CFD methods. The inlet velocity was set to be 2m/s.

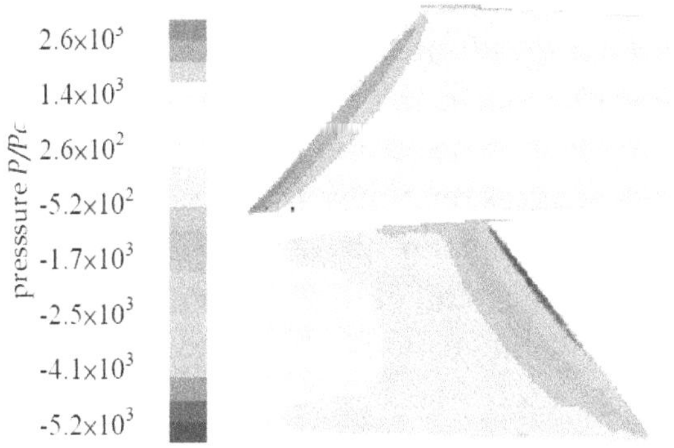

Fig. 20. Pressure distribution chart when angle of attack is 20°

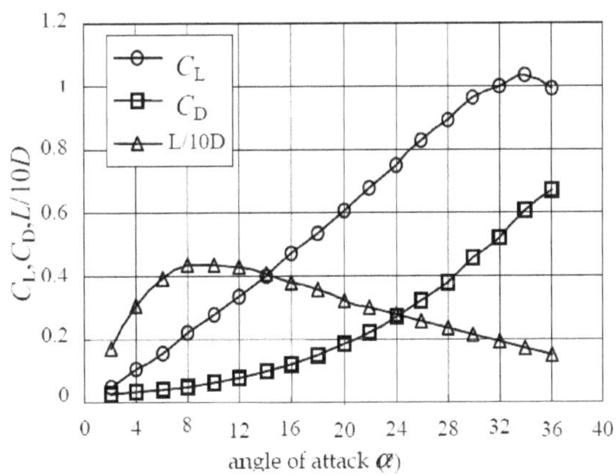

Fig. 21. C_L, C_D and L / 10D variation curve with different angles of attack

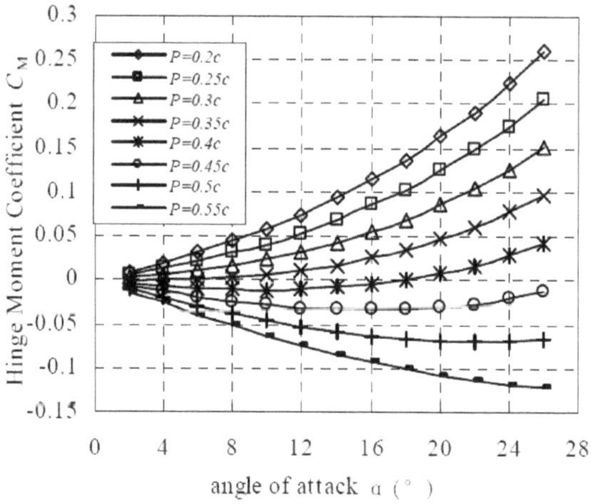

Fig. 22. Hinge moment with different angles of attack

The pressure distribution of the rudder is illustrated in Figure 20, where we can find that there is higher pressure on the front flow face and was local higher pressure area on the back flow face of the tail, that means there exist roundabout flow at the tail of the rudder. Figure 21 shows the relationship between lift, drag and angle of attack. The relationship between *L/D* and angle of attack is also illustrated in the figure 21, the L/D value reduces ten times for the same scale with other two curves. It can be known that the maximal lift to drag ratio was about 8° and the angle of stall about 34°, so the angles of stall of three dimensional rudders are greater than two-dimension section. The hinge moment of rudders with different axis of rudder position is shown in Figure 22, where we can seen that the hinge moment varied with the angle of attack. The hinge moments are little while P = 0.4c for the rudder we design no matter how the angle of attack changed.

4.5 Results and discussion

Aiming at the key problems of the rudder design for autonomous underwater vehicle, the hydrodynamic characteristic of the NACA00xx series section at different angles of attack were simulated when velocity was 2m/s by using the two-dimensional computational fluid dynamics (CFD). For the rudder we design, the stall angle is about 34° for the three dimensional rudders and about 20° for the two-dimensional foil section, so the angle of stall of three dimensional rudders are greater than two-dimension foil section. The area of the rudder of PETREL was calculated using the DNV rules; The hinge moments are little when $P = 0.4c$ for the rudder we design no matter how the angle of attack changed.

5. SHROUD HYDRODYNAMIC EFFECTS ANALYSIS[25]

For the PETREL, the propeller plays a significant role in the vehicle's hydrodynamic performance, so analysis of the hydrodynamic effect of a propeller with a shroud on a winged HUG was performed with Fluent Inc.'s (Lebanon,New Hampshire) CFD software FLUENT6.2.

5.1 Models description

To analyze the effects of the shroud, two simulations were performed, where one model is with the shroud and the other without. Two models are shown in Figure 23.

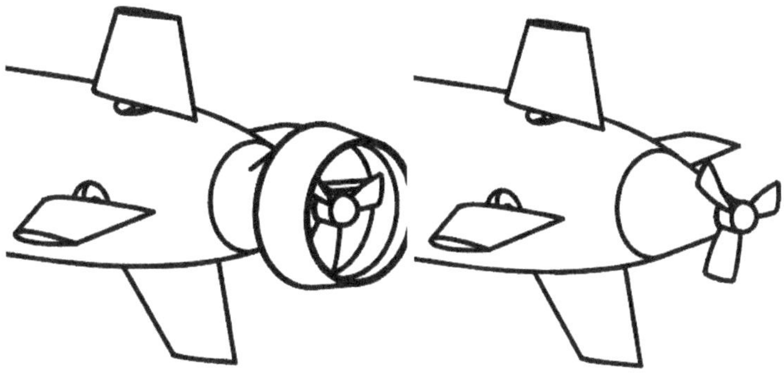

(a) model 1 (b) model 2

Fig. 23. The models studied in the paper

5.2 Effect of shroud on the glide drag

The drag on the vehicle can be expresses as equation (16).

$$D = \frac{1}{2}\rho V^2 C_D A \quad (16)$$

Where, D is the force of drag in Newton, ρ is the density of water in kg/m^3, V is the velocity of the vehicle in m/s, A is the reference area in m^2, C_D is the drag coefficient (dimensionless). The reference area A of the PETREL is 0.096m^2.

Figure 24 shows the overall drag of the two models in the glide mode. The propeller in this mode doesn't rotate. The overall drags of two models are calculated by CFD firstly and then are fitted by the semi-empirical formulae (16). The drag coefficients of two models are respectively 0.32 and 0.26. The average relative error of overall drag between CFD and semi- empirical formulae is 4.7%. The overall drag increase 21%-26% with the propeller shroud compared with the model two according to the CFD computation results, so the shroud greatly increased the drag of the hybrid in glide mode. The drag components of the model1 at the speed of 0.5m/s without angle of attack are shown in Fig. 25. The drag on the body, rudders and wings is mainly viscous forces, while the drags on the propeller, shroud and GPS antenna pole are primarily the pressure forces,. As shown in Figure 26, the propeller and its shroud make up over 30% of total resistance and the percentage will increase with the increment of the velocity. The reason for the high percentage is because of the great pressure drags on the shroud in the glide mode. The local velocity streamline diagram near the shroud of model one shown in the Figure 27. In the Figure, we can see that v_{in} and P_{in} are the velocity and pressure inside the shroud of water, v_{out} and P_{out} are the velocity and pressure outside the shroud of water. Because the propeller doesn't rotate in the glide mode, the velocity of water inside the shroud is slower than that outside the shroud, so there exits $v_{out} > v_{in}$. According to the Bernoulli equation there was $P_{in} > P_{out}$, so a pressure force f is produced by the pressure difference. The percentage of the shroud drag to total resistance is 26%-35% at the different speed due to the pressure force in the glide mode.

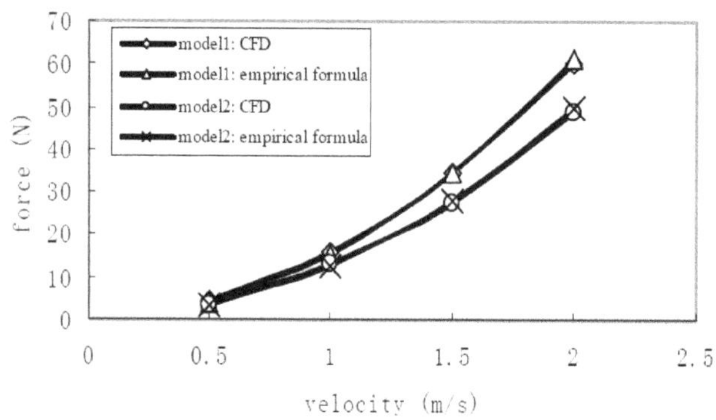

Fig. 24. The overall drags of the two models at difference velocities

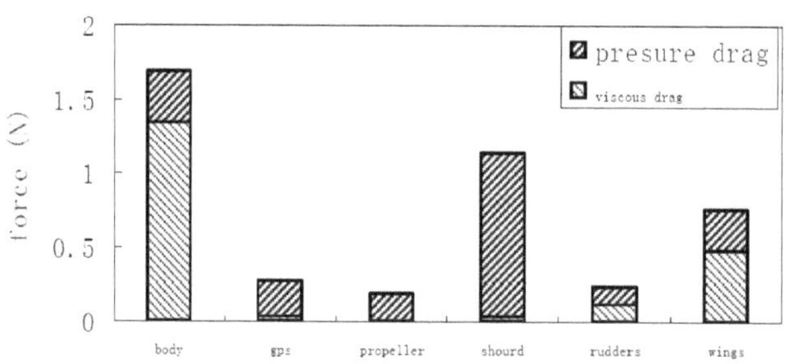

Fig. 25. The drag components of the mode 1 at the speed 0.5m/s

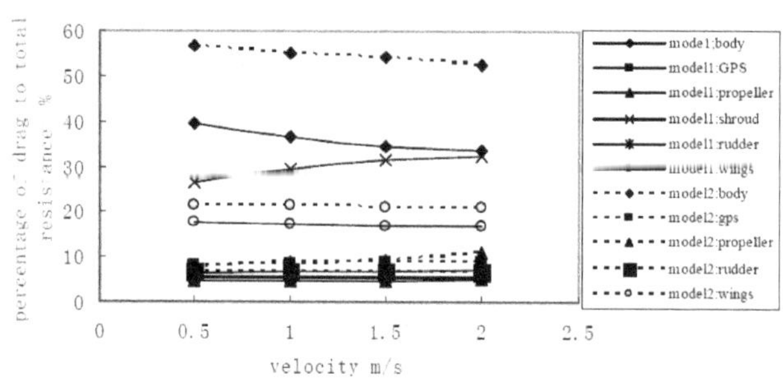

Fig. 26. The drag distribution of vehicle at the different velocities

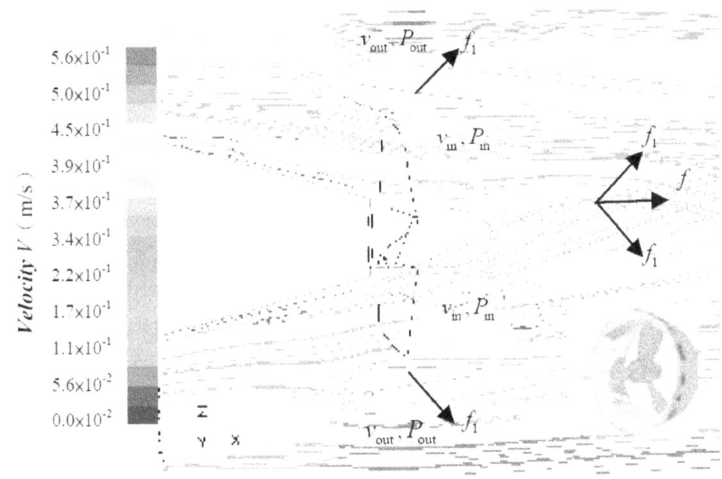

Fig. 27. The local velocity streamline diagram of model1 (v = 0.5m / s)

5.3 Effect of shroud on the glide efficiency

The specific energy consumption can be defined using classical aerodynamics [7] as

$$E_e = \frac{DU}{Bu} = \frac{Bw}{Bu} = \frac{w}{u} = \frac{D}{L} = \frac{C_D}{C_L} \tag{17}$$

Underwater gliders will have a higher glide efficiency when E_e is lower. So the lift to drag ratio *L/D* is a measure of glide efficiency, where bigger values represent higher glide efficiency [7].

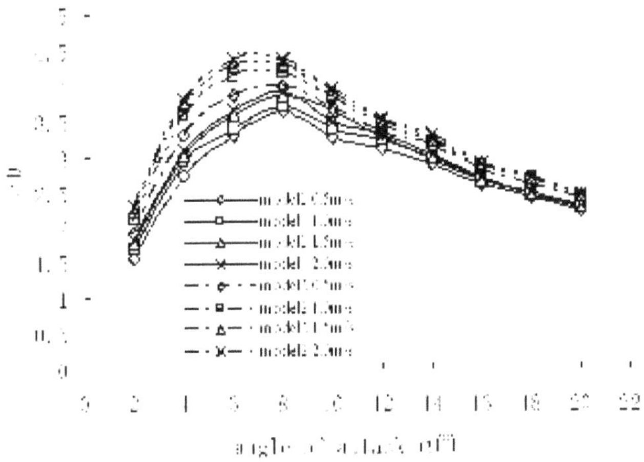

Fig. 28. Lift-to-drag ratio versus angle of attack

The Lift-to-drag ratio versus angle of attack is plotted in Fig. 28, the relations of model one is indicated by the solid lines. The Lift-to-drag ratio of model one is lower than the model two at different angles of attack, that means the vehicle with the shroud will have a lower glide efficient than that without. The Lift-to-drag ratio of model one is less than model two by 20% to 5% for the varied angles of attack within the range from 2°to 20°. The maximum lift-to-drag ratio occurred at the angle of attack 6°-8°for both the models at different speed.

5.4 Effect of shroud on the glide stability

The underwater gliders usually are designed for static stability[17], the dimensionless hydrodynamic moment arm l'_a often used to represent the static stability of the underwater vehicles motion. The equations of the l'_a are shown in equations (7) and (8).

Existing oceanographic gliders are designed to be static stable in steady glides for the easy control and high energy economy. The hybrid-driven underwater glider PETREL was designed as static stability for the high energy economy in the glide mode.

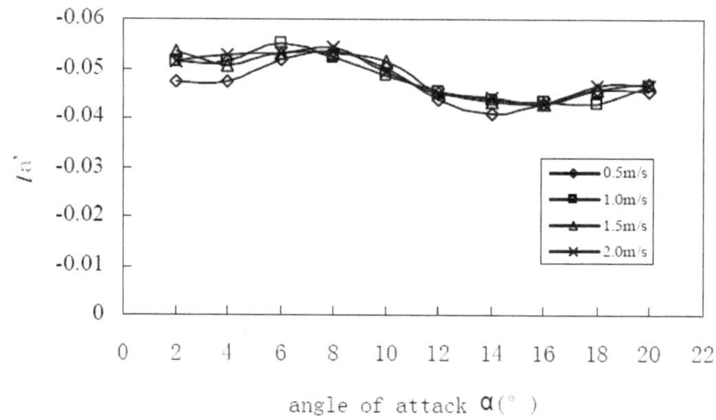

Fig. 29. The static stability coefficient l'_a versus angle of attack of model one

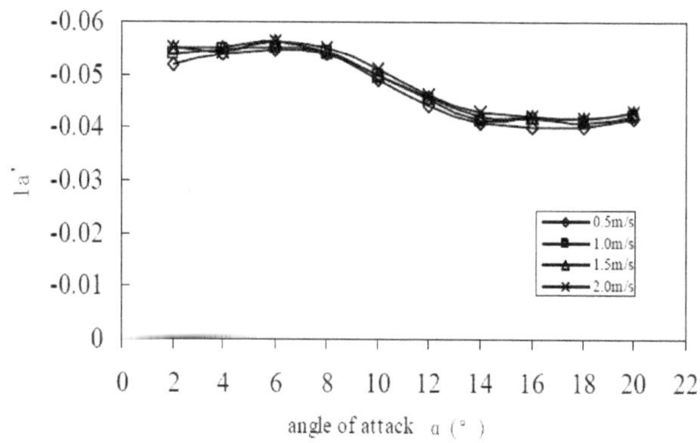

Fig. 30. The static stability coefficient l'_a versus angle of attack of model two

Figure 29 show the static stability coefficient I'_a versus angle of attack of model one and model two. It is static stability for both of the two models in terms of our design intention. The stability decreases when the angle of attack gets bigger than 8°, but the stability slightly increases for model one when the angle of attack is more than 12°. The glide speed has little effect on the stability as shown in the Figure 29 and Figure 30. Figure 31 shows the moment of the shroud versus angle of attack of model one. The values of the moment were positive when the angle of attack is lower than 8° for the v = 0.5 m/s and v = 1 m/s, and the angle of attack is less than 10° for the v = 1.5 m/s and v = 2.0 m/s. The values of the moment were negative when the angle of attack gets higher than those critical angles. So the effect of shroud on the static stability of model one is that, when the angle of attack is lower than the critical angle the shroud will makes the stability decreasing but makes the stability increasing when the angle of attack is higher than the critical angle. as shown in the Figure 31, the action of the shroud makes the stability slightly increased when the attack angle is higher than 12° .

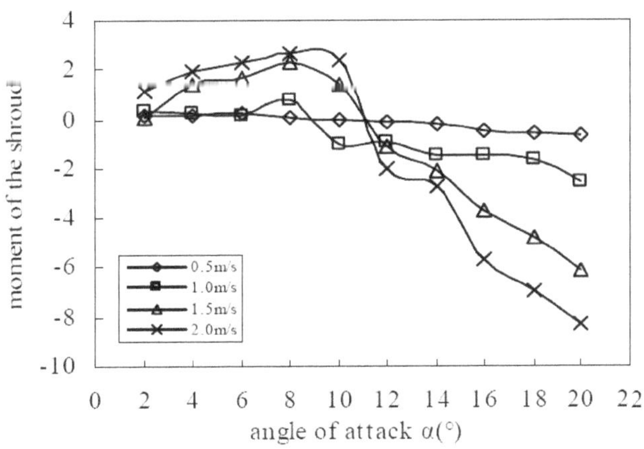

Fig. 31. The moment of the shroud versus angle of attack of model one

5.5 Conclusions

It was found that overall drag increased by 21 to 26 percent for the model with the propeller shroud compared with the one without a shroud, but with the same structure and size, the shroud's resistance is mainly pressure force.

The shroud made the lift-to-drag of the vehicle in glide mode decrease by as much as 20 percent when the angle of attack was 2º. As the angle of attack increased, the shroud's effect was minimized, and the decrease in lift-to-drag ratio ranged down to five percent at an angle of attack of 20º, meaning glide efficiency decreased due to the propeller shroud.

Finally, the shroud decreases the stability of the HUG when the angle of attack is lower than the critical angle, but increases it when the angle of attack is higher than the critical angle. The critical angle is between 8º and 10º for velocities lower than one meter per second, and between 10ºand 12ºfor velocities in the range of one to two meters per second.

These findings indicate that for an underwater glider, the shroud will increase drag and decrease the glide efficiency, but it is good for stability when the angle of attack is larger than 8º. Therefore, the shroud is not a successful design element for the HUG in glide mode, but in propeller mode the shroud can increase the thrust of the vehicle.

Using CFD to analyze the shroud's hydrodynamic effects shows that the vehicle should only be equipped with this feature for activities requiring operation in propeller mode.

6. FLOW FIELD ANALYSIS

6.1 Velocity field

The direct route flow field with the velocity of the hybrid underwater glider PETREL at 0.5m/s、1m/s、1.5m/s and 2m/s was simulated by using CFD ways. The simulation results are shown in Figure 32.

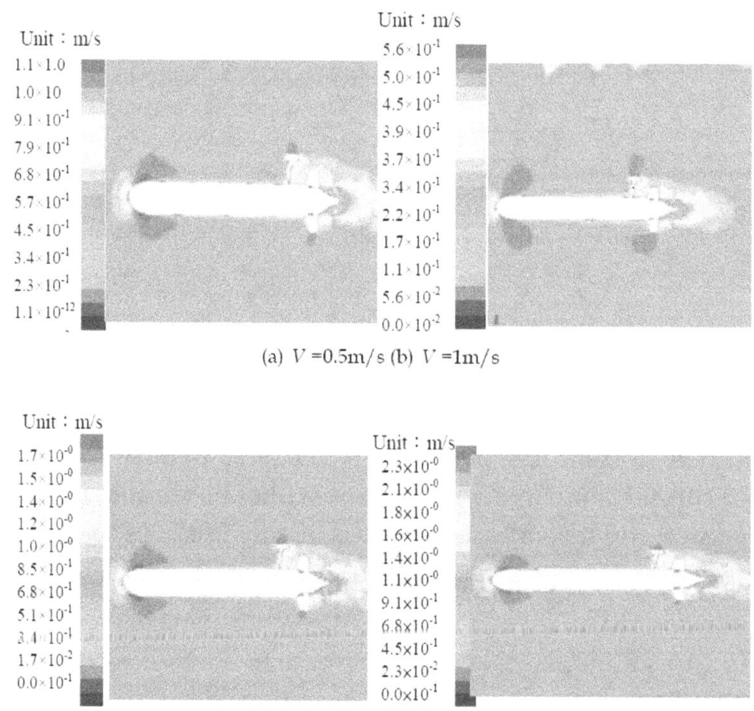

(a) V =0.5m/s (b) V =1m/s

(c) V =1.5m/s (d) V =2.0m/s

Fig. 32. The flow field at different velocity

It is seen that the flow field patterns in the figures are nearly the same. There was high flow rate region near the abrupt curve surfaces of the vehicle head, ballast of the GPS, rudders, while there was also the low flow field domain on the front of those parts and near the tail of the vehicle. The high flow rate region area decreases as the velocity increases. The existence of the mast of GPS makes the flow field behind it disturbed, and makes the flow field asymmetrical. These changes will increase the drag and hydrodynamic moment on the vehicle.

The steady turning flow field in longitudinal vertical and horizontal plane with the velocity of vehicle at 0.5m/s, is shown in figure 33.

It is noted from Figure 33 that the pattern of the steady turning flow field in longitudinal vertical plane and in horizontal plane has notability difference. Due to the rotational speed, the flow field is obviously asymmetric and appears large scale high flow rate region and low flow rate region in the back of the field. An extra hydrodynamic moment is induced because of the asymmetry of the flow field.

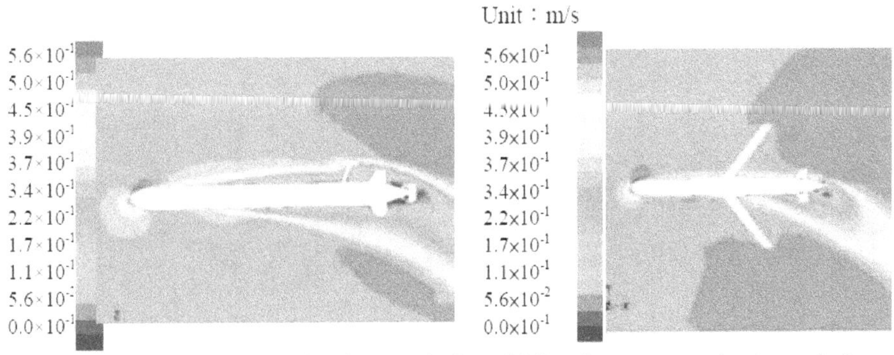

(a) Steady turning in longitudinal vertical plane (b) Steady turning in horizontal plane

Fig. 33. The steady turning flow field

6.2 Pressure distribution

The pressure distributions on the vehicle at the speeds 0.5m/s and 2m/s when the angle of attack α is zero are shown in Figure 34 and Figure 35. There has a tendency that the pressure on the vehicle gradual reduction from head to tail of the vehicle, a high pressure region on the head and a low pressure region on the tail, which induced the pressure drag on the vehicle. The pressure of the high pressure region become higher and the low pressure region become lower with the speed of the vehicle increasing, it means that the pressure drag on the vehicle increase with the speed increasing. It can be known from the pressure distribution on the propeller shroud that pressures drag act on the shroud because of there has higher pressure inside the shroud and lower pressure outside the shroud. The reason for thus pressure distribution is that the propeller doesn't rotating in the glide mode which makes the velocity of flow inside the shroud slower than the outside. So the shroud should be removed or the profile changed to reduce the drag on the vehicle in glide mode.

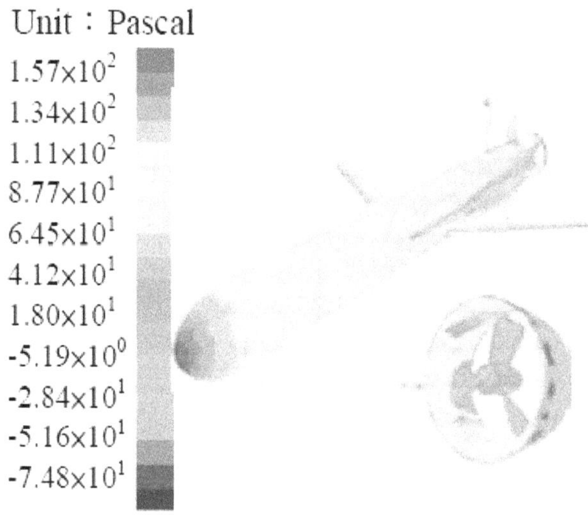

Fig. 34. Pressure distribution (V =0.5m/s)

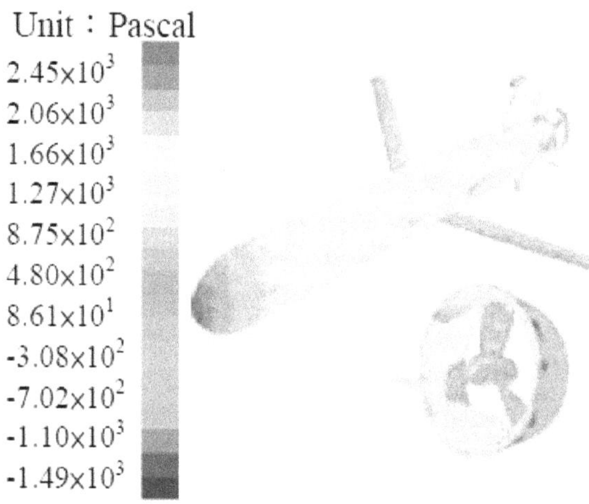

Fig. 35. Pressure distribution (V =2m/s)

The pressure distributions on the vehicle at the speed 0.5m/s when the angle of attack α isn't zero are shown in Figure 36. The pressure distribution on the vehicle isn't symmetry, the pressure of front flow surface higher than back flow surface, when glide with an angle of attack. The wing has the biggest degree of asymmetry of the pressure distribution which makes the wings the main lift generating parts. The asymmetry of the pressure distribution on the vehicle also induces the hydrodynamic moment on the vehicle.

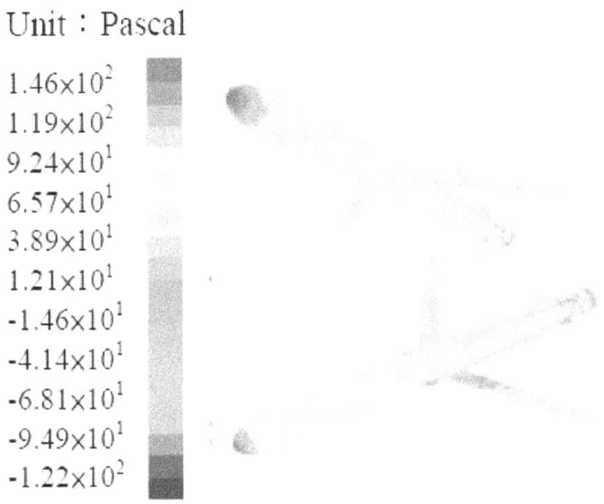

Fig. 36. Pressure distribution (V =0.5m/s, $\alpha = 6$)

7. CONCISE SUMMARY

This chapter examines the hydrodynamic impacts of the primary components of a hybrid-driven underwater glider, particularly during glide mode, and analyzes the simulation results of the three principal hydrodynamic components utilizing computational fluid dynamics (CFD) methodologies. This paper utilized FLUENT 6.2, a CFD software developed by Fluent Inc. in Lebanon, New Hampshire. The primary conclusions are: The glide efficiency is predominantly impacted by the chord length, whereas the vehicle's stability is primarily affected by the sweep angle. Additionally, the position of the wings chiefly impacts glide stability, with minimal effect on glide efficiency. The vehicle achieves the optimal lift-to-drag ratio at an approximate angle of attack of 6°. The endurance of the hybrid glider with the same configuration as PETREL will decrease by 10%~35% compared with the legacy gliders.

For the rudder we design, the angle of stall is about 34° for the three dimensional rudders and about 20° for the two-dimensional foil section, so the angle of stall of three dimensional rudder is greater than two-dimension foil section. The area of the rudder of PETREL was calculated using the DNV rules; The hinge moments are little when P = 0.4c for the rudder we design no matter how the angle of attack changes.

It was found that overall drag increased by 21 to 26 percent for the model with the propeller shroud compared with the one without a shroud, but with the same structure and size, the shroud's resistance is mainly pressure force. The shroud made the lift-to-drag of the vehicle in glide mode decrease by as much as 20 percent when the angle of attack was 2º. As the angle of attack increased, the shroud's effect was minimized, and the decrease in lift-to-drag ratio ranged down to five percent at an angle of attack of 20º, meaning glide efficiency decreased due to the propeller shroud. Finally, the shroud decreases the stability of the HUG when the angle of attack is lower than the critical angle, but increases it when the angle of attack is higher than the critical angle. The critical angle is between 8º and 10º for velocities lower than one meter per second, and between 10ºand 12ºfor velocities in the range of one to two meters per second.

The results demonstrate that the underwater glider's shroud will augment drag, diminish glide efficiency, yet enhance stability when the angle of attack exceeds 8º. Consequently, the shroud is ineffective as a design element for the HUG in glide mode; nevertheless, it can enhance the vehicle's thrust in propeller mode. CFD analysis of the shroud's hydrodynamic effects indicates that the vehicle should be outfitted with this feature solely for operations necessitating propeller mode. The velocity field and pressure distribution of the hybrid glider PETREL were examined, elucidating the impact of these primary components on the vehicle's hydrodynamic characteristics.

REFERENCES

1. C. C. Eriksen, T. J. Osse, R. D. Light, T, et al, (2001)"Sea glider: A long range autonomous underwater vehicle for oceanographic research," *IEEE Journal of Oceanic Engineering,* Vol. 26, 2001, pp. 424–436
2. J. Sherman, R. E. Davis, W. B. Owens, et al , "The autonomous underwater glider "Spray"," *IEEE Journal of Oceanic Engineering,* vol.26, 2001, pp. 437–446
3. D. C. Webb, P. J. Simonetti, C. P. Jones, "SLOCUM, an underwater glider propelled by environmental energy," *IEEE Journal of Oceanic Engineering,* vol.26, 2001, pp. 447– 452
4. R. E. Davis, C. C. Eriksen and C. P. Jones, "Autonomous Buoyancy-driven Underwater Gliders," The Technology and Applications of Autonomous Underwater Vehicles. G.. Griffiths, ed., Taylor and Francis, London, 2002, pp. 37-58
5. R. Bachmayer, N. E. Leonard, J. Graver, E. Fiorelli, P. Bhatta, D. Paley, "Underwater gliders: Recent development and future applications," Proc. *IEEE International Symposium on Underwater Technology* (*UT'04*), Tapei, Taiwan, 2004, pp.195- 200
6. D. L. Rudnick, R. E. Davis, C. C. Eriksen, D. M. Fratantoni, and M. J. Perry, "Underwater gliders for ocean research, " *Marine Technology Society Journal,* vol. 38, Spring 2004, pp. 48–59
7. S. A. Jenkins, D. E. Humphreys, J. Sherman, et al. "Underwater glider system study." *Technical Report,* Office of Naval Research, 2003

8. Marthiniussen, Roar ; Vestgård, Karstein ; Klepaker, Rolf Arne ; Strkersen, Nils, Fuel cell for long-range AUV, *Sea Technology,* vol. 38, pp. 69-73
9. Blidberg, D.Richard ,Solar-powered autonomous undersea vehicles. *Sea Technology,* vol. 38, pp. 45-51
10. Ferguson, J,Pope, A.,Butler, B.1, Verrall, R.I.1 ,Theseus AUV-Two record breaking missions, *Sea Technology,* vol. 40, pp. 65-70
11. Albert M. Bradley, Michael D. Feezor, Member, IEEE, Hanumant Singh, and F. Yates Sorrell, Power Systems for Autonomous Underwater Vehicles, *IEEE JOURNAL OF OCEANIC ENGINEERING,* VOL. 26, NO. 4, OCTOBER 2001,pp.526-538
12. Amit Tyagi*, Debabrata Sen, Calculation of transverse hydrodynamic coefficients using computational fluid dynamic approach, *Ocean Engineering 2006 (33),* pp. 798–809
13. Douglas E.H, Correlation and validation of a CFD based hydrodynamic &dynamic model for a towed undersea vehicle, OCEANS, 2001,Vol3, pp.1755-1760
14. Li Wanping. (2004).*Computational fluid dynamics.* Huazhong University of Science and Technology Press, 978-7-5609-3214-9, Wuha (in Chinease)
15. Wu Wangyi. (1983).*Fluid mechanics (volume I).* Higher Education Press, 7301001991,Beijin. (in Chinease).
16. Wu Ziniu. (2001) .The basic principles of computational fluid dynamics, science press, 7-03-008128-5, Beijin. (in Chinese)
17. Graver J G. *Underwater gliders: dynamics, control and design.* The USA: Princeton University, 2005
18. Wu Jianguo, Chen Chaoyin, Wang Shuxin, et al, Hydrodynamic Characteristics of the Wings of Hybrid-Driven Underwater Glider in Glide Mode, *Journal of Tianjin University,* 2010,Vol43,No.1,pp.84-89, (in Chinease)
19. Jenkins S A, Humphreys D E, Sherman J, et al, Alternatives for enhancement of transport economy in underwater gliders, *IEEE Proceedings of OCEANS,* 2003. 948- 950
20. Wu Baoshan, Xing Fu, Kuang Xiaofeng, et al, Investigation of hydrodynamic characteristics of submarine moving close to the sea bottom with CFD methods, *Journal of Ship Mechanics,* 2005, Vol.9,No.3,pp. 19-28
21. Shi Shengda. 1995. *Submarine Maneuverability.* National Defense Industry Press, 9787118013498,Beijing(in Chinese).
22. Wu Jianguo, Zhang Hongwei, Design and research on the rudder of Mini-type AUV, *ocean technology,* 2009, Vol3, No.2, pp.5-8.

23. P.M.Ostafichuk, *AUV hydrodynamics and modeling for improve control,* Canada: University of British Columbia, 2004

24. Timothy Curtis, B.Eng, The design, conatruction, outfitting, and preliminary testing of the C-SCOUT autonomous underwater vehicle (AUV), Canada, Faculty of *Engineering and Applied Science Memorial University of Newfoundland,* 2001

25. Jianguo Wu, Chaoying Chen and Shunxin Wang, Hydrodynamic Effects of a shroud Design For a Hybrid-Driven Underwater Glider, *Sea Technology,* 2010, Vol.51, No.6,pp.45-47

Chapter 4

REAL-TIME OPTIMAL NAVIGATION AND OBSTACLE EVASION FOR UNMANNED MARINE VEHICLES

1. INTRODUCTION: AN OVERVIEW

The single most important near-term technical challenge of developing an autonomous capability for The purpose of unmanned vehicles is to evaluate and react suitably to proximate objects in their trajectory. For unmanned aerial vehicles (UAVs), the near field may stretch several nautical miles in all directions, whereas for unmanned ground and maritime vehicles, it may only cover a few dozen yards straight front of the vehicle. However, while formulating obstacle avoidance (OA) maneuvers, it is frequently essential to incorporate a level of rational planning that transcends merely adjusting the vehicle's trajectory reactively. For unmanned maritime vehicles (UMVs), the capacity to produce near-optimal operational area trajectories in real time is particularly crucial during sidescan sonar surveys in cluttered environments (e.g., kelp forests or coral reefs), operations in constrained waterways (e.g., rivers or harbors), or executing feature-based, terrain-relative navigation, among others. A principal aim of sidescan sonar surveys is to achieve complete area coverage while preventing harm to the survey apparatus. A real-time trajectory generator should ideally minimize deviations from the predetermined survey geometry while enabling the vehicle to retarget locations overlooked owing to prior operational maneuvers. Likewise, for operations in constrained waterways, efficient OA trajectories must integrate all available environmental data, including terrain, bathymetry, and water currents. In general, this OA capability should be integrated into an onboard planner or trajectory generator that computes optimal (or near-optimal) viable trajectories more rapidly than real time. The planner for unmanned undersea vehicles (UUVs) must be able to generate comprehensive three-dimensional (3D) trajectories; however, certain applications may necessitate restricting the planner's output to two dimensions (2D) for vertical or horizontal operating modes. For unmanned surface vehicles (USVs), the latter scenario is the exclusive operational mode.Consider a typical hardware setup consisting of a UUV augmented with an autopilot (Fig.1). The autopilot not only stabilizes the overall system, but also enables vehicle control at a higher hierarchical level than simply changing a throttle setting δT (t) , or deflecting stern plane δ s (t) or rudder δr (t) angles.

In Fig.1, $\boldsymbol{x}^{WP}$, $\boldsymbol{y}^{WP}$, $\boldsymbol{z}^{WP}$ are the vectors defining *x*, *y*, and *z* coordinates of some points in the local tangent (North-East-Down (NED)) plane for waypoint navigation. Alternatively a typical autopilot may also accept some reference flight-path angle γ (t) (or altitude/depth) command and heading Ψ(t) (or yaw angle ψ (t)), respectively. The motion sensors, accelerometers, and rate gyros measure the components of inertial acceleration, $\ddot{x}_I(t)$, $\ddot{y}_I(t)$ and $\ddot{z}_I(t)$, and angular velocity - roll rate p(t) , pitch rate q(t), and yaw rate r(t).

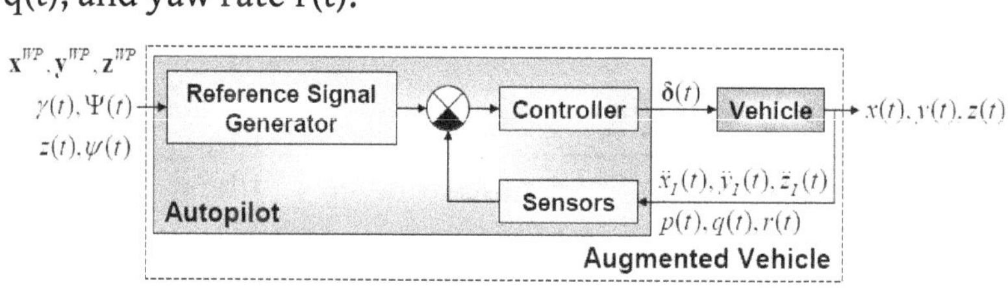

Fig. 1. A. UUV augmented with an autopilot

A trajectory generator would consider an augmented UUV as a new plant (Fig.2) and provide this plant with the necessary inputs based on the mission objectives (final destination, time of arrival, measure of performance, etc.). Moreover, the reference signals, γ (t) and Ψ(t) , are to be computed dynamically (once every few seconds) to account for disturbances (currents, etc.) and newly detected obstacles.

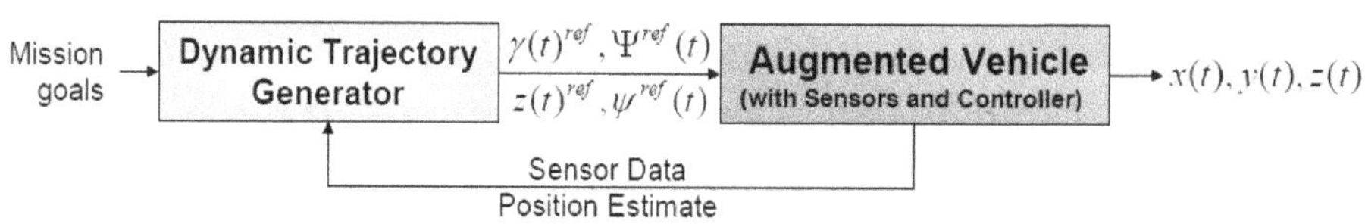

Fig. 2. Providing an augmented UUV with a reference trajectory

Ideally, the trajectory generator software should also produce the control inputs δ^{ref} (t) corresponding to the feasible reference trajectory (Fig.3) (Basset et al., 2008). This enhanced setup assures that the inner-loop controller deals only with small errors. (Of course this setup is only viable if the autopilot accepts these direct actuator inputs.)

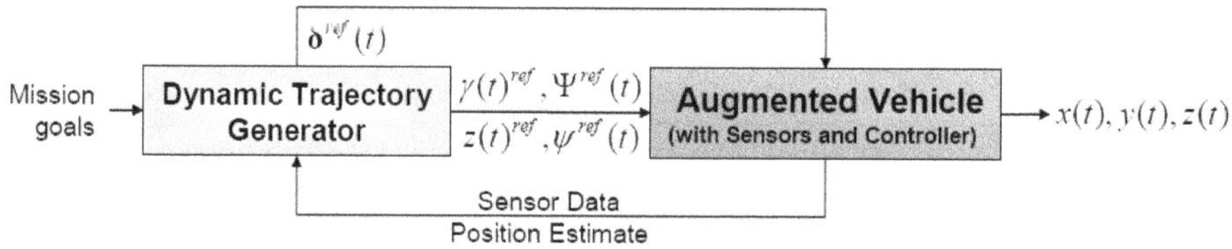

Fig. 3. Providing an augmented UUV with the reference trajectory and reference controls

This chapter aims to elucidate the dynamic trajectory generator created at the Naval Postgraduate School (NPS) for the unmanned marine vehicles (UMVs) of the Center for Autonomous Vehicle Research (CAVR) and to demonstrate the foundational structure of the OA framework based on it. Section 2 delineates a comprehensive viable trajectory generation challenge, whereas Section 3 presents the foundational concepts of the suggested framework for addressing this issue, which employs the inverse dynamics in the virtual domain (IDVD) methodology. Section 4 examines the simplifications that arise from constraining the general spatial problem to two planar subcases. Section 5 delineates the REMUS UUV and SeaFox USV, together with their forward-looking sonar (FLS) systems utilized for oceanographic research at NPS. Section 6 discusses path-following principles and practical implementation details for tracking nonlinear trajectories using standard vehicle autopilots. Section 7 delineates findings from computer simulations and field studies across many scenarios that benefit from accelerated calculation of near-optimal routes.

2. PROBLEM FORMULATION

Let us consider the most general case and formulate an optimization problem for computing collision-free trajectories in 3D (it can always be reduced to a 2D problem by eliminating two states). We will be searching within a set of admissible trajectories described by the state vector

$$\mathbf{z}(t)=\left[x(t),y(t),z(t),u(t),v(t),w(t)\right]^T\in S,\ S=\left\{\mathbf{z}(t)\in Z^6\subset E^6\right\},\ t\in\left[t_0,t_f\right] \tag{1}$$

where the components of the velocity vector – surge u, sway v, and heave w, defined in the body frame {b} – are added to the UUV NED coordinates x, y and z ($z = 0$ at the surface and increases in magnitude with depth). While many UUVs are typically programmed to operate at a constant altitude above the ocean floor, it is still preferable to generate vertical trajectories in the NED local tangent plane because the water surface is a more reliable absolute reference datum than a possibly uneven sea floor. In general, however, it is a trivial matter to convert the resulting depth trajectory z(t) to an altitude trajectory h(t) for vehicles equipped with both altitude and depth sensors. Section 6.2 describes such practical considerations in detail.

The admissible trajectories should satisfy the set of ordinary differential equations describing the UUV kinematics

$$\begin{bmatrix}\dot{x}(t)\\ \dot{y}(t)\\ \dot{z}(t)\end{bmatrix}={}^{u}_{b}R\begin{bmatrix}u(t)\\ v(t)\\ w(t)\end{bmatrix} \tag{2}$$

In (2) ${}^{u}_{b}R$ is the rotation matrix from the body frame {b} to the NED frame {u}, defined using two Euler angles, pitch $\theta(t)$ and yaw $\psi(t)$, and neglecting a roll angle as

$$ {}^{u}_{b}R(t)=\begin{bmatrix}\cos\psi(t)\cos\theta(t) & -\sin\psi(t) & \cos\psi(t)\sin\theta(t)\\ \sin\psi(t)\cos\theta(t) & \cos\psi(t) & \sin\psi(t)\sin\theta(t)\\ -\sin\theta(t) & 0 & \cos\theta(t)\end{bmatrix} \quad (3)$$

Although we are not going to exploit it in this study, admissible trajectories should also obey UUV dynamic equations describing translational and rotational motion. This means that the following linearized system holds for the vector $\varsigma(t)$, which includes speed components *u, v, w* (being a part of our state vector **z**(t)) and angular rates *p, q, r*:

$$\dot{\varsigma}(t)=\mathbf{A}\varsigma(t)+\mathbf{B}\delta(t) \quad (4)$$

Here **A** and **B** are the state and control matrices and $\delta=[\delta_T,\delta_s,\delta_r]^T$ is the control vector (Healey, 2004).

Next, the admissible trajectories (1) should satisfy the initial and terminal conditions

$$\mathbf{z}(t_0)=\mathbf{z}_0,\ \mathbf{z}(t_f)=\mathbf{z}_f \quad (5)$$

Finally, certain constraints should be obeyed by the state variables, controls and their derivatives. For example, in the case of a UUV these can include obvious constraints on the UUV depth:

$$z_{\min}\le z(t)\le z_{\max} \quad (6)$$

Where $z_{max}(x, y)$ describes a programmed operational depth limit. For vehicles programmed to operate at some nominal altitude above the sea floor, the $z_{max}(x, y)$ constraint can be converted into a minimum altitude Section 6.2.

A 3D OA requirement can be formulated as

$$[x(t);y(t);z(t)]\cap\Re=0 \quad (7)$$

where $\Re$ is the set of all known obstacle locations. The constraints are usually imposed not only on the controls themselves $|\delta|\le\delta_{\max}$ but on their time derivatives as well $|\dot{\delta}|\le\dot{\delta}_{\max}$ to account for actuator dynamics. Knowing the system's dynamics (4) (or simply complying with the autopilot specifications), these latter constraints can be elevated to the level of the reference signals, for instance

$$|\theta(t)| \le \theta_{\max} \text{ and } |\dot{\psi}(t)| \le \dot{\psi}_{\max} \quad (8)$$

The objective is to find the best trajectory and corresponding control inputs that minimize some performance index J. Typical performance index specifications include: i) minimizing time of the manoeuvre $t_f - t_0$, ii) minimizing the distance travelled to avoid the obstacle(s), and iii) minimizing control effort or energy consumption. In addition, the performance index may include some "exotic" constraints dictated by a sensor payload. For example, a UUV may require vehicle trajectories which point a fixed FLS at specified terrain features or minimize vehicle pitch motion in order to maintain level, horizontal flight along a survey track line for accurate synthetic aperture sonar imagery (Horner et al., 2009).

Before we proceed with the development of the control algorithm, it should be noted that quite often the UUV surge velocity is assumed to be constant, $u(t) \equiv U_0$, to provide enough control authority in two other channels. This uniquely defines a throttle setting $\delta_T(t)$, and leaves only two control inputs, $\delta_s(t)$ and $\delta_r(t)$, for altering the vehicle's trajectory. It also allows us to consider matrices A and B in (4) to be constant (time- and states-independent). If this assumption is not required, inverting kinematic and dynamic equations will differ slightly from the examples presented in the next section. However, the general ideas of the proposed approach remain unchanged.

3. REAL-TIME NEAR-OPTIMAL GUIDANCE

For the dynamic trajectory generator shown in Figs. 2 and 3, it is advocated to use the direct method-based IDVD (Yakimenko, 2000). The primary rationale is that this approach features several important properties required for real-time implementations: i) the boundary conditions including high-order derivatives are satisfied *a priori*; ii) the resulting control commands are smooth and physically realizable, iii) the method is very robust and is not sensitive to small variations in input parameters, iv) any compound performance index can be used during optimization. Moreover, this specific method uses only a few variable parameters, thus ensuring that the iterative process during optimization converges very fast compared to other direct methods. The IDVD-based trajectory generator consists of several blocks. The goal of the first block, to be discussed next, is to produce a candidate trajectory, which satisfies the boundary conditions.

3.1 Generating a candidate trajectory

Again, consider the most general case of a UUV operating in 3D (as opposed to a USV). Suppose that each coordinate x_i, $i = 1,2,3$ of the candidate UUV trajectory is represented as a polynomial of degree M of some abstract argument τ, the virtual arc

$$x_i(\tau) = \sum_{k=0}^{M} a_{ik}\tau^k \,, \tag{9}$$

(for simplicity of notation we assume $x_1(\tau) \equiv x(\tau)$, $x_2(\tau) \equiv y(\tau)$ and $x_3(\tau) \equiv z(\tau)$). In general, analytic expressions for the trajectory coordinates can be constructed from any combination of basis functions to produce a rich variety of candidate trajectories. For example, a combination of monomials and trigonometric functions was utilized in (Yakimenko & Slegers, 2009).

As discussed in (Yakimenko, 2000; Horner & Yakimenko, 2007) the degree M is determined by the number of boundary conditions that must be satisfied. Specifically, it should be greater or equal to the number of preset boundary conditions but one. In general the desired trajectory includes constraints on the initial and final position, velocity and acceleration: $x_{i0}, x_{if}, x'_{i0}, x'_{if}, x''_{i0}, x''_{if}$. In this case the minimal order of polynomials (9) is 5, because all coefficients in (9) will be uniquely defined by these boundary conditions, leaving the "length" of the virtual arc τ_f as the only varied parameter. For more flexibility in the candidate trajectory, additional varied parameters can be obtained by increasing the order of the polynomials (9). For instance, using seventh-order polynomials will introduce two more varied parameters for each coordinate expression. Rather than varying two coefficients in these extended polynomials directly, we will vary the initial and final jerk, $x''_{i0}{}'$ and xi'''_{if}, respectively. In this case, coefficients a_{ik} in (9) can be determined by solving the obvious system of linear algebraic equations equating polynomials (9) to $x_{i0}, x_{if}, x'_{i0}, x'_{if}, x''_{i0}, x''_{if}, x''_{i0}{}'$ and $x''_{if}{}'$ at two endpoints ($\tau = 0$ and $\tau = \tau_f$) (Yakimenko, 2000, 2008).

By construction, the boundary conditions (5) will be satisfied unconditionally for any value of the final arc τ_f. However, varying τ_f will alter the shape of the candidate trajectory.

Figure 4 demonstrates a simple example whereby a UUV operating 2m above the sea floor at 1.5m/s must perform a pop-up manoeuvre to avoid some obstacle. Even with a single varied parameter, changing the value of τ_f allows the UUV to avoid obstacles of different heights. Similar trajectories could be produced solely in the horizontal plane or in all three dimensions. It should be pointed out that even at this stage infeasible candidate trajectories will be ruled out. (In Fig.4 the trajectory requiring the UUV to jump out of the water is infeasible because it violates the constraint (6).)

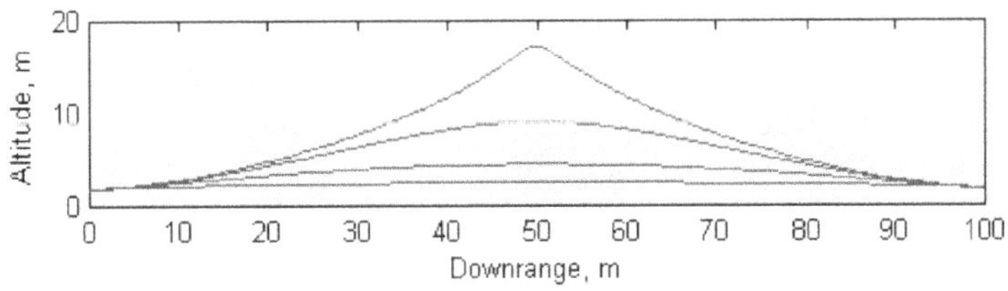

Fig. 4. Varying the candidate trajectory while changing the value of τ_f

With six free parameters, which in our case are components of the initial and final jerk ($x''_{i0}{}'$ and $xi'''_{if}, i = 1,2,3$) the trajectory generator can change the overall shape of the trajectory even further. To this end, Fig. 5 illustrates candidate trajectories for a UUV avoiding a 10m obstacle located between its initial and final points. These trajectories were generated by varying just two components of the jerk, x'''_{30} and x'''_{3f}, and minimizing τ_f. This additional flexibility can produce trajectories which satisfy operational constraints (6), as well as OA constraints (7).

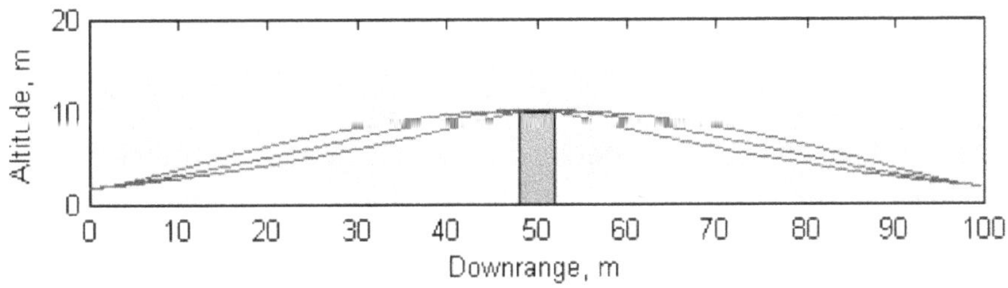

Fig. 5. Candidate trajectories obtained by varying the terminal jerks

The selection of a specific trajectory will be based upon whether the trajectory is feasible (satisfies constraints (8)) and if so, whether it assures the minimum value of the performance index calculated using the values of the vehicle states (and controls) along that trajectory. As an example, Fig. 6 presents collision-free solutions for two different locations of a 10m-tall obstacle when five varied parameters, $x''_{10}{}'$, $x'''1_f$, x'''_{30}, x'''_{3f} and τ_f, are optimized to assure feasible minimum-path-length trajectories

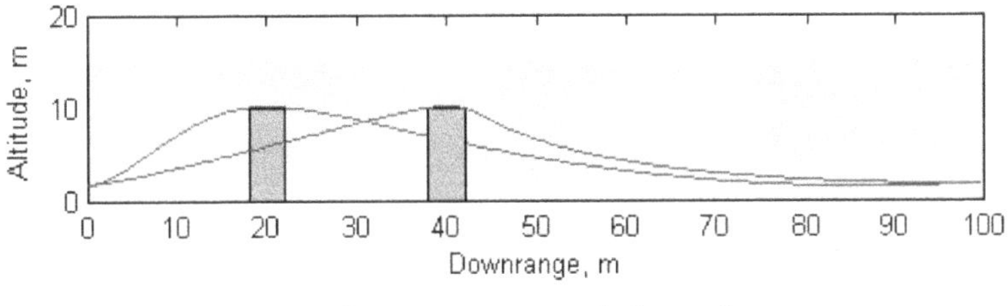

Fig. 6. Examples of minimum-path-length trajectories

Now, let us address the reason for choosing some abstract parameter τ as an argument for the reference functions (9) rather than time or path length, which are commonly used.

Assume for a moment that $\tau \equiv t$. In this case, once we determine the trajectory we unambiguously define a speed profile along this trajectory as well, since

$$V(t) = \sqrt{u(t)^2 + v(t)^2 + w(t)^2} = \sqrt{\dot{x}(t)^2 + \dot{y}(t)^2 + \dot{z}(t)^2} \tag{10}$$

Obviously, we cannot allow this to happen because we want to vary the speed profile independently. With the abstract argument τ this becomes possible via introduction of a speed factor λ such that

$$\lambda(\tau) = \frac{d\tau}{dt}. \tag{11}$$

Now instead of (10) we have

$$V(\tau) = \lambda(\tau)\sqrt{x'(\tau)^2 + y'(\tau)^2 + z'(\tau)^2} \tag{12}$$

and by varying $\lambda(\tau)$ we can achieve any desired speed profile.

The capability to satisfy higher-order derivatives at the trajectory endpoints, specifically at the initial point, allows continuous regeneration of the trajectory to accommodate sudden changes like newly discovered obstacles. As an example, Fig. 7 demonstrates a scenario whereby a UUV executing an OA manoeuvre discovers a second obstacle and must generate a new trajectory beginning with the current vehicle states and control values (up to the second-order derivatives of the states). The suggested approach enables this type of continuous trajectory generation and ensures smooth, non-shock transitions.

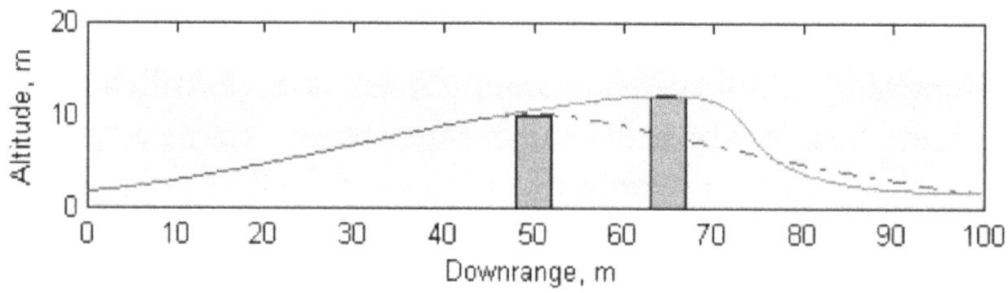

Fig. 7. Example of dynamic trajectory reconfiguration

3.2 Inverse dynamics

The second key block inside the dynamic trajectory generator in Figs. 2 and 3 accepts the candidate trajectory as an input and computes the components of the state vector and control signals required to follow it. In this way we can ensure that each candidate trajectory does not violate any constraints (including those of (8)).

First, using the following relation for any parameter ζ,

$$\dot{\zeta}(\tau)=\frac{d\zeta}{d\tau}\frac{d\tau}{dt}=\zeta'(\tau)\lambda(\tau) \tag{13}$$

we convert kinematic equations (2) into the τ domain

$$\lambda(\tau)\begin{bmatrix} x'(\tau) \\ y'(\tau) \\ z'(\tau) \end{bmatrix} = {}^{u}_{b}R \begin{bmatrix} U_0 \\ v(\tau) \\ w(\tau) \end{bmatrix} \tag{14}$$

Next, we assume the pitch angle to be small enough to let sinθ (t) ≈ 0 and cosθ (t) ≈ 1, so that the rotation matrix (3) becomes

$${}^{u}_{b}R(\tau)=\begin{bmatrix} \cos\psi(\tau) & -\sin\psi(\tau) & 0 \\ \sin\psi(\tau) & \cos\psi(\tau) & 0 \\ 0 & 0 & 1 \end{bmatrix} \tag{15}$$

While this step is not required, it simplifies the expressions in the following development. Inverting (14) via the rotation matrix (15) yields

$$\begin{bmatrix} U_0 \\ v \\ w \end{bmatrix} = \lambda \begin{bmatrix} \cos\psi & \sin\psi & 0 \\ -\sin\psi & \cos\psi & 0 \\ 0 & 0 & 1 \end{bmatrix}\begin{bmatrix} x' \\ y' \\ z' \end{bmatrix} \tag{16}$$

Hereafter each variable's explicit dependence on τ will be omitted from the notation.

Now the three equations of system (16) must be resolved with respect to three unknown parameters, *v*, *w* and ψ. While the last one readily yields

$$w=\lambda z' \tag{17}$$

the first two require more rigorous analysis.

Consider Fig. 8. Geometrically, a scalar product of two vectors on the right-hand-side of the first equation in (16) represents the length of the longest side of the shaded rectangle. Similarly, the second equation expresses the length of the shortest side of this rectangle. From here it follows that the square of the length of the diagonal vector can be expressed in two ways: $v^2\lambda^{-2}+U_0^2\lambda^{-2}=x'^2+y'^2$. This yields

$$v = \sqrt{\lambda^2(x'^2 + y'^2) - U_0^2} \tag{18}$$

From the same figure it follows that

$$\psi = \Psi - \tan^{-1}\frac{v\lambda^{-1}}{U_0\lambda^{-1}} = \Psi - \tan^{-1}\frac{v}{U_0}, \quad \Psi = \tan^{-1}\frac{y'}{x'} \tag{19}$$

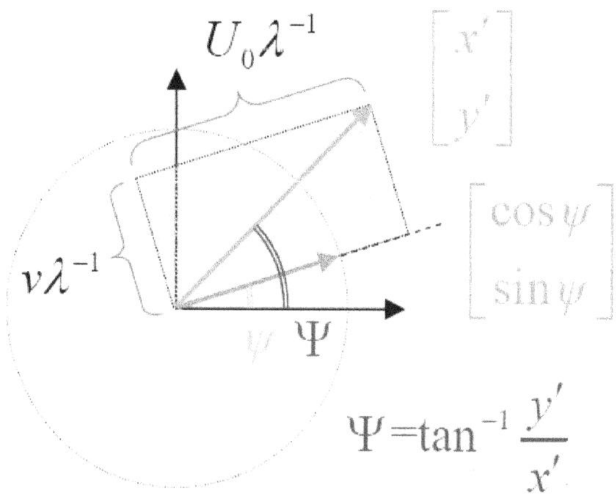

Fig. 8. Kinematics of horizontal plane parameters

Now, using these inverted kinematic equations, we can check whether each candidate trajectory obeys the constraints imposed on it (constraints (8)).

3.3 Discretization

We proceed with computing the remaining states along the reference trajectory over a fixed set of *N* points (for instance, N=100) spaced evenly along the virtual arc $[0;\tau_f]$ with the interval

$$\Delta\tau = \tau_f(N-1)^{-1} \tag{20}$$

so that

$$\tau_j = \tau_{j-1} + \Delta\tau, \quad j = 2,\dots,N, \; (\tau_1 = 0) \tag{21}$$

In order to determine coefficients for polynomials (9) we will have to guess on the values of the varied parameters τ_f, x''_{i0}, x'''_{i0}, x'''_{if}, and x'''_{if}. These guesses will be used along with the known or desired boundary conditions x_{i0}, x'_{i0}, x''_{i0}, x_{if}, x'_{if}, and x''_{if}. The boundary conditions on coordinates x_{i0} and x_{if} come directly from (5). According to (14), the given boundary conditions on surge, sway, and heave velocities define the first-order time derivatives of the coordinates as

$$\begin{bmatrix} x'_{0;f} \\ y'_{0;f} \\ z'_{0;f} \end{bmatrix} = \lambda_{0;f}^{-1}\,{}^{u}_{b}R_{0;f} \begin{bmatrix} U_0 \\ v_{0;f} \\ w_{0;f} \end{bmatrix} \tag{22}$$

They also define the initial and final pitch and yaw angles used to compute ${}^{u}_{b}R_{0;f}$ in (22) as

$$\theta_{0;f} = \gamma_0 + \tan^{-1}\frac{-w_{0;f}}{\sqrt{U_0^2 + v_{0;f}^2}} \text{ and } \psi_{0;f} = \Psi_0 - \tan^{-1}\frac{v_{0;f}}{U_0} \tag{23}$$

In equation (22) we may use any value for the initial and final speed factor λ, for example, $\lambda_{0;f} = 1$. This value simply scales the virtual domain; the higher the values for λ, the larger the values for τ_f. This follows directly from equations (11) and (12) that $\lambda_{0;f}\,\tau^{-1}_f = U_0\, s^{-1}_f$, where s_f is the physical path length.

Finally, initial values for the second-order derivatives are provided by the UUV motion sensors (see Figs. 1-3) (after conversion to the τ domain), while final values for the second- order derivatives are usually set to zero for a smooth arrival at the final point. Having an analytical representation of the candidate trajectory (9) defines the values of x_{ij}, and x'_{ij}, $i = 1,2,3$, $j = 1,\dots, N$.

Now, for each node $j = 1,\dots, N$ we compute

$$\lambda_j = \Delta\tau\Delta t_{j-1}^{-1} \tag{24}$$

where

$$\Delta t_{j-1} = \frac{\sqrt{(x_j - x_{j-1})^2 + (y_j - y_{j-1})^2 + (z_j - z_{j-1})^2}}{\sqrt{U_0^2 + v_{j-1}^2 + w_{j-1}^2}} \tag{25}$$

and then use (17)-(19) to compute w, v, ψ and Ψ at each timestamp. The vertical plane parameters, flight path angle γ and pitch angle θ, can be computed using the following relations:

$$\gamma_j = \tan^{-1}\frac{-z'_j}{\sqrt{x_j'^2 + y_j'^2}}, \quad \theta_j \approx \gamma_j + \tan^{-1}\frac{-w_j}{\sqrt{U_0^2 + v_j^2}} \tag{26}$$

In order to check the yaw rate constraints (8) we must first numerically differentiate the expression for Ψ in (19).

3.4 Optimization

When all parameters (states and controls) are computed in each of N points, we can compute the performance index J and the penalty function. For example, we can combine constraints (6) and (8) into the joint penalty

$$\Delta = \left[k^{z_{\min}}, k^{z_{\max}}, k^{\theta}, k^{\psi} \right] \begin{bmatrix} \min_j(0; z_j - z_{\min})^2 \\ \max_j(0; z_j - z_{\max})^2 \\ \max_j(0; |\theta_j| - \theta_{\max})^2 \\ \max_j(0; |\dot{\psi}_j| - \dot{\psi}_{\max})^2 \end{bmatrix} \quad (27)$$

with k^{zmin}, k^{zmax}, k^{θ} and k^{ψ} being scaling (weighting) coefficients. Now the problem can be solved using numerical methods such as the built-in *fmincon* function in the Mathworks' MATLAB development environment. Alternatively, by combining the performance index J with the joint penalty Δ we may exploit MATLAB's non-gradient *fminsearch* function. For real-time applications, however, the authors prefer to use a more robust optimization routine based on the gradient-free Hooke-Jeeves pattern search algorithm (Yakimenko, 2011).

4. PLANAR CASES

This section presents two simplified cases for a vehicle manoeuvring exclusively in either the horizontal or vertical plane.

4.1 Horizontal plane guidance

For the case of a UUV manoeuvring in the horizontal plane or a USV, the computational procedure is simplified. The trajectory is represented by only two reference polynomials for coordinates x1 and x2. Hence, we end up having only five varied parameters, which are: τ_f, x'''_{10}, x'''_{20}, x'''_{if} and x'''_{2f}. The remaining kinematic formulas are identical to those presented above with $z \equiv 0$, $z' \equiv 0$ and $\gamma \equiv 0$. Figure 9 shows an example of a planar scenario in which a USV has to compute a new trajectory twice. First, after detecting an obstacle blocking its original path, a new trajectory is generated to steer right and pass safely in front of the object (dotted line). Second, while executing the first avoidance manoeuvre the USV detects that the object has moved south into its path. It therefore generates a new trajectory to steer left and pass safely behind the object's stern. The complete trajectory is shown as a solid line.

4.2 Vertical plane guidance

For the case of a UUV manoeuvring in the vertical plane, the 3D algorithm can be reduced to the 2D case in a manner similar to the horizontal case. Specifically, using five varied parameters, τ_f, $x''_{10}{}'$, x'''_{30}, x'''_{1f} and x'''_{3f}, we can develop reference trajectories for x_1 and x_3, and then use the same general equations developed in Section 3, assuming $y \equiv 0$, $y' \equiv 0$, and $\Psi \equiv 0$.

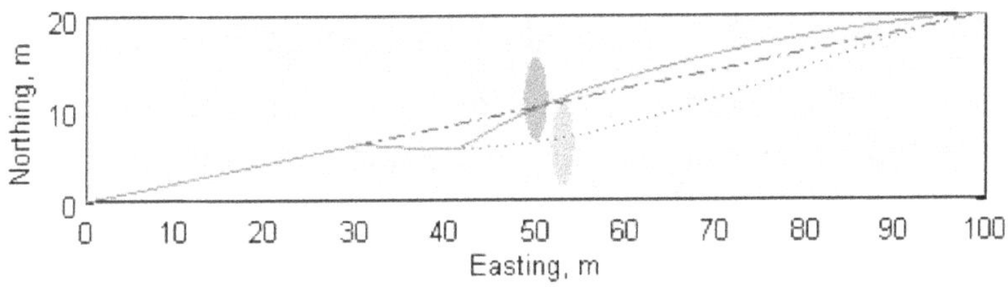

Fig. 9. Moving obstacle avoidance in a horizontal plane

Alternatively, we can use a single reference polynomial to approximate just x_3 and then integrate the third equation of (4) to get the heave velocity *w*. That allows computation of the time period Δt_{j-1} using instead of (25).

$$\Delta t_{j-1} = (z_j - z_{j-1}) w_{j-1}^{-1} \tag{28}$$

Another way of dealing with vertical plane manoeuvres is to invert the dynamic equations (4) (Horner & Yakimenko, 2007). After developing the reference functions for two coordinates, x_1 and x_3, the stern plane δs control input is computed subject to five variable parameters: τ_f, x'''_{10}, x'''_{30}, x'''_{1f}, and x'''_{3f}.

In this case, the corresponding time period Δt_{i-1} is computed similarly to (28):

$$\Delta t_{j-1} = t_j - t_{j-1} = \frac{z_j - z_{j-1}}{w_{j-1}\cos\theta_{j-1} + u_0 \sin\theta_{j-1}} \approx \frac{z_j - z_{j-1}}{w_{j-1}} \tag{29}$$

and the heave velocity is calculated using the third equation of system (4) as

$$\begin{aligned} x'_{j-1} &= \lambda_{j-1}^{-1}(w_{j-1}\sin\theta_{j-1} + U_0\cos\theta_{j-1}) & x_j &= x_{j-1} + \Delta\tau x'_{j-1} \\ w'_{j-1} &= \lambda_{j-1}^{-1}\left(A_{33}w_{j-1} + A_{35}q_{j-1} + B_{32}\delta_{s;j-1}\right) & w_j &= w_{j-1} + \Delta\tau w'_{j-1} \end{aligned} \tag{30}$$

The next step involves computing the pitch angle, pitch rate and pitch acceleration as

$$\theta_j = \cos^{-1}\left(\lambda_j \frac{u_0 x'_j + w_j z'_j}{w_j^2 + U_0^2}\right),\ q_j = \dot{\theta}_j \approx \frac{\theta_j - \theta_{j-1}}{\Delta t_{j-1}},\ \dot{q}_j = \ddot{\theta}_j \approx \frac{q_j - q_{j-1}}{\Delta t_{j-1}} \tag{31}$$

Finally, we can compute the dive plane deflection required to follow the trajectory using the 5th equation of system (4) as

$$\delta_{s;j} = (\dot{q}_j - A_{53}w_j - A_{55}q_j)B_{52}^{-1} \quad (32)$$

In this case the last two terms in the joint penalty Δ , similar to that of (27) but developed for the new controls, enforce $|\delta_s| \le \delta_{s\max}$ and $|\dot{\delta}_s| \le \dot{\delta}_{s\max}$.

5. TEST VEHICLES AND SENSING ARCHITECTURE

The aforementioned trajectory generation framework has been executed on multiple UMVs. Prior to presenting theoretical and experimental results involving specific maritime vehicle platforms, we will first discuss two vehicles currently utilized at CAVR: the REMUS UUV and the SeaFox USV. Both vehicles apply FLS to identify and locate obstacles in their surroundings and utilize the proposed direct technique to produce real-time OA trajectories. This section delineates system-level descriptions of both platforms, encompassing their sensors, and suggests a methodology for constructing the OA framework atop the trajectory generation framework, therefore further augmenting the architecture seen in Figs. 2 and 3.

5.1 REMUS UUV and Sea Fox USV

Remote Environmental Monitoring Units (REMUS) are unmanned underwater vehicles (UUVs) created by the Woods Hole Oceanographic Institution and commercially marketed by Hydroid, LLC (Hydroid, 2011). The NPS CAVR possesses and runs two REMUS 100 vehicles to serve many navy-sponsored research initiatives. The REMUS 100 is a modular, 0.2-meter diameter unmanned underwater vehicle designed for operations in coastal areas at depths of up to 100 meters. Standard configurations have a length of less than 1.6m and a weight under 45kg, facilitating global transit and deployment by a two-person crew (Fig.10a). Primarily engineered for hydrographic surveys, REMUS is outfitted with sidescan sonar and sensors for the acquisition of oceanographic data, including conductivity, temperature, depth, and optical backscatter. The REMUS 100 system employs dual external transponders for long baseline acoustic localization or ultra-short baseline terminal homing, in conjunction with an Acoustic Doppler Current Profiler/Doppler Velocity Log (ADCP/DVL). The ADCP/DVL quantifies vehicle altitude, velocity relative to the ground or water, and current velocity profiles in body-fixed coordinates.

To support ongoing CAVR research into sonar-based OA, terrain-relative navigation, and multi-vehicle operations in cluttered environments, each NPS REMUS vehicle has been modified to incorporate a FLS, multi-beam bathymetric sonar, acoustic communications modem, navigation-grade inertial measurement system, and fore/aft horizontal/vertical cross-body thrusters for hovering or precise manoeuvring. (Figure 10b provides a close up of the NPS REMUS FLS arrays with nose cap removed.) To maximize the REMUS system's utility as a research platform, Hydroid developed the RECON communications interface so that sensor and computer payloads can interact with the REMUS autopilot. Using this interface, NPS payloads receive vehicle sensor data and generate autopilot commands based on NPS sonar processing, trajectory generation, and path-following algorithms.

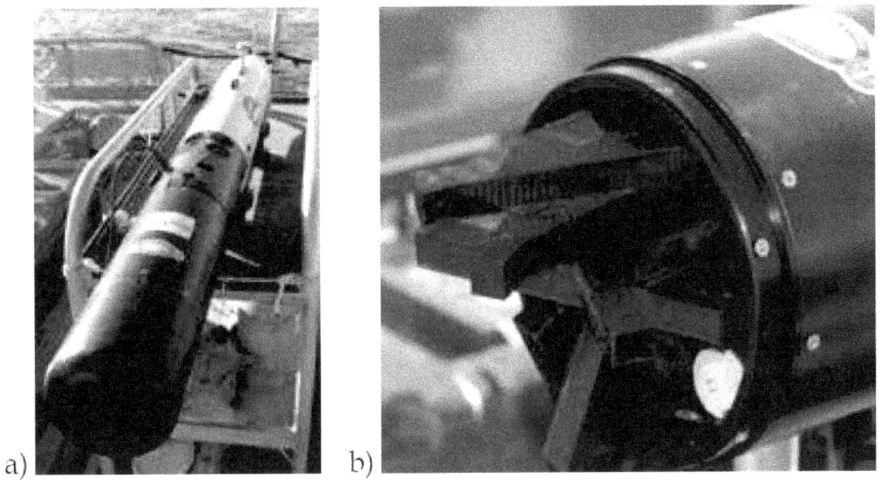

Fig. 10. NPS REMUS 100 UUV (a) and FLS arrays (b)

The SeaFox USV was designed and manufactured by Northwind Marine (Seattle, WA) as a remote-controlled platform for intelligence, surveillance, reconnaissance, anti-terrorist force protection, and maritime interdiction operations (Northwind Marine, 2011). SeaFox is a 4.88m long, aluminium, rigid-hulled inflatable boat with a 1.75m beam; 0.25m draft; fold- down communications mast; and fully-enclosed electronics and engine compartments. SeaFox's water jet propulsion system is powered by a JP5-fueled, 185-HP V-6 Mercury Racing engine, and can deliver a top speed of 74km/h. Standard sensing systems include three daylight and three low light navigation cameras for remote operation, as well as twin daylight and infrared gyro-stabilized camera turrets for video surveillance. All video is accessible over a wireless network via two onboard video servers.

The NPS SeaFox was modified to enable fully-autonomous operations by integrating a payload computer with the primary autopilot (Fig.11). Meanwhile, the original remote control link was retained to provide an emergency stop function. NPS algorithms running on the payload computer generate rudder and throttle commands that are sent directly to the SeaFox autopilot. Recent navigational upgrades include a satellite compass that uses differential Global Positioning System (GPS) navigation service for accurate heading information, a tactical-grade inertial measurement unit for precise attitude estimation, and an optional ADCP/ DVL for water velocity measurements. To support ongoing CAVR research into autonomous riverine navigation, the NPS SeaFox was further upgraded to deploy a retractable, pole-mounted FLS system for underwater obstacle detection and avoidance (Gadre et al., 2009). Figure 12 shows the SeaFox USV operating on a river with its sonar system deployed below the waterline.

5.2 Sonar system

The NPS REMUS and SeaFox vehicles rely on FLS to detect and localize obstacles in their environment. Both platforms utilize commercial blazed array sonar systems manufactured by BlueView Technologies (BlueView Technologies, 2011). These sonar systems comprise one or more pairs of arrays grouped into sonar "heads." Each sonar head generates a 2D cross-sectional image of the water column in polar coordinates, typically plotted as the image plane's field of view angle vs. range. Due to the sonar arrays' beam width, the resulting FLS imagery has a 12-degree out-of-plane ambiguity. The REMUS FLS system consists of two fixed sonar heads, which provide a 90-degree horizontal field of view (FOV) and a 45-degree vertical FOV. Similarly, the SeaFox FLS system is comprised of twin sonar heads mounted on port and starboard pan/tilt actuators, providing each side with a 45- degree FOV image at an adjustable mounting orientation that can be swept through the water column for increased sensor coverage.

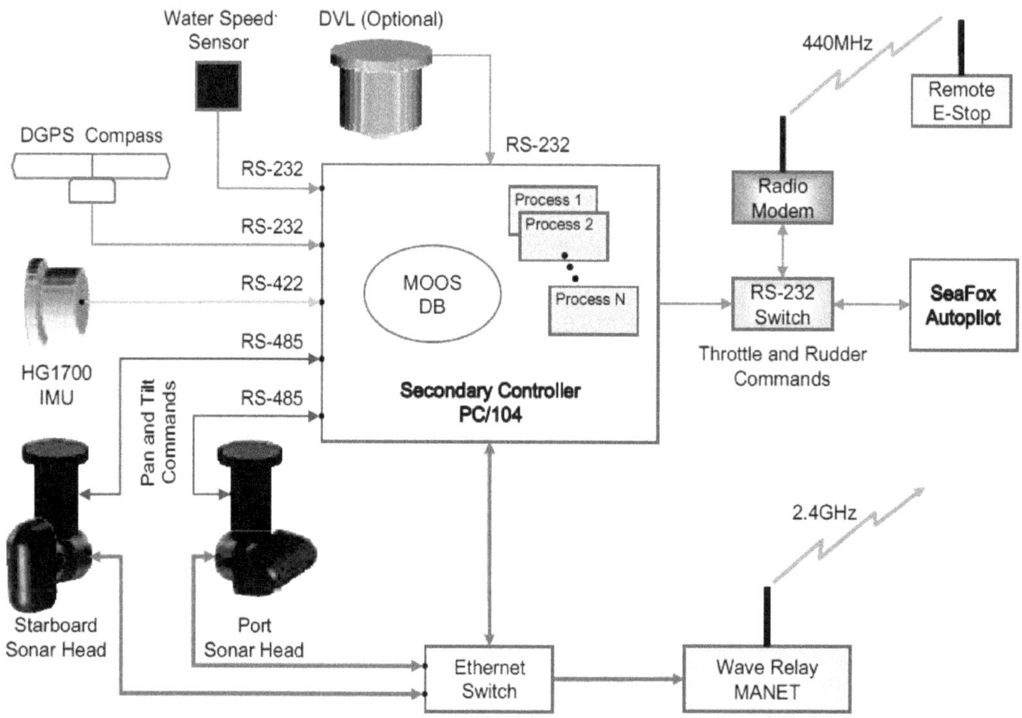

Fig. 11. SeaFox sensors and control architecture

Fig. 12. SeaFox USV navigating on the Sacramento River near Rio Vista, CA

5.3 Obstacle avoidance framework

The proposed OA framework built into the architecture of Figs. 2 and 3 is shown in Fig.13. It consists of an environmental map, a planning module, a localization module, sensors and actuators (Horner & Yakimenko, 2007). The environmental map can include a priori knowledge, such as the positions of charted underwater obstacles, but also incorporate unexpected threats discovered by sonar. The positions of all obstacles are eventually resolved in the vehicle-centred coordinate frame with the help of the localization module. The planning module is responsible for generating collision-free trajectories the vehicle should follow. This reference trajectory, possibly with reference controls, is then used to excite actuators.

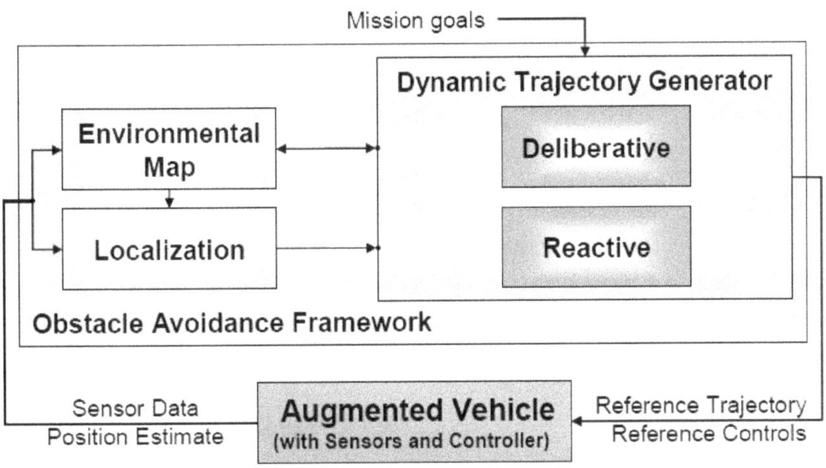

Fig. 13. Components of the NPS OA framework

The proposed OA framework supports both deliberative and reactive obstacle avoidance behaviours. Deliberative OA involves the ability to generate and follow a trajectory that avoids all known obstacles between an arbitrary start location and some desired goal location, whereas reactive OA involves the ability to avoid any previously unknown obstacles detected while following this trajectory. Since the sonar system continuously resamples the environment, this reactive behaviour can be achieved by a deliberative planner as long as i) it executes fast enough to incorporate all new obstacle information from the sonar, and ii) it generates feasible trajectories which begin with the vehicle's current state vector. Specifically, since the REMUS and SeaFox FLS have limited range and limited fields of view in both image planes, new trajectories must be generated continuously (e.g. on some fixed time interval or upon detection of a new obstacle) during execution of the current manoeuvre to ensure reactive avoidance of new obstacles.

For instance, consider a REMUS vehicle utilizing sidescan sonar to map a minefield prior to a mine clearance operation as an illustration of deliberative operational autonomy. The objective of this mission is delineated by the series of waypoints constituting a standard lawn-mowing survey pattern. Upon detection of an obstacle along a designated track line, the optimal OA maneuver for this mission should also minimize the total deviation from the track line, as we aim for complete sensor coverage of the survey region. Therefore, deliberative OA entails the optimization of a specific performance metric. Similarly, although computerized nautical charts or past vehicle inspections may facilitate the identification of certain impediments in advance, this information is often incomplete or obsolete. Vehicles must possess the ability to retain in memory the positions of any undiscovered obstacles encountered throughout their mission, enabling future trajectories to circumvent them, even when these obstacles fall outside the sonar's present range of view. Consequently, deliberative OA also involves the development and upkeep of obstacle maps.

5.4 Obstacle detection and mapping

Detecting obstacles from sonar imagery is challenging because several factors affect the intensity of sonar reflections off objects in the water column. These factors include the size, material, and geometry of an object relative to the sonar head; interference from other acoustic sensors; and the composition of the acoustic background (e.g. bottom type, amount of sediment, etc.) to name a few (Masek, 2008). Once an obstacle has been detected, other image processing algorithms must measure its size and compute its location within the navigational reference frame. While localizing obstacles via the range and bearing data embedded in the sonar imagery is straightforward, computing their true size is very difficult. First, for the REMUS FLS, an obstacle's height and width can be measured directly by both sonar heads only when it is located within a narrow 12-degree by 12-degree "window" directly ahead of the vehicle. Due to this narrow beam width, most obstacles are not imaged by both the horizontal and vertical sonar at the same time. Moreover, FLS images do not contain information in the region behind an obstacle's ensonified leading edge; this portion of the image is occluded. Therefore, the true horizontal and vertical extent of each obstacle must be deduced from multiple views of the same object. For a vehicle with a fixed sensor like the REMUS, this may be accomplished by deliberately inducing vehicle motion to vary the sonar angle (Furukawa, 2006) or by generating trajectories that will image the object from a different location at a later time. For these scenarios, it is desirable to balance OA behaviours with exploration behaviours in order to maximize sensor coverage and generate more complete obstacle maps. In this way, the proposed trajectory generation framework can be adapted to produce exploratory trajectories which more accurately measure the size and extent of detected obstacles (Horner et al., 2009). Nevertheless, due to the uncertainty in sonar images arising from environmental factors, sensor geometry, or obstacle occlusion, it is prudent to make conservative assumptions about an obstacle's boundaries until other information becomes available.

For the remainder of this section, we highlight different representations for incorporating obstacle size, location, and uncertainty into an obstacle map for efficient collision detection during the trajectory optimization phase. These representations can be tailored to the working environment. For operations in a kelp forest, for example, kelp stalks often appear as point-like features in horizontal-plane sonar imagery (Fig.14) but seldom appear in vertical-plane images. By making the reasonable assumption (for this environment) that these obstacles extend vertically from the sea floor to the surface, it may be simpler to perform horizontal-plane OA through this type of obstacle field. Nevertheless, when building an obstacle map comprised primarily of point features, mapping algorithms must account for the uncertainty inherent in sonar imagery. One simple but effective technique adds spherical (3D) or circular (2D) uncertainty bounds to each point feature stored in the obstacle map. Candidate OA trajectories which penetrate these boundaries violate constraint (7). Under this construct, collision detection calculations are reduced to a simple test to determine whether line segments in a discretized trajectory intersect with the uncertainty circle (2D) or sphere (3D) for each obstacle in the map. In general, when checking for line segment intersections with a circle or sphere there are five different test cases to consider (Bourke, 1992). Our application, however, requires only two computationally efficient tests to determine: i) which line segment along a discretized trajectory contains the closest point of approach (CPA) to an obstacle, and ii) whether this CPA is located inside the obstacle's uncertainty bound.

Most objects appear in sonar imagery not as point features, but as complex shapes. Unlike point features, it is difficult and computationally expensive to determine exhaustively whether or not a candidate vehicle trajectory will collide with these shapes. Instead, we can bound an arbitrary shape with a minimal area rectangle (or box, in 3D) aligned with the shape's principle axes (Fig.15). This type of object, called an oriented bounding box, is widely used in collision-detection algorithms for video games. One technique, based on the Separating Axis Theorem from complex geometry, results in an extremely fast test for line segment intersections with an oriented bounding box (Kreuzer, 2006). With slight modification, this test can also be used to detect when a trajectory passes directly above a bounding box. In our application, we use the OpenCV computer vision library to generate a bounding box around each object detected in the horizontal image plane. Subsequently, we calculate the centroid, dimensional extent, and orientation of each box in relation to the vehicle's navigation frame. Owing to occlusion, the width derived from this rectangle fails to adequately represent the actual dimensions of the barrier; hence, we presume a constant value for this parameter. To construct a 3D (specifically 2.5D) bounding box around the item, we calculate its vertical dimensions from vertical sonar images. The current assumption is that barriers extend from the ocean floor to their measured height above the bottom; however, this method can be modified to include obstructions hanging in the water column or extending from the surface to a measured depth (e.g., ships in a harbor).

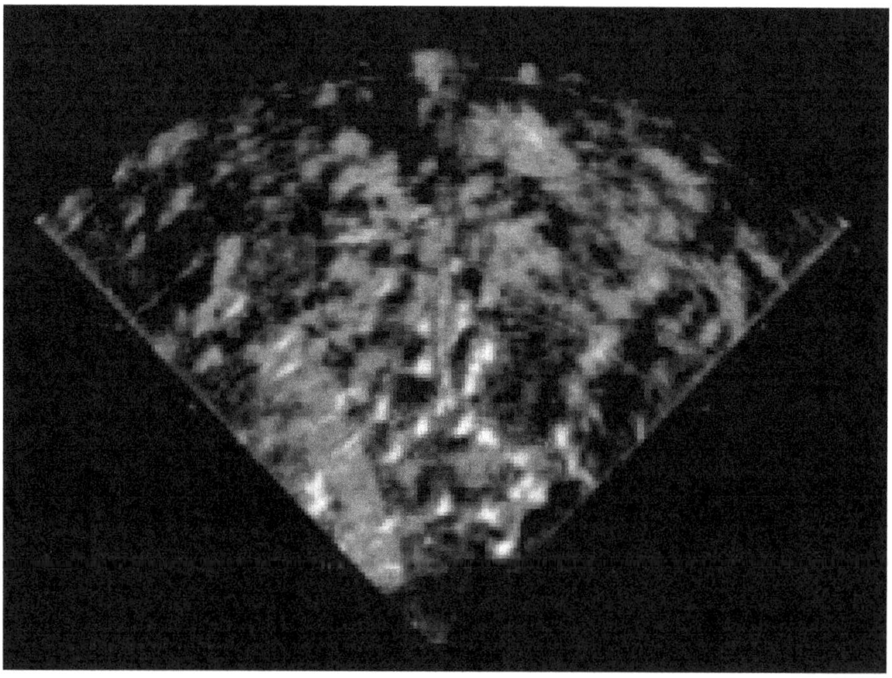

Fig. 14. Horizontal FLS image of a kelp forest

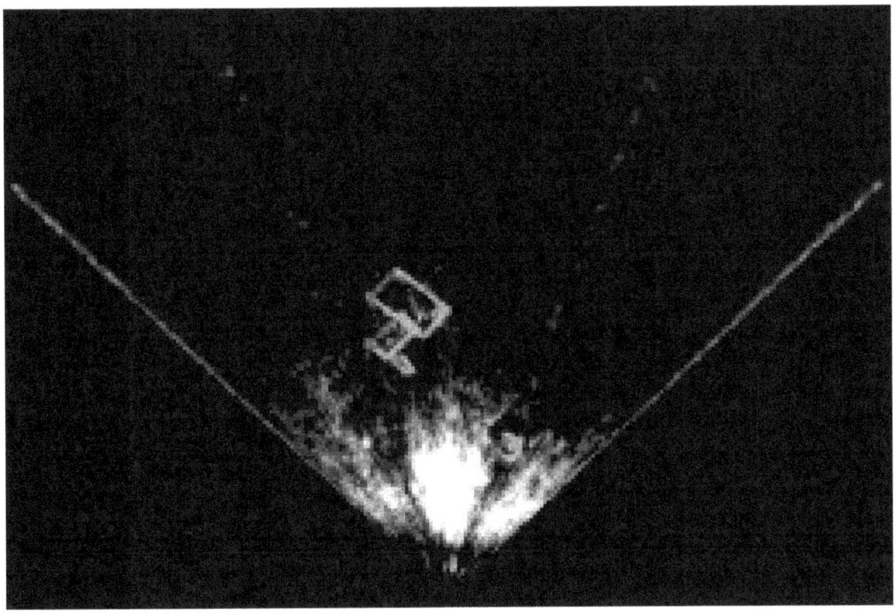

Fig. 15. Example of the bounding boxes used in conservative collision detection calculations

Oriented bounding boxes are effective for mapping discrete impediments in open-water situations; nevertheless, they necessitate an additional image processing step and are not well adaptable to operations in constrained waterways. In these contexts, a probabilistic occupancy grid is advantageous for effectively mapping extensive continuous obstacles (e.g., harbor breakwaters) or natural landscapes (e.g., riverbanks). Occupancy grids partition the surroundings into a grid of cells, assigning a probability to each cell indicating the likelihood of it being occupied by an obstacle. Utilizing a probabilistic sensor model, Bayes' Theorem calculates the likelihood of a specific cell being occupied, depending on the present sensor data. An estimate of each cell's occupancy state can be continuously refined by an iterative method that integrates all prior data (Elfes, 1989). Figure 16a illustrates an occupancy grid map of a river produced by the SeaFox FLS system. In this graphic, each pixel represents a 1-meter square grid cell, with color indicating the cell's likelihood of occupancy (red) or vacancy (green). The inset section of the occupancy grid map is superimposed with an obstacle map of oriented bounding boxes in Fig. 16b. Utilizing discrete bounding boxes to depict an extensive, continuous shoreline rapidly becomes unmanageable as the number of required boxes increases. The occupancy grid architecture serves as a more efficient depiction of obstacle maps for extensive operations in constrained waterways.

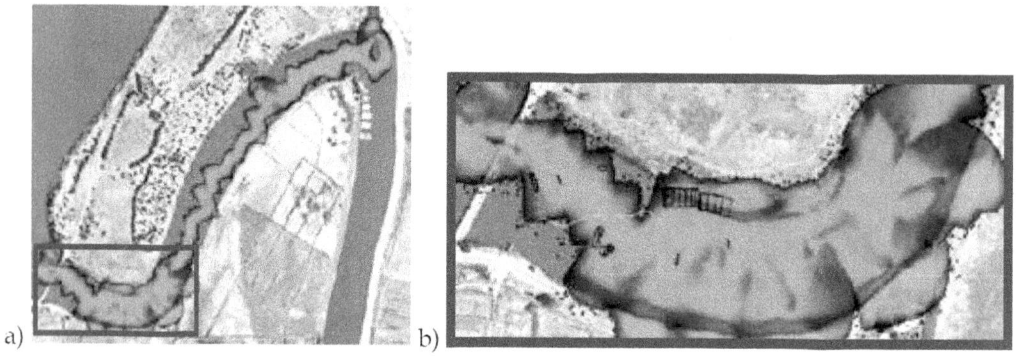

Fig. 16. Occupancy grid for a river as generated by the SeaFox FLS system

NPS has developed probabilistic sonar models for the BlueView FLS and has successfully combined separate 2D occupancy grids in order to reconstruct the 3D geometry of an obstacle imaged by the REMUS UUV's horizontal and vertical sonar arrays (Horner et al., 2009). Using this occupancy grid framework, each candidate trajectory's risk of obstacle collision is computed using the occupancy probabilities (a direct lookup operation) of the grid cells it traverses. Trajectory optimization for OA entails minimizing the cumulative risk of collision along the entire trajectory.

6. PATH FOLLOWING

While the REMUS UUV and SeaFox USV are both commercial vehicles with proprietary autopilots, both provide communications interfaces that allow experimental sensor and computer payloads to override the primary autopilot via high-level commands. The REMUS RECON interface, for example, closely resembles the augmented autopilot depicted in Figs. 2 and 3 (although direct actuator inputs are only available for propeller and cross-body thrusters settings). For full overriding control, a payload module must periodically send valid commands containing all of the following: i) desired depth or altitude, ii) desired vehicle or propeller speed, and iii) desired heading, turn rate, or waypoint location. The developed trajectory generator (described in Section 3) outputs reference trajectories as parameterized expressions for each coordinate in a spatial curve plus a speed factor to use while traversing that curve. Using these expressions as reference trajectories, the 3D path following controller developed earlier (Kaminer et al., 2007) can compute turn rate and pitch rate commands required to drive a vehicle onto (and along) the desired trajectory. The RECON interface, however, does not accept pitch rate commands (for vehicle safety reasons). Therefore, in order to use the aforementioned path following controller to track 3D trajectories with the REMUS UUV, controller outputs must be partitioned into horizontal (turn rate) and vertical (depth or altitude) commands as described in the following section (obviously, the SeaFox USV only uses the turn rate commands).

6.1 Horizontal plane

Consider the 2D problem geometry depicted in Fig.17, which defines an inertial {I} frame, Serret-Frenet {F} error frame and body-fixed reference frame {b}. The kinematic model of the vehicle (2)-(3) reduces to

$$\begin{bmatrix} \dot{x}(t) \\ \dot{y}(t) \end{bmatrix} = \begin{bmatrix} U_0 \cos\psi(t) \\ U_0 \sin\psi(t) \end{bmatrix} \tag{33}$$

with dynamics described by

$$\dot{\psi} = r \tag{34}$$

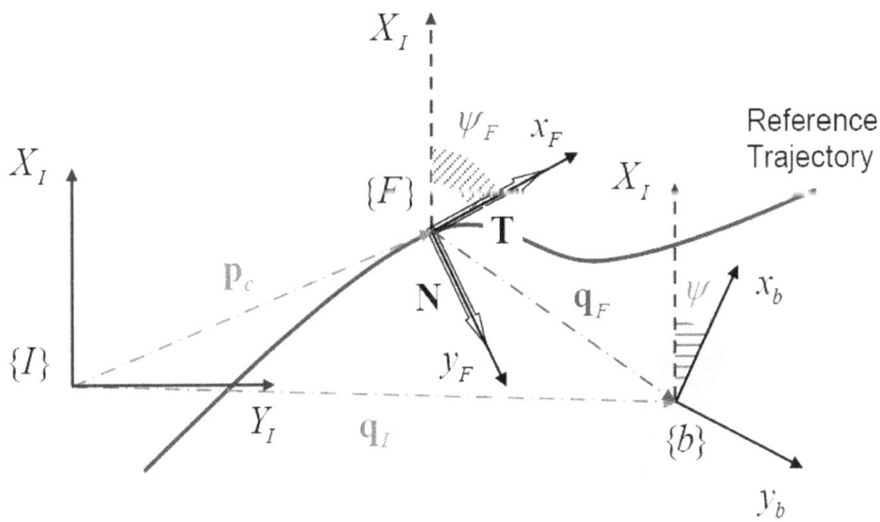

Fig. 17. Horizontal path-following kinematics

By construction, the local trajectory planner produces an analytic expression for each component of the spatial trajectory as a function of virtual arc length, pc (τ). We can also compute analytic expressions for $\mathbf{p}'_c(\tau)$ and $\mathbf{p}''_c(\tau)$, the first and second derivatives, respectively, of the spatial trajectory. Using the relationships in Fig.17, the errors can be expressed in the Serret-Frenet frame {F} as

$$\mathbf{q}_F = \begin{bmatrix} x_F \\ y_F \end{bmatrix} = {}_I^F R(\mathbf{q}_I - \mathbf{p}_c) \tag{35}$$

where ${}_I^F R = [\mathbf{T},\mathbf{N}, \mathbf{B}]^T$ is a rotation matrix constructed from the tangent, normal, and binormal vectors of the Serret-Frenet error frame {F}. The tangent vector is computed from the expression for the trajectory's first derivative as:

$$\mathbf{T} = \frac{\mathbf{p}_c'(\tau)}{\left\|\mathbf{p}_c'(\tau)\right\|} \tag{36}$$

For the 2D problem, the normal vector components can be computed directly from the tangent vector components: $N_x = -T_y$ and $N_y = T_x$. Additionally, the signed curvature of the trajectory can be computed using the expressions for the trajectory's first and second derivatives as:

$$\kappa(\tau) = \frac{p_{cy}''(\tau)p_{cx}'(\tau) - p_{cx}''(\tau)p_{cy}'(\tau)}{\left\|\mathbf{p}_c'(\tau)\right\|^3} \tag{37}$$

Taking the time derivative of **q**F, we obtain the following state space representation for the error kinematics (i.e. the position and heading of the vehicle's body-fixed frame {b} relative to the Serret-Frenet frame {*F*}, which follows the desired trajectory):

$$\begin{aligned} \dot{x}_F &= -\dot{l}(1-\kappa y_F) + U_0 \cos\psi_e \\ \dot{y}_F &= -\dot{l}(\kappa x_F) + U_0 \sin\psi_e \\ \dot{\psi}_e &= \dot{\psi} - \dot{\psi}_F = u_\psi - \dot{\psi}_F = u_\psi - \kappa\dot{l} \end{aligned} \tag{38}$$

where l is the path length of the desired spatial curve and l describes the speed at which a virtual target travels along this curve.

The goal is to drive the vehicle's position error ($\mathbf{q}_F$) and heading error (Ψe) to zero. This will drive the vehicle to the commanded trajectory location ($\mathbf{p}_c$) and align its velocity vector with the trajectory's tangent vector (**T**). The control signal uψ must now be chosen to asymptotically drive the vehicle position and velocity vectors onto the commanded trajectory. We choose the candidate Lyapunov function

$$V = \frac{1}{2}\left(x_F^2 + y_F^2 + (\psi_e - \delta_\psi)^2\right) \tag{39}$$

where δ_ψ is a shaping function that controls the manner in which the vehicle approaches the path

$$\delta_\psi = \sin^{-1}\left(\frac{-y_F}{\left|y_F\right| + d}\right) \tag{40}$$

with $d > 0$ an arbitrary constant.

Using some algebra, we choose the following control laws to ensure that $\dot{V} < 0$:

$$\begin{aligned} \dot{l} &= K_1 x_F + U_0 \cos\psi_c \\ u_\psi &= \kappa \dot{l} + \dot{\delta}_\psi + K_2(\psi_c - \delta_\psi) - \frac{\sin\psi_c - \sin\delta_\psi}{\psi_c - \delta_\psi} U_0 y_F \end{aligned} \quad (41)$$

In these expressions K_1, K_2, and d can be used as gains to tune the closed-loop performance of the path following controller.

6.2 Vertical plane

Now consider the REMUS UUV manoeuvring in the vertical plane using altitude commands. For survey operations, REMUS is typically programmed to follow a lawnmower pattern at a constant altitude above the sea floor determined by the desired sidescan sonar range. Since the ADCP/DVL sensor continuously measures vehicle altitude above the bottom, this operating mode ensures safe operation over undulating bottoms with slopes of up to 45 degrees (Healey, 2004). Obstacle avoidance manoeuvres are required to safely negotiate steeper slopes, step-like terrain features (e.g. sand bars or coral heads), or objects proud of the ocean floor. As described earlier, since the REMUS FLS is mounted in a fixed orientation, it may detect new obstacles while executing a manoeuvre to avoid the current obstacle threat. Periodic or detection-based replanning can handle these situations. This scenario was illustrated conceptually in Fig.7.

When negotiating a step-up ridge or sand bar whose extent is not known due to sonar occlusion at the time of detection, it may not be desirable to follow the planned vertical trajectory to its completion. Between planning iterations, a simple yet safer approach is to revert back to constant altitude control once the vehicle reaches a position directly above the detected object boundaries. This condition can be checked using a 2.5D version of the 3D bounding box intersection test described above. Figure 18 illustrates a simulation whereby the REMUS FLS detects the leading edge of a ridge at maximum sonar range. Image processing algorithms compute range to the object (80m) as well as its width (W, into the page) and its height above the seafloor (5.5m) but cannot determine the length of the ridge since it is occluded by its own leading edge. Therefore, the obstacle detection algorithm generates a 3D bounding box measuring W m wide x 1.0m long (assumed) x 5.5m high.

While the IDVD-method planner generates a vertical trajectory in NED coordinates, in shallow water it is safer to operate the vehicle in an altitude control mode. Therefore, the vertical coordinate trajectory is converted from a depth plan into an altitude plan by assuming constant water depth over the planning horizon and using the relationships

$$\begin{aligned} D &= h_{nom} + z(0) \qquad h_{plan}(\tau) = D - z_{plan}(\tau) \\ \Delta h_{plan}(\tau) &= h_{plan}(\tau) - h_{nom} \qquad h_{cmd}(\tau) = h_{nom} + \Delta h_{plan}(\tau) \end{aligned} \tag{42}$$

where D is the water depth computed at planner initialization time, h_{nom} is the nominal altitude set point, and h_{cmd} (τ) is the altitude command sent to the autopilot via the RECON interface. The resulting altitude plan Δh_{plan} (τ) is shown in Fig.18 as a deviation from the nominal mission segment altitude. As seen, sending this altitude command directly to the vehicle autopilot would cause an undesirable jump in the altitude profile once the ADCP/DVL sensor measures the vehicle's true altitude above the ridge. Instead, switching the altitude command to the nominal mission segment altitude once the vehicle reaches the ridge will produce the desired altitude profile). Note that even though Fig.18 depicts a sudden drop in altitude when the vehicle passes the ridge while commanding the nominal mission segment altitude, in practice the vehicle dynamics will ensure that the UUV executes a smooth transition back to its nominal survey altitude.

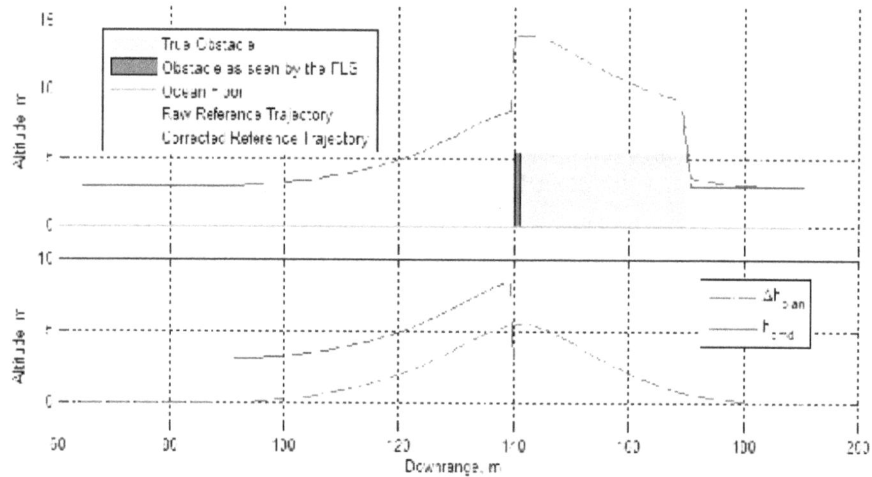

Fig. 18. Simulation results for vertical OA using UUV altitude control mode

7. COMPUTER SIMULATIONS AND SEA TRIALS

The suggested path-planning approach can be customized for a particular vehicle or operational environment by adjusting the performance index J to include vehicle or actuator dynamics (feasibility restrictions) and mission objectives such as OA or underwater rendezvous. This section provides simulated and in-water experimental outcomes for four distinct applications utilizing the proposed trajectory optimization framework for unmanned marine vehicle guiding. i) Underwater docking of an unmanned underwater vehicle (UUV) with a mobile underwater recovery system (MURS); ii) Optimal utilization of a terrain-relative feature map to enhance UUV self-localization precision; iii) Two-dimensional or three-dimensional obstacle avoidance in cluttered environments; and iv) Specific unmanned surface vehicle (USV) applications for sonar-based obstacle avoidance in riverine operations.

7.1 Underwater recovery

The goal of underwater recovery (Yakimenko et al., 2008) is to be able to compute a rendezvous trajectory from any point on the UUV holding pattern to any point on the MURS holding pattern as shown in Fig.19 (note hereinafter that depth values are shown as negative numbers).

While the stochastic simulation shown in Fig.19 employs circular race tracks, in practice the MURS would establish a race track that allowed it to travel back and forth along two long track legs (see Fig.20). These legs are needed to allow sufficient time to contact the UUV (which is assumed to be in its holding pattern somewhere within communication range) and allow it to transit from its holding pattern to the rendezvous point. The proposed sequence of events is to have the MURS (at position 1 in Fig.20) signal the UUV (at position 2) and command it to proceed to a rendezvous point by a certain time. The UUV computes the trajectory required to comply with the command. If the commanded rendezvous is feasible, the UUV sends an acknowledgement message. Otherwise (i.e. the request violated some constraint) the UUV sends a denial message (stage A in Fig.20) and requests that the MURS command a different rendezvous point or time. The final point of the trajectory is located in the approximate location of the MURS docking station at a given time. Knowing the geometry of the MURS allows the planner to construct a "keep out" zone corresponding to the MURS propeller and aft control surfaces. The UUV rendezvous trajectory must avoid this area. Once the rendezvous plan has been agreed upon and acknowledged, both the UUV and the MURS proceed to position 3 for rendezvous (stage B). Finally, at position 4 the recovery operation (stage C) is completed.

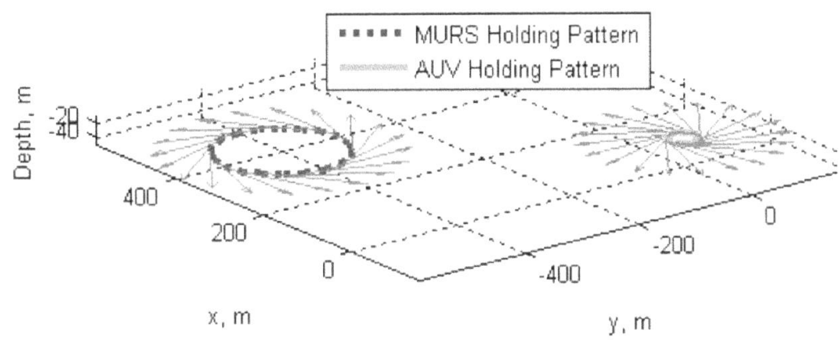

Fig. 19. Manifold of initial and final conditions

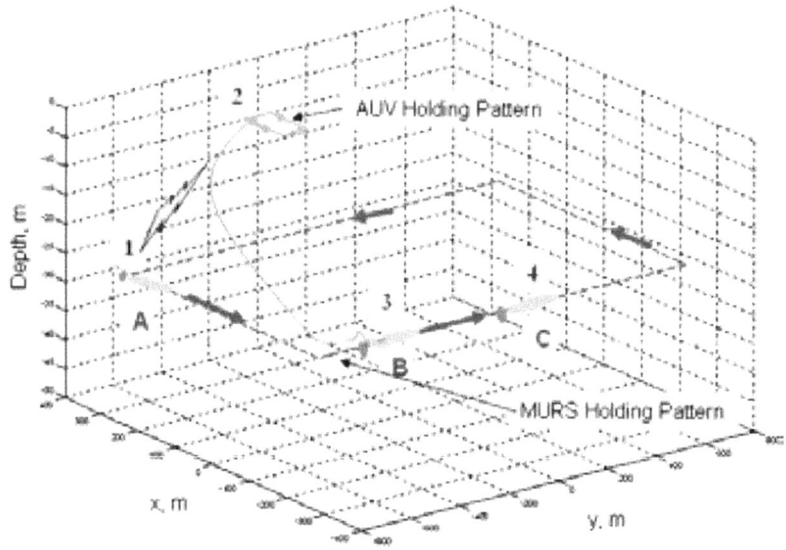

Fig. 20. Proposed rendezvous scenario

The simulated rendezvous scenario assumes three stages: communication (A), execution (B), and recovery (C), respectively. From the trajectory generation standpoint we are primarily concerned with optimizing the path that would bring the UUV from its current position (point 2) to a certain rendezvous state (point 3) in the preset time T_r proposed by the MURS, while obeying all possible real-life constraints and avoiding the MURS keep out zone.

Figures 21 and 22 present a computer simulation in which a MURS is moving due east at 1m/s (1.94kn) with the docking station at a depth of 15m. A UUV is located 800 meters away. The MURS wishes to conduct a rendezvous operation T_r minutes later and sends the corresponding information to the UUV. This information includes the proposed final position x_f, y_f, z_f rendezvous course, speed, and time. Figure 21 shows several generated trajectories, which meet the desired objectives for this scenario and also avoid an obstacle located along the desired path to MURS. These trajectories differ by the arrival time T_r.

During handshaking communications with the MURS, the UUV determines whether the suggested T_r is feasible. Of the four trajectories shown, the trajectory generated for $T_r = 450$s happens to be infeasible (the constraints on controls are violated). The solution of the minimum-time problem for this scenario yielded 488 seconds as the soonest possible rendezvous time.

The other three trajectories shown in Fig.21 are feasible. That means that the boundary conditions are met (by construction) and all constraints including OA are satisfied (via optimization). As an example, Fig.22 shows the time histories for the yaw rate $\dot{\psi}_C$ and flight path angle γ_c vehicle control parameters as well as the UUV's speed as it followed the trajectory for $T_r = 600s$.

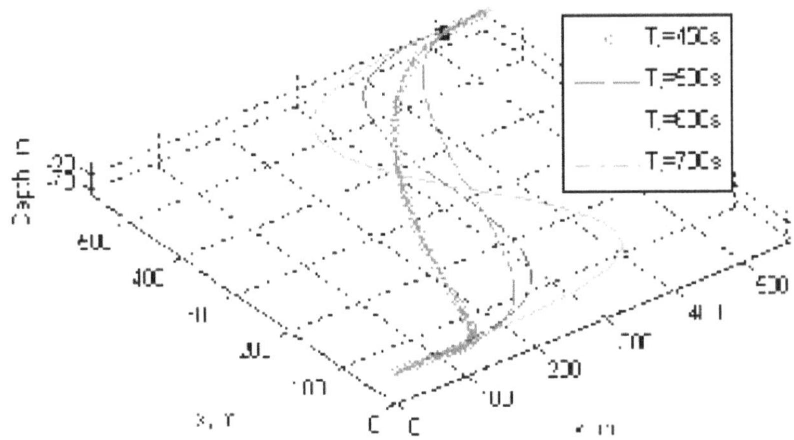

Fig. 21. Examples of rendezvous trajectories

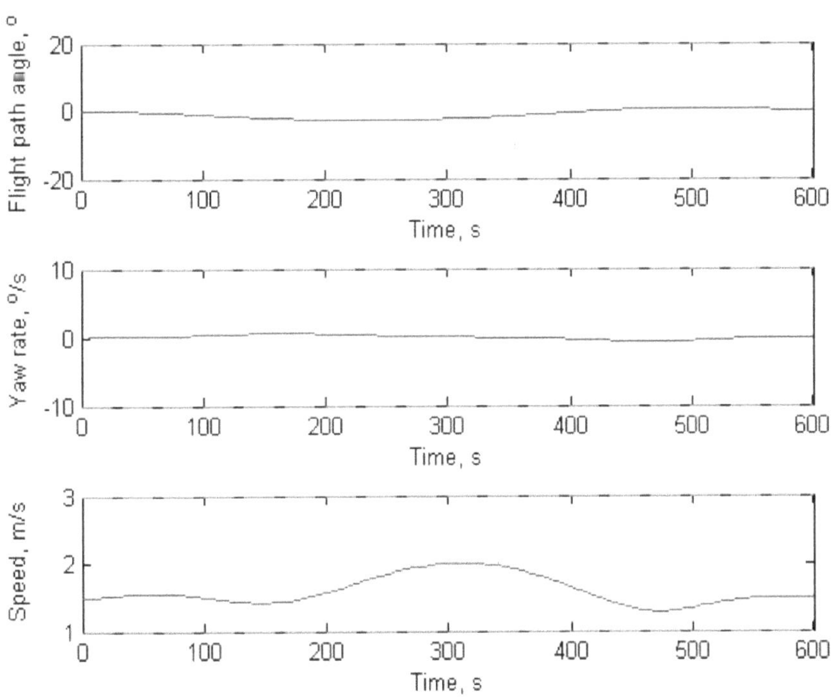

Fig. 22. Constrained vehicle parameters for $T_r = 600s$

Stochastic simulations of the manifolds shown in Fig.21 illustrate that a successful rendezvous can take place in all cases as long as T_r is greater than a certain value. Furthermore, they show that minimization of the performance index using the IDVD method ensures that a smooth, realizable trajectory is calculated in just a few seconds, regardless of the initial guess. Converting code to an executable file in lieu of using an interpretative programming language reduces execution time down to a fraction of a second.

7.2 Feature-based navigation

Over the past decade, numerous UUVs have been created to execute diverse underwater operations. Survey-class vessels are equipped with precise navigational and sonar instruments for ocean bottom mapping, rendering them rather costly. Vehicles that do not possess these payloads might execute several beneficial tasks at a significantly reduced cost; nevertheless, their performance will deteriorate with time due to faulty self localization unless external navigation aids are utilized. Consequently, it is intriguing to examine collaborative operations through a fleet of vehicles for optimal usefulness at a fair expense. The NPS CAVR has been examining a concept of operations known as feature-based navigation. This method enables vehicles equipped solely with a GPS receiver and inexpensive image sonar to utilize a precise sonar map produced by a survey truck. This map consists of landscape or bottom object features that serve as potential navigational references. The sonar map is uploaded to the economical follow-on vehicles prior to launch. Commencing from an initial GPS position fix acquired at the surface, these vehicles thereafter navigate underwater by comparing current sonar imaging with the sonar features from the survey vehicle's map.

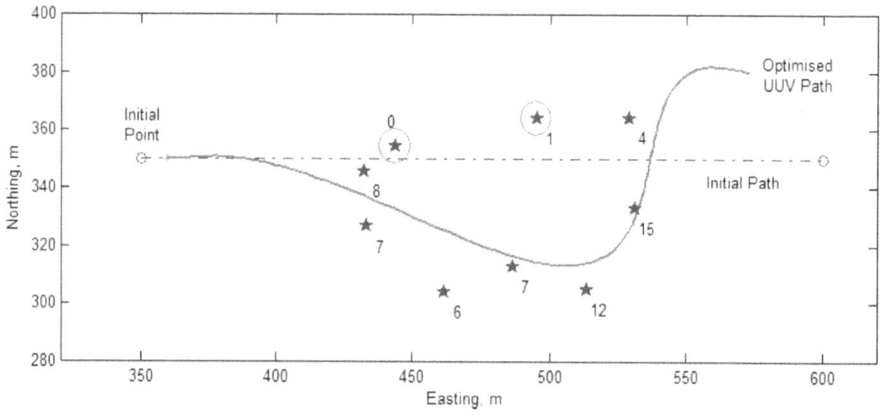

Fig. 23. Simulation results for a feature-based navigation application

The localization precision of vehicles utilizing feature-based navigation can be enhanced by optimizing the frequency of navigational reference detections by imaging sonar. This simulation illustrates the customization of the IDVD trajectory generating framework for this application. A straightforward geometric model of a FLS with a range of 60m, a 30-degree horizontal field of view, and a notional ping rate of 1Hz was utilized to create a novel performance index that prioritizes potential trajectories directing the sonar towards navigational references in the a priori feature map. In this instance, we aimed to identify trajectories capable of acquiring a minimum of three sonar photos for each feature on the map. Figure 23 illustrates the outcomes of a computer simulation, with annotations indicating the frequency of sonar imaging for each object. The resultant trajectory is viable (i.e., adheres to turn rate limits) and produces three or more sonar photos of all but two targets.

7.3 Obstacle avoidance in cluttered environments

Another application which benefits from the aforementioned trajectory generation algorithm is real-time OA in a highly cluttered environment. Figure 24 illustrates simulated trajectories for avoiding a field of point-like objects in the 2D horizontal plane (e.g. a kelp forest) and in all three dimensions (e.g. a mine field). In both simulations, the performance index was designed to minimize deviations from a predefined survey track line while avoiding all randomly generated obstacles via a CPA calculation. Terminal boundary conditions for the OA manoeuvre were chosen to ensure the UUV rejoined the desired track line before reaching the next waypoint (i.e. the manoeuvre terminated at a position 95% along the track segment). Initial boundary conditions were chosen to simulate a random obstacle detection which triggers an avoidance manoeuvre after the UUV has completed about 10% of the predefined track segment. For illustration purposes, Fig.24 includes several candidate trajectories evaluated during the optimization process although the algorithm ultimately converged to the trajectory depicted with a thicker (red) line (CPA distances to each obstacle appear as dashed lines).

Figure 25 shows the results from an initial sea trial of 3D OA that took place in Monterey Bay on 9 December 2008. This experiment tested periodic trajectory generation and replanning on the REMUS UUV using a simulated obstacle map comprised of oriented bounding boxes. As seen in Fig.25, initially the REMUS UUV follows a predefined track segment (dash-dotted line) at 4 meters altitude. At some point the vehicle's FLS simulator "detects" an obstacle (i.e. the current REMUS position and orientation place the virtual obstacle within the range and aperture limits of the FLS). This activates the OA mode, and the planner generates an initial trajectory (green) from the current vehicle position to the final waypoint. REMUS follows this trajectory until the next planning cycle (4 seconds later) when the vehicle generates a new trajectory and continues this path planning-path following cycle.

7.4 Obstacle avoidance in restricted waterways

The NPS CAVR in collaboration with Virginia Tech (VT) is developing technologies to enable safe, autonomous navigation by USVs operating in unknown riverine environments. This project involves both surface (laser) and subsurface (sonar) sensing for obstacle detection, localization, and mapping as well as global-scale (wide area) path planning, local- scale trajectory generation, and robust vehicle control. The developed approach includes a hybrid receding horizon control framework that integrates a globally optimal path planner with a local, near-optimal trajectory generator (Xu et al., 2009).

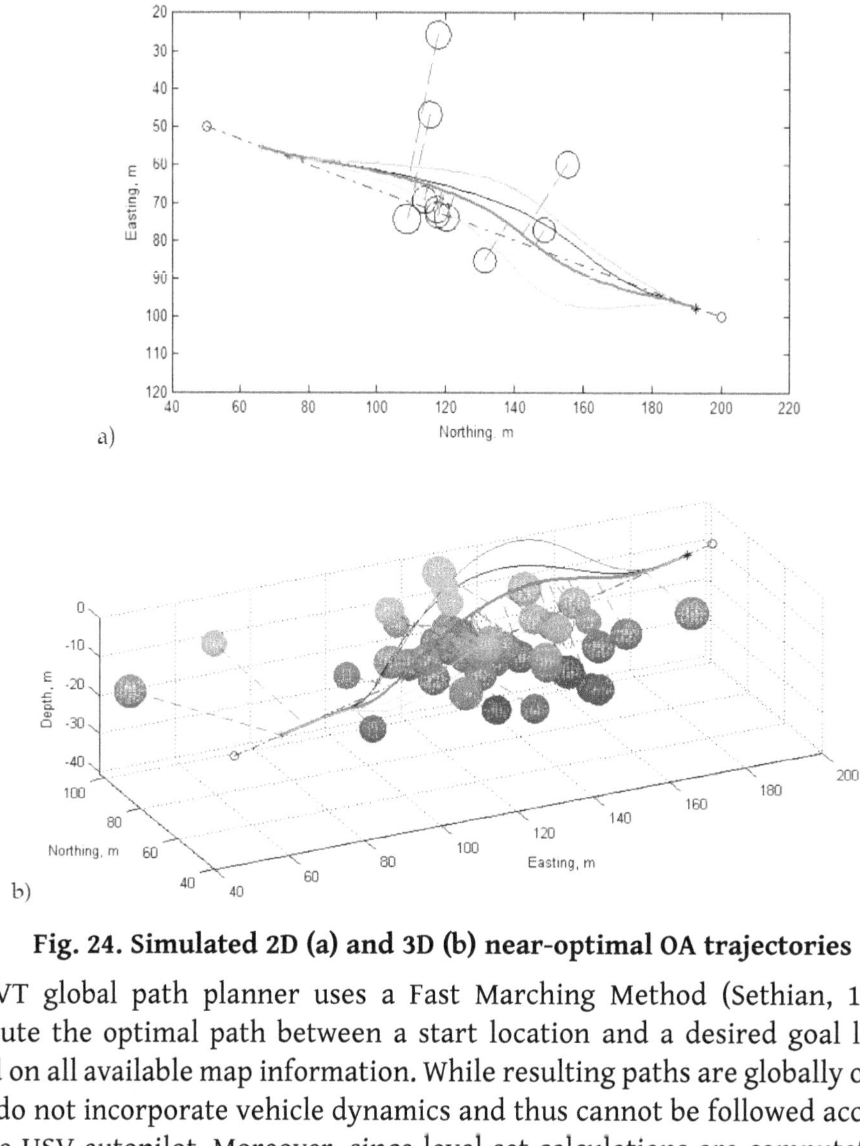

Fig. 24. Simulated 2D (a) and 3D (b) near-optimal OA trajectories

The VT global path planner uses a Fast Marching Method (Sethian, 1999) to compute the optimal path between a start location and a desired goal location based on all available map information. While resulting paths are globally optimal, they do not incorporate vehicle dynamics and thus cannot be followed accurately by the USV autopilot. Moreover, since level set calculations are computationally expensive, global plans are recomputed only when necessary and thus do not always incorporate recently detected obstacles. Therefore, a complimentary local path planner operating over a short time horizon is required to incorporate current sensor information and generate feasible OA trajectories. The IDVD- based trajectory generator described above is ideally suited for this purpose. VT has developed a set of matching conditions which guarantee the asymptotic stability of this framework. When these matching conditions are satisfied, the sequence of local trajectories will converge to the global path's goal location. If the local trajectories no longer satisfy these conditions (usually because the global path is no longer compatible with recently detected obstacles), the global path is recomputed.

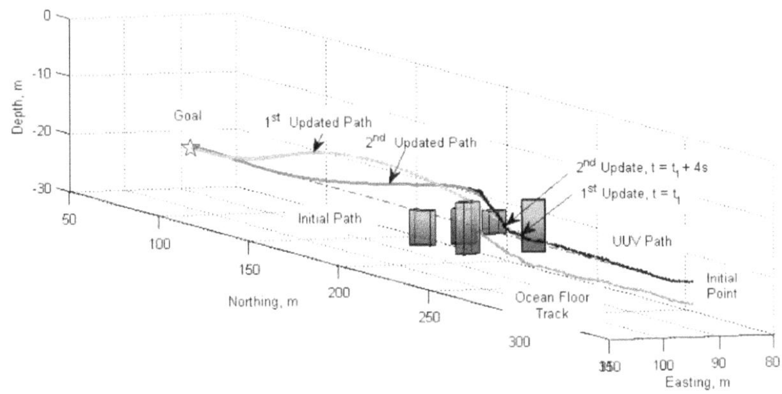

Fig. 25. REMUS sea trial results demonstrating periodic planning and path following

Simulation results demonstrate the need for local trajectories that incorporate vehicle dynamics and real-time sensor data (Fig.26). For this simulation, an initial level set map was computed using an occupancy grid created by masking land areas as occupied and water areas as unoccupied in an aerial image of the Sacramento River operating area. Performing gradient descent on the level set from the USV's initial position produces an optimal path shown in blue. To simulate local trajectory generation with a stale global plan, the initial level set map was not updated during the entire simulation. Meanwhile, to simulate access to real- time sensor data, the local planner was provided with a complete sonar map generated during a previous SeaFox survey of the area. In Fig.26, this sonar map has been overlaid on the *a priori* map with red and green colour channels representing the probability that a cell is occupied or unoccupied, respectively. Black pixels represent cells with unknown status. A short green line segment depicts the USV's orientation when the local planner is invoked, and the resulting trajectory is shown in yellow. The first simulation (Fig.26a) shows a local trajectory which deviates from the stale global plan to avoid a sand bar detected with sonar. In the second simulation (Fig.26b) the USV is initially heading in a direction opposite from the global path, but the local planner generates a dynamically feasible trajectory to turn around and rejoin the global path later.

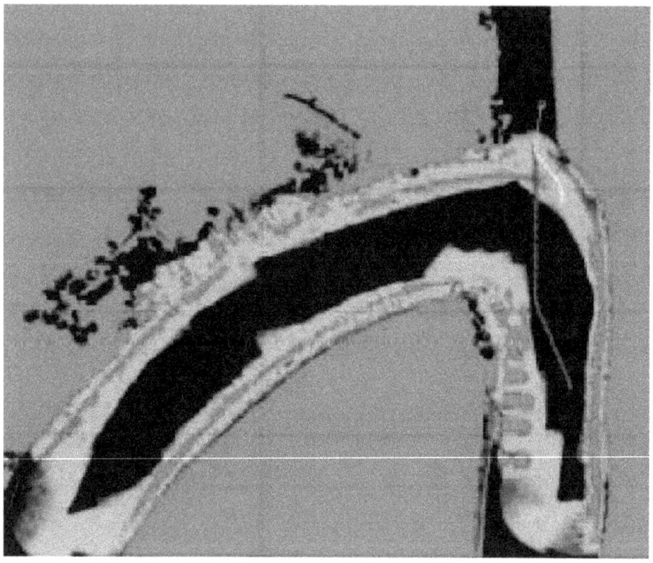

a)

b)

Fig. 26. Simulated local OA trajectories

To track these local trajectories, the 2D controller described in Section 6.1 was implemented on the SeaFox USV by mapping the controller's turn rate commands into rudder commands understood by the SeaFox autopilot. After validating the turn rate controller design during sea trials on Monterey Bay, the direct method trajectory generator and closed-loop path following controller were tested on the Pearl River in Mississippi on 22 May 2010 (Fig.27). For this test, the local planner used a sonar map of the operating area to generate the trajectory (the cyan line) from an initial orientation (depicted by the yellow arrow) to a desired goal point (depicted by a circle). The SeaFox USV then followed it almost precisely (the magenta line). Figure 27 illustrates that the trajectory generator was activated at a random position while the USV executed a clockwise maneuver. As the USV was directed to return to its initial position after completing this maneuver, the pink line encompasses a segment of its return trajectory; otherwise, the actual USV path would closely resemble the reference trajectory in this plot.

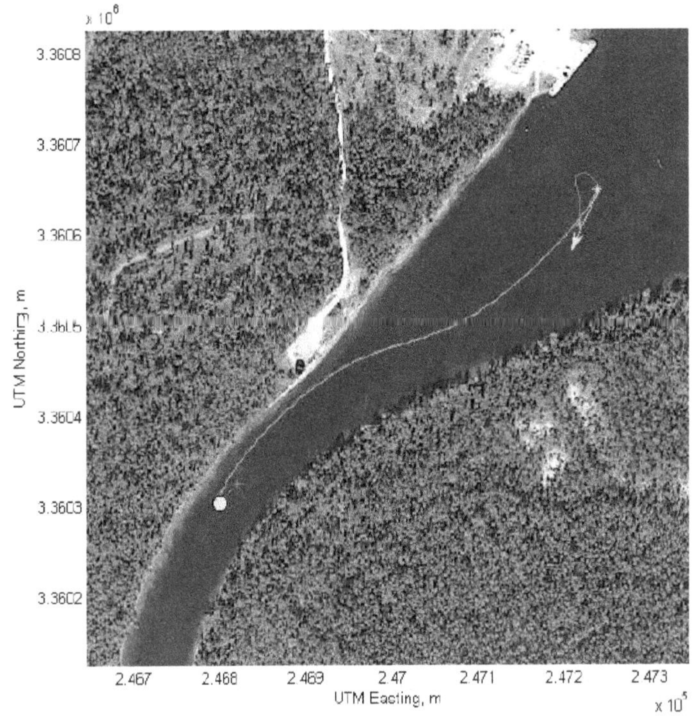

Fig. 27. Path-following controller test on the Pearl River

8. CONCISE SUMMARY

A trajectory planner for onboard use, utilizing the Inverse Dynamics in the Virtual Domain direct technique discussed in this chapter, effectively enhances an unmanned maritime vehicle's autopilot by providing smooth, viable paths and associated controls. It also enables the integration of advanced sensors, such as forward-looking sonar, enabling both proactive and reactive obstacle evasion. This methodology has been applied to both unmanned underwater and surface vehicles, exhibiting significant potential. The suggested approach not only computes near-optimal collision-free trajectories significantly quicker than real time but also accommodates any practically viable compound performance index. The created control architecture is both universal and user-friendly, as evidenced by several simulations and initial sea trials. This chapter revealed results from a limited number of preliminary sea trials. Subsequent study will advance the proposed trajectory architecture to facilitate additional tactical scenarios.

REFERENCES

Basset, G., Xu, Y. & Yakimenko, O. (2010). Computing short-time aircraft maneuvers using direct methods," *Journal of Computer and Systems Sciences International*, 49(3), 145-176

BlueView Technologies, Inc. (2011). 2D Imaging sonar webpage. Available from: www.blueview.com/2d-Imaging-Sonar.html

Bourke, P. (1992). Intersection of a line and a sphere (or circle). Professional webpage. Available from: http://paulbourke.net/geometry/sphereline

Elfes, A. (1989). Using occupancy grids for mobile robot perception and navigation. *Computer*, 22(6), 46-57

Furukawa, T. (2006). *Reactive obstacle avoidance for the REMUS underwater autonomous vehicle using a forward looking sonar.* MS Thesis, NPS, Monterey, CA, USA

Gadre, A., Kragelund, S., Masek, T., Stilwell, D., Woolsey, C. & Horner, D. (2009). Subsurface and surface sensing for autonomous navigation in a riverine environment. In: *Proceedings of the Association of Unmanned Vehicle Systems International (AUVSI) Unmanned Systems North America convention, Washington, DC, USA*

Healey, A. J. (2004). Obstacle avoidance while bottom following for the REMUS autonomous underwater vehicle. In: *Proceedings of the IFAC conference, Lisbon, Portugal*

Horner, D. & Yakimenko, O. (2007). Recent developments for an obstacle avoidance system for a small AUV. In: *Proceedings of the IFAC conference on Control Applications in Marine Systems, Bol, Croatia*

Horner, D., McChesney, N., Kragelund, S. & Masek, T. (2009). 3D reconstruction with an AUV-mounted forward-looking sonar. In: *Proceedings of the International symposium on Unmanned Untethered Submersible Technology (UUST09), Durham, NH, USA*

Hydroid, Inc. (2011). REMUS 100 webpage. Available from: www.hydroidinc.com/remus100.html

Kaminer, I., Yakimenko, O., Dobrokhodov, V., Pascoal, A., Hovakimyan, N., Cao, C., Young, A. & Patel, V. (2007). Coordinated path following for time-critical missions of multiple UAVs via L1 adaptive output feedback controllers. In:

Proceedings of the AIAA Guidance, Navigation, and Control conference, Hilton Head, SC, USA

Kreuzer, J. (2006). 3D programming – weekly: Bounding boxes. Collision detection tutorial webpage. Available from: www.3dkingdoms.com/weekly/weekly.php?a=21

Masek, T. (2008). *Acoustic image mModels for navigation with forward-looking sonars.* MS Thesis, NPS, Monterey, CA, USA

Northwind Marine. (2011). SeaFox webPage. Available from: www.northwindmarine.com/military-boats

Sethian, J. (1999). Fast marching method. *SIAM Review,* 41(2), 199-235

Xu, B., Kurdila, A. J. & Stilwell, D. J. (2009). A hybrid receding horizon control method for path planning uncertain environments. In: *Proceedings of the IEEE/RSJ International conference on Intelligent Robots and Systems, St. Louis, MO, USA*

Yakimenko, O. & Slegers, N. (2009). Optimal control for terminal guidance of autonomous parafoils. In: *Proceedings of the 20th AIAA Aerodynamic Decelerator Systems Technology conference, Seattle, WA, USA*

Yakimenko, O. (2000). Direct method for rapid prototyping of near optimal aircraft trajectories. *Journal of Guidance, Control, and Dynamics,* 23(5), 865-875

Yakimenko, O. (2011). *Engineering computations and modeling in MATLAB/Simulink.* AIAA Education Series, ISBN 978-1-60086-781-1, Arlington, VA, USA

Yakimenko, O. A. (2008). Real-time computation of spatial and flat obstacle avoidance trajectories for AUVs. In: *Proceedings of the 2nd IFAC workshop on Navigation, Guidance and Control of Underwater Vehicles (NGCUV'08), Killaloe, Ireland*

Yakimenko, O.A., Horner, D.P. & Pratt, D.G. (2008). AUV rendezvous trajectories generation for underwater recovery, In: *Proceedings of the 16th Mediterranean conference on Control and Automation, Corse, France*

5 Chapter

DECENTRALIZED CONTROL FUNCTIONS FOR AUV FORMATION GUIDANCE

1. INTRODUCTION: AN OVERVIEW

Autonomous Underwater Vehicles (AUVs) represent the most intricate category of unattended marine systems, characterized by mobility, complicated dynamics, and non-holonomic kinematics. They are progressively acknowledged as a fundamental technology for advancing human scientific and economic interests in the deep ocean (Papoulias et al., 1989). A recent analysis by Bildberg (2009) concludes, according to several prominent researchers, that AUVs are swiftly advancing towards maturity. The autonomy of AUVs constitutes their primary capability. They independently investigate oceanic phenomena pertinent to human scientific and commercial interests. Expertly designed autonomous control enables them to operate reliably and predictably in response to waves, currents, wind, sea state, and several other natural disturbances and operational situations. Consequently, they are currently assuming a prominent role in representing human presence and interests in the Ocean over a broad spectrum of themes.

- Physical oceanography (Plueddemann et al., 2008; Tuohy, 1994),
- Marine biology, conservationist biology, marine ecology management, biological oceanography (Farrell et al., 2005; Pang, 2006; Pang et al., 2003),
- Geology, petrology, seismology, hydrography (for the benefit of e.g. the oil and gas industry, maritime civil engineering etc.),
- Maritime and naval archaeology, submerged cultural heritage protection and management,
- Marine traffic management, search and rescue, hazardous material and waste management, emergencies and catastrophes management and first responding (Carder et al., 2001; Pang, 2006)
- Maritime security, customs enforcement, border protection and defense (Allen et al., 1997; 2004; Clegg & Peterson, 2003; Curtin et al., 1993; Eisman, 2003; US Navy, 2004).

Autonomous Underwater Vehicles (AUVs) represent the most intricate category of unattended marine systems, characterized by mobility, complicated dynamics, and non-holonomic kinematics. They are progressively acknowledged as a fundamental technology for advancing human scientific and economic interests in the deep ocean (Papoulias et al., 1989). A recent analysis by Bildberg (2009) concludes, according to several prominent researchers, that AUVs are swiftly advancing towards maturity. The autonomy of AUVs constitutes their primary capability. They independently investigate oceanic phenomena pertinent to human scientific and commercial interests. Expertly designed autonomous control enables them to operate reliably and predictably in response to waves, currents, wind, sea state, and several other natural disturbances and operational situations. Consequently, they are currently assuming a prominent role in representing human presence and interests in the Ocean over a broad spectrum of themes.

1.1 The cooperative paradigm

Since the 1970s, robotics and control engineers have examined the cooperative paradigm. Cooperative control comprises comprehensive, terminating algorithms and machine-executed strategies that enable several individual agents to accomplish a certain job optimally. This optimality arises from the agents utilizing each other's resources (e.g., maneuvering capabilities) to more efficiently minimize a cost function that quantifies the overall task's "budget," in contrast to what each agent could achieve independently (without the advantages of collaboration).In the marine environment, such "social" leveraging is beneficial in several ways. Firstly, deployment of more AUVs significantly reduces the time needed to survey a given theater of operations. This has enormous economic repercussion in terms of conserved hours or days of usually prohibitively expensive ship-time (for the vessel that is rendering operational support to the AUV fleet). Secondly, deployment of a larger number of AUVs *diversifies the risk to operations.* In a group scenario, loss of a (small) number of AUVs doesn't necessarily preclude the achievement of mission goals. Lastly, if each of the group AUVs are furnished with adaptive-sampling algorithms, such as in the chemical plume-tracing applications (Farrell et al., 2005; Pang, 2006; Pang et al., 2003), deployment of multiple vehicles guarantees much faster convergence to the points of interest.

Cooperative control frameworks are categorized into centralized and decentralized strategies. The objective of a centralized cooperative control system is to ascertain the actions of each agent based on a comprehensively known data set of the problem, encompassing the state vectors of all agents involved. The centralized system generates a globally optimal solution by evaluating the current resource allocation of the entire group, along with the comprehensive, if potentially imperfect, understanding of the environment. The status data are typically gathered by polling all agents via a communication network. Subsequent to the polling cycle, the centralized system transmits low-level guiding commands to individual agents. This method facilitates the attainment of a global optimum in decision-making based on all available information, although it is significantly reliant on fault-tolerant, quality-assured, high-bandwidth connectivity. In a decentralized framework, as outlined in this chapter, each agent has incomplete state and perception information on all other agents and the observable environment, and independently determines its own course of action. The primary challenge in decentralized cooperative control is attaining consensus among independently thinking autonomous agents.

2. THE FRAMEWORK OF VIRTUAL POTENTIALS

A solution utilizing virtual or artificial potentials is offered to tackle the challenge of reactive formation guiding for multiple AUVs traveling in a sea environment. The virtual potentials mitigate several prominent issues faced by competing reactive formation guidance techniques, which are susceptible to the following challenges:

- Reliance on the *perfect knowledge* of a map of the waterspace,
- Lack of *reaction* to the decentralized, agent-local process of accumulating or perfecting knowledge of the environment on top of the initially imperfect situational awareness of each individual agent,
- Trajectory planning that is sub-optimal, or optimal based on a *hard-coded* criterion, without possibility of adjusting or restating that criterion at run-time, because the cost function is implicit in the choice of mathematical tools (such as a distinct set of curve formulations used for trajectories etc.).

Stemming from these considerations, we propose a scheme where each AUV in a 2D formation imbedded in the "flight ceiling" plane as previously discussed maintains a local imperfect map of the environment. Every possible map only ever consists of a finite number of instantiations of any of the three types of features:

- A way-point that is commanded for the entire formation, $w \in \mathbb{R}^2$
- Obstacles which need to be circumnavigated in a safe and efficient manner, (Oi), $\forall i = 1 \ldots n_{obs}$ $\mathbb{O}_i \subset \mathbb{R}^2$,
- Vertices of the *characteristic cell* of the chosen formation geometry, covered in more detail in sec. 2.3.3 and 3.

With this in mind, let the virtual potential be a real, single valued function $P : \mathbb{R}^2 \to \mathbb{R}$, mapping almost every attainable position of an AUV on the "flight ceiling" to a real. Let P-s total differential exists almost wherever the function itself is defined. P can be said to live on the subspace of the full-rank state-space of the AUVs, $C = \mathbb{R}^6 \times \mathbb{SE}^3$. The state-space of the AUV is composed of the Euclidean 6-space R6 spanned by the angular and linear velocities, $\{[\mathbf{v}^T\ \omega^T]^T\} = \{[u\ v\ w \mid p\ q\ r]^T\} \equiv \mathbb{R}^6$ and a full 3D, 6DOF configuration-space $\{[x^T\ \Theta^T]^T\} = \{[x\ y\ z \mid \varphi\ \vartheta\ \psi]^T\}$ which possesses the topology of the *Special Euclidean group of rank* 3, $\mathbb{SE}^3$. Function *P* therefore maps to a real scalar field over that same *C*.

Furthermore, this framework will be restricted to only those *P* that can be expressed in terms of a sum of finitely many terms:

$$\exists n \in \mathbb{N} \mid P_\Sigma = \sum_{i=1}^{n} P_i \tag{1}$$

Where P_i is of one of a small variety of considered function forms. Precisely, we restrict our attention to three function forms with each one characteristic of each of the three mentioned *types of features* (way-point, obstacle, vertices of formation cells).

The critical issue in the guidance problem at hand is Euclidean 2D distance (within the "flight ceiling") between pairs of AUVs in the formation, and each AUV and all obstacles. Therefore, our attention is further restricted to only such $\{P_i\} \subset \mathcal{L}(C \to \mathbb{R})$ with $\mathcal{L}$ being the space of all functions mapping C to R whose total differential exists almost wherever each of the functions is defined on *C*, which can be represented as the composition $P_i \equiv p_i \circ di$, $pi : \mathbb{R}^+_0 \to \mathbb{R}$, and $d_i : C \to \mathbb{R}^+_0$ a Euclidean 2D metric across the "flight ceiling". Consequently, P_i is completely defined by the choice of $p_i(d)$, the *isotropic potential contour generator.* Choices and design of p_i (d)-s will be discussed in sec. 2.3.

With all of the above stated, a *decentralized total control function* $f : \mathbb{Z} \to \mathbb{R}^2$ is then defined as a sampling, repeated at sample times $k \in \mathbb{Z}_0$, of the 2D vector field $E : \mathbb{W}_i \to \mathbb{R}^2$ over a subspace $\mathbb{W}_i \subseteq \mathbb{R}^2 \subset C$, the *navigable waterspace*:

$$\forall x \in \mathbb{W}_i \subseteq \mathbb{R}^2 \subset C,\ \mathbf{E}(x) = -\nabla P_\Sigma(x) \tag{2}$$

Where $\mathbb{W}_i = \mathbb{R}^2 \setminus \left(\cup_i \mathbb{O}_i \cup \cup_j \mathbb{O}_j^{(ag)}\right)$ contains all of $\mathbb{R}^2$ to the exclusion of closed connected subsets of $\mathbb{R}^2$ that represent interiors of obstacles, $\{\mathbb{O}_i\}$ and those that represent safety areas around all the j-th AUVs ($j \neq$ i) other than the i-th one considered. $\mathbb{W}_i$ is an open, connected subset of $\mathbb{R}^2$, the "flight ceiling", inheriting its Euclidean-metric-generated topology and always containing the way-point w. Sampling **E** at the specific $x_i(k) \in \mathbb{W}_i$, the location of the i-th AUV, results in $f_i(k)$, the *total decentralized control function* for the i-th AUV at time *k* and location x_i.

2.1 Passivity

The decentralized total control function *f* is used as the forcing signal of an idealized dimensionless charged particle of unit mass, modeled by a *holonomic 2D double integrator.* If any AUV were able to behave in this manner, the AUV would follow an *ideal conservative trajectory* given by:

$$x_i(t) = \iint_{\tau=0}^{t} \mathbf{E}\left[x_i(\tau)\right] d\tau^2 + x_i^{(0)} \tag{3}$$

This ideal conservative trajectory, while stable in the BIBO sense, is in general not asymptotically stable, nor convergent by construction. The simplest case when this doesn't hold is when **E**(x) is an *irrotational*[1] vector field whose norm is affine in the $\|x - w\|$ 2D Euclidean distance:

$$\|\mathbf{E}(x)\| = e\|x - w\| + E_0;\ e \in [0, \infty) \tag{4}$$

And whose direction is always towards *w*:

$$\forall x,\ \mathbf{E}(x) \cdot (x - w) \stackrel{\text{id}}{=} e\|x - w\|^2 + E_0\|x - w\| \tag{5}$$

In that case (3) can be regarded as a linear second or third order system with two of the poles in ±i. Such a system exhibits borderline-stable oscillation – a hallmark of its *conservativeness.* An example of such BIBO-stable non-convergent oscillation is given in figure 1.

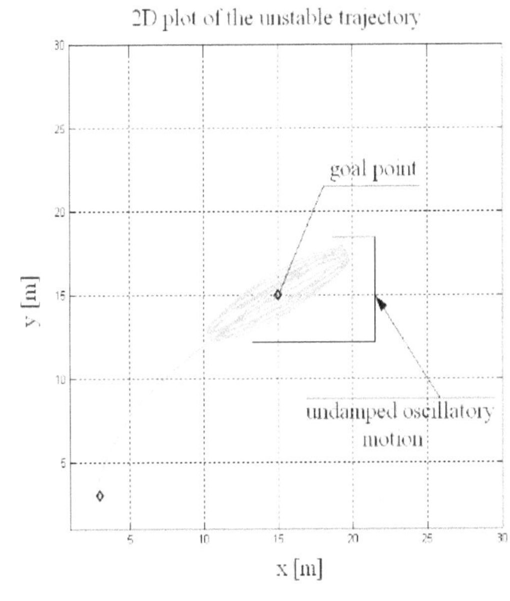

Fig. 1. An example of an oscillatory trajectory due to the conservativeness of the virtual potential system.

Note that this analysis is irrespective of the initial condition $\dot{x}_0$ as long as (4, 5) approximate **E**(x) sufficiently well in some open ε-ball centered on w. However, AUVs are in general not able to actuate as ideal holonomic 2D double integrators. The introduction of *any* finite non-zero lag in the above discussion, which is sure to exist from first physical principles in a real AUV, is sufficient to cause dissipation and as a consequence passivity and convergence to *w*.

2.2 Local minima

In addition to the problem of passivity, the virtual potential approach suffers from the existence of local minima. Without further constraints, the nature of $\mathbf{E}(\mathbb{W}_i)$ so far discussed doesn't preclude a dense, connected, closed state-subspace $c_{i0}^{(j)} \subseteq \mathrm{C}$, containinguncountablymanyinitialvectors $\{x_0^{(l)} \mid l \in \mathbb{R}\}$ of"related"trajectories(with the indexing by AUV denoted by i, and the enumeration of the distinct points of convergence other than the way-point by j omitted for clarity) that *do not* converge to the way-point w or a finitely large orbit around it, but rather to another point $x^{(j)}_{\infty}$ (or a finitely large orbit around *it*). Therefore, for each of these uncountably many "nearby" trajectories (to be visualized as a "sheaf" of trajectories emanating from a distinct, well defined neigbourhood in Wi for some range of initial linear and angular velocities) there exists a lower bound t_l after $\|x_i(t > t_l) - x_\infty^{(l)}\| \le \|x_i(t > t_l) - w\|$ almost always. The set $\{t_l\}$ is also dense and connected.

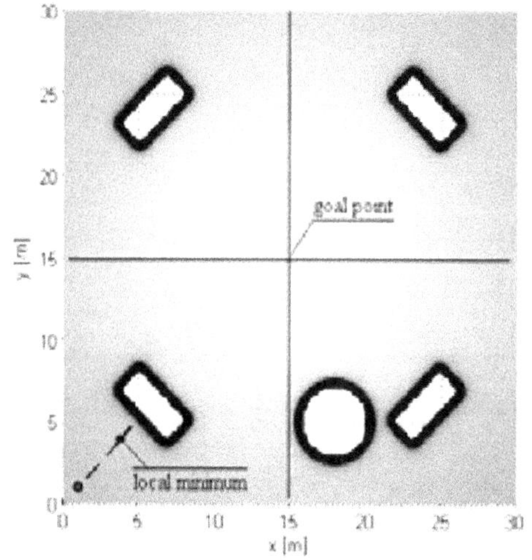

Fig. 2. Example of a local minimum occurring in virtual potential guidance in a 2D waterspace.

Furthermore, there is no prejudice as to the number of such $C^{(j)}_{i0}$-s, i.e. there exists $C_{i0}^{\Sigma} \subseteq C, C_{i0}^{\Sigma} = \cup_j C_{i0}^{(j)}$. There may be multiple disjoint dense, connected, closed sets of initial conditions of the trajectory of the i-th AUV which all terminate in the same, or distinct local minima. The enumerator j may even come from $\mathbb{R}$ (i.e. there may be *uncountably many distinct local minima,* perhaps arranged in dense, connected sets – like curves or areas in $\mathbb{R}^2$).

An example of an occurrence of a local minimum is depicted in figure 2. In order to resolve local minima, an intervention is required that will ensure that either one of the following conditions is fulfilled:

- The set $\mathcal{C}^{\Sigma}_{i0}$ is empty by construction.
- A halting P-complete algorithm is introduced that for every $x0 \in \mathcal{C}^{\Sigma}_{i0}$, triggering at t_0 with $\varepsilon(t_0) = \sup || x(t > t_0)\ x_\infty | x_0 ||$ characterizing a ε-ball centered on the particular x_∞ and containing all $x(t > t_0)$, to intervene in $\mathbf{E}(\mathbb{W}_i)$ *guaranteeing* that this entire ball is outside (a possibly existing) new $\mathcal{C}^{\Sigma'}_{i0}$ (with $x'_0 \leftarrow x(t_0)$).

Out of the two listed strategies for dealing with local minima, the authors have published extensively on strategy 2 (Barisic et al., 2007a), (Barisic et al., 2007b). However, strategy 1 represents a much more robust and general approach. A method guaranteeing $\mathcal{C}^{\Sigma}_{i0} \stackrel{\text{id}}{=} \emptyset$ by designing in *rotors* will be described in sec. 2.4.

2.3 Possible contour generators and decentralized control mechanisms

As for the potential contour generators $p_i(\text{d}) : \mathbb{R}^+_0 \to \mathbb{R}$, their definition follows from the global goals of guidance for the formation of AUVs. Bearing those in mind, the potential contour generators of each *feature type*, *p(o,w,c)* (for obstacle, way-point and formation cell vertex, accordingly) are specified below.

2.3.1 Obstacles

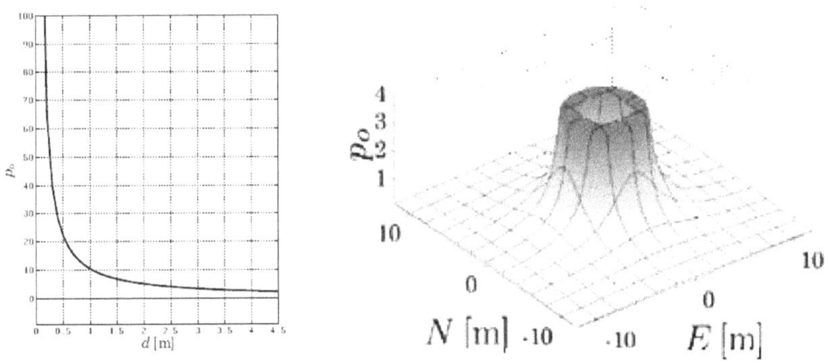

(a) Graph of $p_o(d) : \mathbb{R}^+ \to \mathbb{R}$. (b) Graph of $p_o(d(x)) : \mathbb{R}^+ \to \mathbb{R} \circ \mathbb{R}^2 \to \mathbb{R}^+$.

Fig. 3. The potential contour generator of an obstacle, $p_o(d(x))$.

$$p_o(d) = \exp\left(\frac{A^+}{d}\right) - 1; \quad \lim_{d\to\infty} p_o(d) = 0; \quad \lim_{d\to 0^+} p_o(d) = \infty \tag{6}$$

$$\frac{\partial}{\partial d} p_o(d) = -\frac{A^+}{d^2}\exp\left(\frac{A^+}{d}\right); \quad \lim_{d\to\infty} \frac{\partial}{\partial d} p_o(d) = 0; \quad \lim_{d\to 0^+} \frac{\partial}{\partial d} p_o(d) = \infty \tag{7}$$

Where:

$p_o(d) : \mathbb{R}^+ \to \mathbb{R}$ is the potential contour generator of obstacles, a strictly monotonously decreasing smooth single-valued Lebesgue-integrable function mapping a non-negative real to a real,

$A^{+} \in \mathbb{R}^{+} \setminus \{0\}$ is a positive real independent parameter dictating the scale of the acceleration away from the obstacle.

2.3.2 Way-points

$$p_w(d) = \begin{cases} d \leq d_0 : & \frac{A_p^-}{2} d^2 \\ d > d_0 : & A_0^-(d - d_0) + p_0 \end{cases}; \quad d_0 \overset{\text{id}}{=} \frac{A_0^-}{A_p^-}; \quad p_0 \overset{\text{if}}{=} \frac{A_0^{-2}}{2A_p^-} \tag{8}$$

$$\therefore p_w(d, d > d_0) = A_0^- d - \frac{A_c^{-2}}{2A_p^-}; \quad \lim_{d \to \infty} p_w(d) = \infty; \quad \lim_{d \to 0^+} p_w(d) = 0 \tag{9}$$

$$\frac{\partial}{\partial d} p_w(d) = \max(A_p^- d, A_0^-); \quad \lim_{d \to \infty} \frac{\partial}{\partial d} p_w(d) = A_0^-; \quad \lim_{d \to 0^+} \frac{\partial}{\partial d} p_w(d) = 0 \tag{10}$$

Where:

$pw(d) : \mathbb{R}^{+} \to \mathbb{R}$ is the potential contour generator of way-points, a strictly monotonously increasing smooth single-valued Lebesgue-integrable function mapping a non-negative real to a real,

$A_p^- \in \mathbb{R}^+ \setminus \{0\}$ is a positive real independent parameter dictating the scale of acceleration towards the way-point in the area of proportional attraction,

$A_0^- \in \mathbb{R}^+ \setminus \{0\}$ is a positive real independent parameter dictating the constant acceleration towards the way-point outside the area of proportional attraction,

$d_p \in \mathbb{R}^+ \setminus \{0\}$ is a positive real independent parameter dictating the radius of the open ball centered on the way-point that constitutes the area of proportional acceleration.

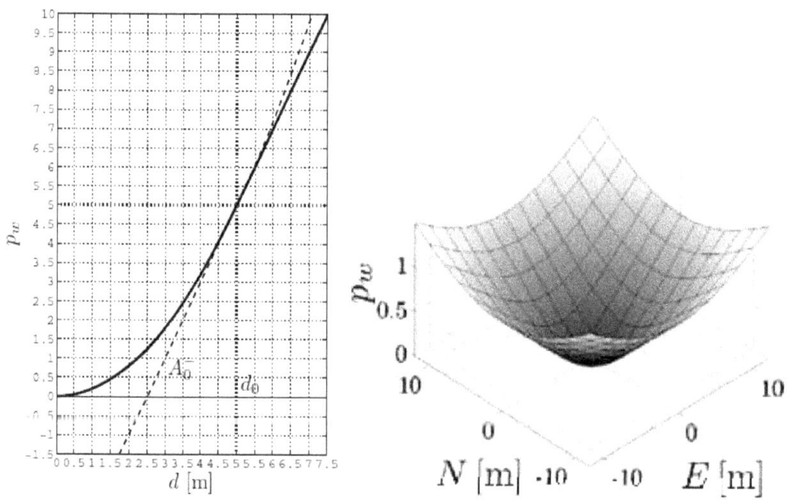

(a) Graph of $p_w(d) : \mathbb{R}^+ \to \mathbb{R}$. (b) Graph of $p_w(d(x)) : \mathbb{R}^+ \to \mathbb{R} \circ \mathbb{R}^2 \to \mathbb{R}^+$.

Fig. 4. The potential contour generator of a way-point, $p_w(d(\mathbf{x}))$.

2.3.3 Cell vertices

An effective contender for a potential contour generator of formation cell vertices, exhibiting behavior akin to a function with local support, is the normal distribution curve, modified for beauty (i.e., with an inverted sign).

$$p_c(d) = -A_c^- d_c \exp\left(1 - \frac{d^2}{2d_c^2}\right); \quad \lim_{d\to\infty} p_c(d) = 0; \quad \lim_{d\to 0^+} p_w(d) = -A_c^- \tag{11}$$

$$\frac{\partial}{\partial d} p_c(d) = \frac{A_c^-}{d_c} d \exp\left(1 - \frac{d^2}{2d_c^2}\right); \quad \frac{\partial}{\partial d} p_c(d)\big|_0 = 0; \quad \lim_{d\to\infty} \frac{\partial}{\partial d} p_c(d) = 0 \tag{12}$$

$$\frac{\partial^2}{\partial d^2} p_c(d) = \frac{A_c^-}{d_c}\left(1 - \frac{d^2}{d_c^2}\right) \exp\left(1 - \frac{d^2}{2d_c^2}\right) \tag{13}$$

$$\therefore\ d_{\max} = \arg\left\{\frac{\partial^2}{\partial d^2} p_c(d) \stackrel{!}{=} 0\right\} \tag{14}$$

$$d_{\max} \stackrel{\text{id}}{=} \pm d_c \tag{15}$$

$$\frac{\partial}{\partial d} p_c(d)\Big|_{d_{\max} \stackrel{\text{id}}{=} d_c} = A_c^- \tag{16}$$

Where:

$pc\,(\mathrm{d}) : \mathbb{R}^+ \to \mathbb{R}$ is the potential contour generator of cell vertices of *characteristic cells* of a formation, a strictly monotonously increasing smooth single-valued Lebesgue-integrable function mapping a non-negative real to a real,

$A_c^- \in \mathbb{R}^+ \setminus \{0\}$ is a positive real independent parameter dictating the scale of acceleration towards the cell vertex at the distance of maximum acceleration towards the vertex (equivalent to the valuation of $A_c^- \cdot \mathcal{N}(\pm\sigma)$ on a Gaussian normal distribution curve),

$d_c \in \mathbb{R}^+ \setminus \{0\}$is a positive real independent parameter dictating the radius of a sphere at which the inflection in the potential contour generator occurs, i.e. the distance at which maximum acceleration towards the vertex occurs (taking the place of σ in (12), which is analogous to a Gaussian normal distribution curve).

The potential of a square formation cell surrounding an agent that figures as an obstacle is represented in figure 5.

2.3.4 Reformulation in terms of decentralized control functions

The monotonicity of (6, 8, 11) ensures that the direction of the gradient of the potential, $\nabla P(x)/\|\nabla P(x)\| \in SO^2$,is always $\pm n_i = (x - x_i)/\|x - x_i\|$. Therefore, since (1, 2) are linear, (2) can be solved analytically for any finite sum of terms of the form specified by (6, 8, 11) up to the values of the independent parameters (A^+, A^-_p, A^-_0, A^-_c, d_c). The procedure follows:

$$-\nabla P_\Sigma(x) = -\nabla \sum_i P_i(x)$$
$$= \sum_i (-\nabla p_i(d_i(x))) \quad (17)$$
$$= -\sum_i \frac{\partial}{\partial d_i(x)} p_i[d_i(x)] \cdot \boldsymbol{n}_i(x) \quad (18)$$

Equation (18) can be summarized by designating the terms in (7, 10, 12) as $a^{(o,w,c)}$ respectively. The terms $a_i^{(o,w,c)} \cdot \boldsymbol{n}_i$, can likewise be denoted $\boldsymbol{a}_i^{(o,w,c)}$, respectively, and represent the *potential-based decentralized control functions* due to the *i*-th feature.

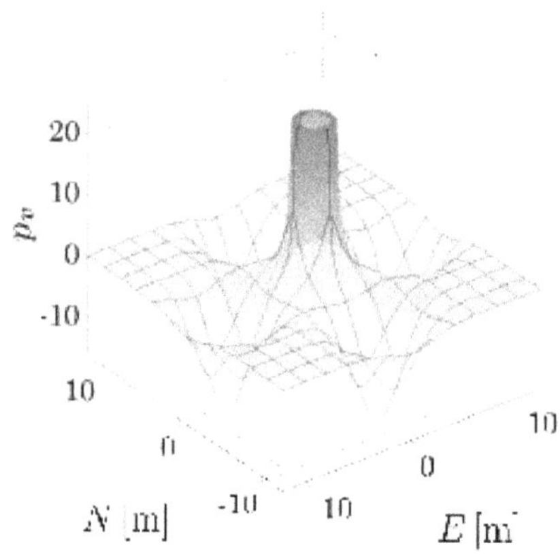

Fig. 5. The potential contour generator of a formation agent.

$$-\nabla P_\Sigma = \underbrace{\sum_i^{\text{obstacles}} a_i^{(o)}(x) \cdot \boldsymbol{n}_i(x)}_{a_i^{(o)}(x)} + \underbrace{a_i^{(w)}(x) \cdot \boldsymbol{n}_w(x)}_{a^{(w)}(x)} + \underbrace{\sum_i^{\text{vertices}} a_i^{(c)}(x) \cdot \boldsymbol{n}_i(x)}_{a_i^{(c)}(x)} \quad (19)$$

$$= \sum_i^{\text{obstacles}} \boldsymbol{a}_i^{(o)}(x) + \boldsymbol{a}^{(w)}(x) + \sum_i^{\text{vertices}} \boldsymbol{a}_i^{(c)}(x) \quad (20)$$

$$= \sum_i^{\text{obstacles}} \frac{A^+}{d_i(x)^2} \exp\left[\frac{A^+}{d_i(x)}\right] \boldsymbol{n}_i(x) + \min[A_p^- d(x),\, A_c^-] \frac{w - x}{\|w - x\|}$$
$$+ \sum_i^{\text{vertices}} \frac{A_c^-}{d_c} d_i(x) \exp\left(1 - \frac{d_i(x)^2}{2d_c^2}\right) \quad (21)$$

2.4 Rotor modification

As stated in section 2.2, the virtual potential method for guidance is highly vulnerable to the emergence of local minima. A strong and straightforward method is required to ensure the avoidance of local minima.In terms of the vector field introduced by (2), the analytical solution of which is presented in (20 – 21), stable local minima occur due to the *irrotationality* of the field, rot $\mathrm{E}(x) \overset{\text{id}}{=} 0$.

In order to avoid irrotationality, and thereby local minima, decentralized control functions proposed in (7, 10, 12) are redesigned, adding *a rotor* component:

$$a_i^{(s)} \leftarrow a_i \qquad a'_i \xleftarrow{\text{redef}} a_i^{(s)} + a_i^{(r)} \tag{22}$$

Where:

a'_i is the redefined total decentralized control function due to the i-th feature (the dash will hereafter be omitted),

$a_i^{(s)}$ is the stator decentralized control function as introduced in the preceding section, denoted with the superscript (s) to contrast it with the newly introduced $a_i^{(r)}$,

$a_i^{(r)}$ is the *rotor decentralized control function,* all of which are continuous real 2D vector fields over the Euclidean 2-space (mapping $\mathbb{R}^2$ to itself) such that they Jacobians exist wherever each of them is defined.

The introduction of $a_i(r)$ establishes a non-zero rot(E) by design, as follows:

$$\begin{aligned} \text{rot } \mathrm{E}(x) &= \sum_i a_i(x) \neq 0 \\ &= \underbrace{\text{rot} \sum_i a_i^{(s)}(x)}_{\overset{\text{id}}{=}0} + \text{rot} \sum_i a_i^{(r)}(x) \\ &= \text{rot} \sum_i a_i^{(r)}(x) \end{aligned} \tag{23}$$

With respect to the way-point, its potential influence on an AUV in this framework must not be prejudiced in terms of the direction of approach. If a decentralized control function of a way-point were augmented with a rotor part, the direction of a_w would deviate from line-of-sight. The same is true of formation cell vertices. Therefore, the only non-zero rotor decentralized control functions are those of *obstacles.* As a result, (23) can be further simplified to:

$$\text{rot } \mathrm{E}(x) = \text{rot} \sum_{i}^{\substack{\text{obstacles,}\\ \text{w.p.,}\\ \text{vertices}}} a_i^{(r)}(x) = \text{rot} \sum^{\text{obstacles}} a_i^{(r)}(x) \tag{24}$$

An individual *obstacle rotor* decentralized control function is defined below:

$\forall_i$ = enum(obstacles)

$$a(x) = a_r(x)\hat{a}_r(x) \tag{25}$$

$$a_r(x) = \frac{A_i^{(r)}}{d_i(x)^2} \exp\left(\frac{A_i^{(r)}}{d_i(x)}\right) \tag{26}$$

$$\hat{a}_r(x) = \begin{bmatrix} 1 & 0 & 0 \\ 0 & 1 & 0 \\ (0 & 0 & 1) \end{bmatrix} \cdot \left(r_i(x) \times [n_i(x)|(0)]^{\mathrm{T}}\right) \tag{27}$$

$$r_i(x) = \left[\frac{w-x_i}{\|w-x_i\|}\middle|(0)\right] \cdot [n_i(x)|(0)]^{\mathrm{T}} \tag{28}$$

$$r_i(x) = \begin{cases} r_i = 1: & n_i(x) \times \left[\frac{v}{\|v\|} - (\frac{v}{\|v\|} \cdot n_i)n_i\middle|(0)\right]^{\mathrm{T}} \\ 0 \le r_i < 1: & \left[\frac{w-x_i}{\|w-x_i\|}\middle|(0)\right]^{\mathrm{T}} \times [n_i(x)|(0)]^{\mathrm{T}} \\ \text{otherwise}: & \vec{0} \end{cases} \tag{29}$$

Where:

$A_i^{(r)} \in \mathbb{R} \setminus \{0\}$ is a positive real independent parameter dictating the scale of acceleration perpendicular to the direction of fastest flight from the obstacle,

$A_i^{(r)} \in \mathbb{R}^+$ is the magnitude of the rotor decentralized control function,

$\hat{a}_i^{(r)} \in SO^2$ is the direction of the rotor decentralized control function,

$x_i \in \mathbb{R}^2$ is the center of the i-th obstacle,

$n_i(x) \in \mathbb{SO}^2$ is the unit vector in the direction of fastest flight from the *i*-th obstacle,

$r_i(x)$ is the *unit rotor direction generator,* such that $\hat{a}_i^{(r)}(x) \stackrel{\text{id}}{=} r_i \times n_i(x)$,

$v \in \mathbb{R}^2$ is the current true over-ground velocity of the AUV (including possible sideslip) projected onto the "flight ceiling".

The rotor decentralized control function and the total decentralized control function consisting of the superposition of the rotor and stator parts, are displayed in figure 6.

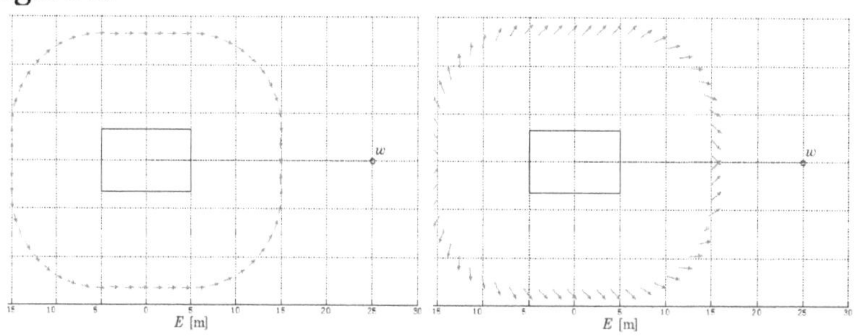

(a) A 2D display of $\hat{a}^{(r)} \in \mathbb{R}^2$ (b) A 2D display of $\hat{a} \in \mathbb{R}^2$

Fig. 6. Direction of the rotor decentralized control function $a_i^{(r)}$ and the two-term $a_i = a_i^{(s)} + a_i^{(r)}$ decentralized control function.

3. POTENTIAL FRAMEWORK OF FORMATIONS

The formation introduced by the proposed framework is the *line graph occurring at the tile interfaces of the square tessellation* of $\mathbb{R}^2$, represented in figure 7. Due to a non-collocated nature of AUV motion planning, an important feature of candidate tessellations is that they be periodic and regular, which the square tessellation is.

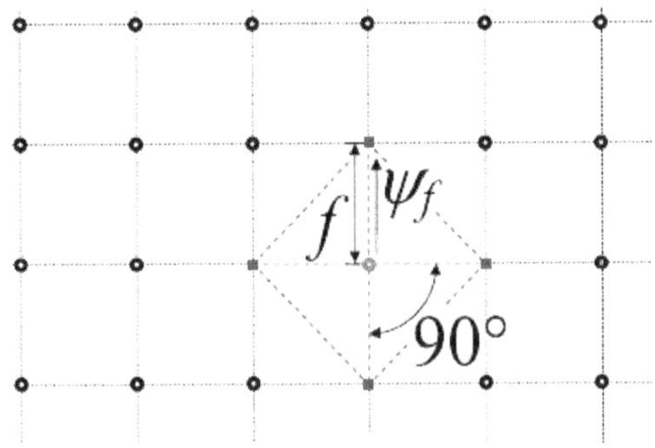

Fig. 7. The square tiling of the plane.

Each AUV whose states are being estimated by the current, i-th AUV, meaning j-th AUV, j ≠ i) is considered to be a center of a formation cell. The function of the presented framework for potential-based formation keeping is depicted in figure 8. In an unstructured motion of the cooperative group, only a small number of cell vertices attached to *j*-th AUVs $\forall j \neq i$, if any, are partially masked by nodes. The *i*-th AUV is attracted strongest to the closest cell vertex, in line with how attractiveness of a node varies with distance expressed in (11). In the structured case in 8.b), presenting an ideal, undisturbed, non-agitated and stationary formation, all the *j*-th AUVs in formation are masking each the attractiveness (w.r.t. the i-th AUV) of the vertex they already occupy. At the same time they reinforce the attractiveness of certain unoccupied vertices at the perimeter of the formation. The vertices that attract the *i*-th AUV the strongest thus become those that result in the most compact formation. Notice in figure 8.b) how certain vertices are colored a deeper shade of blue than others, signifying the lowest potential.

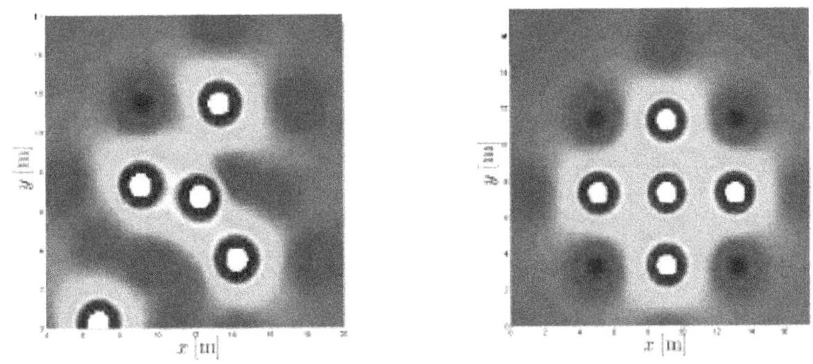

(a) Disordered arrangement (b) Formation arrangement

Fig. 8. The potential masking of agents out of and in formation.

The square formation cell is a cross figure appearing at the interstice of four squares in the tessellation, comprised of the j-th AUV and the four cell vertices attached to it, in the sense that their position is completely determined based on the i-th AUV's local estimation of j-th AUV's position, $(\hat{x}_j^{(i)})$, as in figure 7. The cell vertices are uniquely determined by $\hat{x}_j^{(i)}$ and an independent positive real scaling parameter f.

4. THE PLATFORM – A LARGE ARIES-PRECURSOR AUV

The vehicle whose dynamic model will be used to demonstrate the developed virtual potential framework is an early design of the NPS2 Aries autonomous underwater vehicle which was resized during deliberations preceding actual fabrication and outfitting. The resulting, smaller Aries vehicle has been used in multiple venues of research, most notably (An et al., 1997; Marco & Healey, 2000; 2001). As Marco & Healey (2001) describe, the vehicle whose model dynamics are used has the general body plan of the Aries, displayed in figure 9, albeit scaled up. The body plan is that of a chamfered cuboid-shaped fuselage with the bow fined using a nose-cone. The modeled Aries-precursor vehicle, the same as the Aries itself, as demonstrated in the figure, combines the use of two stern-mounted main horizontal thrusters with a pair of bow- and stern-mounted rudders (four hydrofoil surfaces in total, with dorsal and ventral pairs mechanically coupled), and bow- and stern-mounted elevators.

Healey & Lienard (1993) have designed sliding mode controllers for the Aries-precursor vehicle, considering it as a full-rank system with states $x = [\mathbf{v}^T \ \omega^T \ \mathbf{x}^T \ \Theta] = [u \ v \ w | p \ q r | x \ y \ z | \varphi \vartheta \ \psi]^T$, relying on the actuators:

$$\mathbf{u}(t) = [\delta_r(t) \ \delta_s(t) \ n(t)]^T \tag{30}$$

Fig. 9. The Aries, demonstrating the body-plan and general type of the model dynamics.

Where:

$\delta_r(t)$ is the stern rudder deflection command in radians,

$\delta_s(t)$ is the stern elevator planes' command in radians,

$n(t)$ is the main propellers' revolution rate in rad/s.

4.1 Model dynamics of the vehicle

The dynamics published by Healey and Lienard are used in the HILS[3] in the ensuing sections, and were developed on the grounds of hydrodynamic modelling theory (Abkowitz, 1969; Gertler & Hagen, 1967), exploited to great effect by Boncal (1987). The equations for the six degrees of freedom of full-state rigid-body dynamics for a cuboid-shaped object immersed in a viscous fluid follow, with the parameters expressed in Table 1.

4.1.1 Surge

$$
\begin{aligned}
& m\left[\dot{u} - vr + wq - x_G(q^2 + r^2) + y_G(pq - \dot{r}) + z_G(pr + \dot{q})\right] = \frac{\rho}{2}L^4\left[X_{pp}p^2 + X_{qq}q^2 + X_{rr}r^2\right. \\
& \left.+ X_{pr}pr\right] + \frac{\rho}{2}L^2\left[X_{\dot{u}}\dot{u} + X_{wq}wq + X_{vp}vp + X_{vr}vr + uq\left(X_{q\delta_s}\delta_s + X_{q\delta_b/2}\delta_{bp} + X_{q\delta_b/2}\delta_{bs}\right)\right. \\
& \left.+ X_{r\delta_r}ur\delta_r\right] + \frac{\rho}{2}L^2\left[X_{vv}v^2 + X_{ww}w^2 + X_{v\delta_r}uv\delta_r + uw\left(X_{w\delta_s}\delta_s + X_{w\delta_b/2}\delta_{bs} + X_{w\delta_b/2}\delta_{bp}\right)\right. \\
& \left.+ u^2\left(X_{\delta_s\delta_s}\delta_s^2 + X_{\delta_b\delta_b/2}\delta_b^2 + X_{\delta_r\delta_r}\delta_r^2\right)\right] - (W - B)\sin\vartheta + \frac{\rho}{2}L^3X_{\delta_s n}uq\delta_s\epsilon(n) + \frac{\rho}{2}L^2\left[X_{w\delta_s n}uw\delta_s\right. \\
& \left.+ X_{\delta_s\delta_s n}u^2\delta_s^2\right]\epsilon(n) + \frac{\rho}{2}L^2u^2X_{prop}
\end{aligned} \tag{31}
$$

4.1.2 Sway

$$
\begin{aligned}
& m\left[\dot{v} + ur - wp + x_G(pq + \dot{r}) - y_G(p^2 + r^2) + z_G(qr - \dot{p})\right] = \frac{\rho}{2}L^4\left[Y_{\dot{p}}\dot{p} + Y_{\dot{r}}\dot{r} + Y_{pq}pq\right. \\
& \left.+ Y_{qr}qr\right] + \frac{\rho}{2}L^3\left[Y_{\dot{v}}\dot{v} + Y_p up + Y_r ur + Y_{vq}vq + Y_{wp}wp + Y_{wr}wr\right] + \frac{\rho}{2}L^2\left[Y_v uv + Y_{vw}vw\right. \\
& \left.+ Y_{\delta_r}u^2\delta_r\right] - \frac{\rho}{2}\int_{x_{nose}}^{x_{tail}}\left[C_{dy}h(x)(v + xr)^2 + C_{dz}b(x)(w - xq)^2\right]\frac{v + xr}{U_{cf}(x)}dx + (W - B)\cos\vartheta\sin\varphi
\end{aligned} \tag{32}
$$

4.1.3 Heave

$$
\begin{aligned}
& m\left[\dot{w} - uq + vp + x_G(pr - \dot{q}) + y_G(qr + \dot{p}) - z_G(p^2 + q^2)\right] = \frac{\rho}{2}L^4\left[Z_{\dot{q}}\dot{q} + Z_{pp}p^2 + Z_{pr}pr\right. \\
& \left.+ Z_{rr}r^2\right] + \frac{\rho}{2}L^3\left[Z_{\dot{w}}\dot{q} + Z_q uq + Z_{vp}vp + Z_{vr}vr\right] + \frac{\rho}{2}L^2\left[Z_w uw + Z_{vv}v^2 + u^2\left(Z_{\delta_s}\delta_s + Z_{\delta_b/2}\delta_{bs}\right.\right. \\
& \left.\left.+ Z_{\delta_b/2}\delta_{bp}\right)\right] + \frac{\rho}{2}\int_{x_{tail}}^{x_{nose}}\left[C_{dy}h(x)(v + xr)^2 + C_{dx}b(x)(w - xq)^2\right]\frac{w - xq}{U_{cf}(x)}dx \\
& + (W - B)\cos\vartheta\cos\varphi + \frac{\rho}{2}L^3Z_{qn}uq\epsilon(n) + \frac{\rho}{2}L^2\left[Z_{wn}uw + Z_{\delta_s n}u\delta_s\right]\epsilon(n)
\end{aligned} \tag{33}
$$

4.1.4 Roll

$$I_y\dot{q} + (I_x - I_z)pr - I_{xy}(qr + \dot{p}) + I_{yz}(pq - \dot{r}) + I_{xz}(p^2 - r^2) + m\left[x_G(\dot{w} - uq + vp)\right.$$
$$\left. -z_G(\dot{v} + ur - wp)\right] + \frac{\rho}{2}L^5\left[K_{\dot{p}}\dot{p} + K_{\dot{r}}\dot{r} + K_{pq}pq + K_{qr}qr\right] + \frac{\rho}{2}L^4\left[K_{\dot{v}}\dot{v} + K_p up + K_r ur\right.$$
$$\left. +K_{vq}vq + K_{wp}wp + K_{wr}wr\right] + \frac{\rho}{2}L^3\left[K_v uv + K_{vw}vw + u^2\left(K_{\delta_b/2}\delta_{bp} + K_{\delta_b/2}\delta_{bs}\right)\right]$$
$$+(y_GW - y_BB)\cos\vartheta\cos\varphi - (z_GW - z_BB)\cos\vartheta\sin\varphi + \frac{\rho}{2}L^4K_{pn}up\epsilon(n)$$
$$+\frac{\rho}{2}L^3u^2K_{prop} \tag{34}$$

4.1.5 Pitch

$$I_x\dot{p} + (I_z - I_y)qr + I_{xy}(pr - \dot{q}) - I_{yz}(q^2 - r^2) - I_{xz}(pq + \dot{r}) + m\left[y_G(\dot{w} - uq + vp)\right.$$
$$\left. -z_G(\dot{u} - vr + wq)\right] + \frac{\rho}{2}L^5\left[M_{\dot{q}}\dot{q} + M_{pp}p^2 + M_{pr}pr + M_{rr}r^2\right] + \frac{\rho}{2}L^4\left[M_{\dot{w}}\dot{w} + M_q uq\right.$$
$$\left. +M_{vp}vp + M_{vr}vr\right] + \frac{\rho}{2}L^3\left[M_{uw}uw + M_{vv}v^2 + u^2\left(M_{\delta_s}\delta_s + M_{\delta_b/2}\delta_{bs} + M_{\delta_b/2}\delta_{bp}\right)\right]$$
$$-\frac{\rho}{2}\int_{x_{tail}}^{x_{nose}}\left[C_{dy}h(x)(v + xr)^2 + C_{dz}b(x)(w - xq)^2\right]\frac{w + xq}{U_{cf}(x)}x\,dx - (x_GW - x_BB)\cdot$$
$$\cdot\cos\vartheta\cos\varphi - (z_GW - z_BB)\sin\vartheta + \frac{\rho}{2}L^4M_{qn}qn\epsilon(n) + \frac{\rho}{2}L^3\left[M_{wn}uw\right.$$
$$\left. +M_{\delta_s n}u^2\delta_s\right]\epsilon(n) \tag{35}$$

4.1.6 Yaw

$$I_z\dot{r} + (I_y - I_x)pq - I_{xy}(p^2 - q^2) - I_{yz}(pr + \dot{q}) + I_{xz}(qr - \dot{p}) + m\left[x_G(\dot{v} + ur - wp)\right.$$
$$\left. -y_G(\dot{u} - vr + wq)\right] + \frac{\rho}{2}L^5\left[N_{\dot{p}}\dot{p} + N_{\dot{r}}\dot{r} + N_{pq}pq + N_{qr}qr\right] + \frac{\rho}{2}L^4\left[N_{\dot{v}}\dot{c} + N_p up\right.$$
$$\left. +N_r ur + N_{vq}vq + N_{wp}wp + N_{wr}wr\right] + \frac{\rho}{2}L^3\left[N_v uv + N_{vw}vw + N_{\delta_r}u^2\delta_r\right] - \frac{\rho}{2}\int_{x_{tail}}^{x_{nose}}\left[C_{dy}\cdot\right.$$
$$\left.\cdot h(x)(v + xr)^2 + C_{dz}b(x)(w - xq)^2\right]\frac{w + xq}{U_{cf}(x)}x\,dx + (x_GW - x_BB)\cos\vartheta\sin\varphi + (y_GW - y_BB)\cdot$$
$$\cdot\sin\vartheta + \frac{\rho}{2}L^3u^2N_{prop} \tag{36}$$

4.1.7 Substitution terms

$$U_{cf}(x) = \sqrt{(v + xr)^2 + (w - xq)^2} \tag{37}$$

$$X_{prop} = C_{d0}(\eta\,|\eta| - 1); \quad \eta = 0.012\frac{n}{u}; \quad C_{d0} = 0.00385 \tag{38}$$

$$\epsilon(n) = -1 + \frac{\text{sign}(n)}{\text{sign}(u)}\cdot\frac{\sqrt{C_t + 1} - 1}{\sqrt{C_{t1} + 1} - 1} \tag{39}$$

$$C_t = 0.008\frac{L^2\eta|\eta|}{2.0}; \quad C_{t1} = 0.008\frac{L^2}{2.0} \tag{40}$$

4.2 Control

The *Aries*-precursor's low-level control encompasses three separate, distinctly designed decoupled control loops:

- Forward speed control by the main propeller rate of revolution,
- Heading control by the deflection of the stern rudder,
- Combined control of the pitch and depth by the deflection of the stern elevator plates.

All of the controllers are sliding mode controllers, and the precise design procedure is presented in (Healey & Lienard, 1993). In the interest of brevity, final controller forms will be presented in the ensuing subsections.

4.2.1 Forward speed

The forward speed sliding mode controller is given in terms of a signed squared term for the propeller revolution signal, with parameters (α, β) dependent on the nominal operational parameters of the vehicle, and the coefficients presented in table 1:

$$n(t)|n(t)| = (\alpha\beta)^{-1}\left[\alpha u(t)|u(t)| + \dot{u}_c(t) - \eta_u \tanh\frac{\tilde{u}(t)}{\phi_u}\right] \tag{41}$$

$$\alpha = \frac{\rho L^2 C_d}{2m + \rho L^3 X_{\dot{u}}}; \quad C_d = 0.0034$$

$$\beta = \frac{n_0}{u_0}; \quad n_0 = 52.359\ \frac{\text{rad}}{\text{s}}; \quad u_0 = 1.832\ \frac{\text{m}}{\text{s}} \tag{42}$$

Table 1. Parameters of the Model Dynamics

W = 53.4 kN	B = 53.4 kN	L = 5.3 m	I_x = 13587 Nms^2
I_x = 13587 Nms^2	I_{yz} = -13.58 Nms^2	I_{xz} = -13.58 Nms^2	I_y = 13587 Nms^2
I_x = 2038 Nms^2	x_G = 0.0 m	x_B = 0.0 m	y_G = 0.0 m
y_B = 0.0 m	z_G = 0.061 m	z_B = 0.0 m	g = 9.81m/s^2
ρ = 1000.0 kg/m^2	m = 5454.54 kg		
$X_{pp} = 7.0 \cdot 10^{-3}$	$X_{qq} = -1.5 \cdot 10^{-2}$	$X_{rr} = 4.0 \cdot 10^{-3}$	$X_{pr} = 7.5 \cdot 10^{-4}$
$X_{\dot{u}} = -7.6 \cdot 10^{-3}$	$X_{wq} = -2.0 \cdot 10^{-1}$	$X_{vp} = -3.0 \cdot 10^{-3}$	$X_{vr} = 2.0 \cdot 10^{-2}$
$X_{q\delta_s} = 2.5 \cdot 10^{-2}$	$X_{q\delta_b}/2 = -1.3 \cdot 10^{-3}$	$X_{r\delta_r} = -1.0 \cdot 10^{-3}$	$X_{vv} = 5.3 \cdot 10^{-2}$
$X_{ww} = 1.7 \cdot 10^{-1}$	$X_{v\delta_r} = 1.7 \cdot 10^{-3}$	$X_{w\delta_s} = 4.6 \cdot 10^{-2}$	$X_{w\delta_b}/2 = 0.5 \cdot 10^{-2}$

$X_{\delta_s \delta_s} = -1.0 \cdot 10^{-2}$	$X_{\delta_b \delta_b/2} = -4.0 \cdot 10^{-3}$	$X_{\delta_r \delta_r} = -1.0 \cdot 10^{-2}$	$X_{q\delta_s n} = 2.0 \cdot 10^{-3}$
$X_{w\delta_s n} = 3.5 \cdot 10^{-3}$	$X_{\delta_s \delta_s n} = -1.6 \cdot 10^{-3}$		
$Y_{\dot{p}} = 1.2 \cdot 10^{-4}$	$Y_{\dot{r}} = 1.2 \cdot 10^{-3}$	$Y_{pq} = 4.0 \cdot 10^{-3}$	$Y_{qr} = -6.5 \cdot 10^{-3}$
$Y_{\dot{v}} = -5.5 \cdot 10^{-2}$	$Y_{p} = 3.0 \cdot 10^{-3}$	$Y_{r} = 3.0 \cdot 10^{-2}$	$Y_{vq} = 2.4 \cdot 10^{-2}$
$Y_{wp} = 2.3 \cdot 10^{-1}$	$Y_{wr} = -1.9 \cdot 10^{-2}$	$Y_{v} = -1.0 \cdot 10^{-1}$	$Y_{vw} = 6.8 \cdot 10^{-2}$
$Y_{\delta_r} = 2.7 \cdot 10^{-2}$			
$Z_{\dot{q}} = -6.8 \cdot 10^{-3}$	$Z_{pp} = 1.3 \cdot 10^{-4}$	$Z_{pr} = 6.7 \cdot 10^{-3}$	$Z_{rr} = -7.4 \cdot 10^{-3}$
$Z_{\dot{w}} = -2.4 \cdot 10^{-1}$	$Z_{q} = -1.4 \cdot 10^{-1}$	$Z_{vp} = -4.8 \cdot 10^{-2}$	$Z_{vr} = 4.5 \cdot 10^{-2}$
$Z_{w} = -3.0 \cdot 10^{-1}$	$Z_{vv} = -6.8 \cdot 10^{-2}$	$Z_{\delta_s} = -7.3 \cdot 10^{-2}$	$Z_{\delta_b/2} = -1.3 \cdot 10^{-2}$
$Z_{qn} = -2.9 \cdot 10^{-3}$	$Z_{wn} = -5.1 \cdot 10^{-3}$	$Z_{\delta_s n} = -1.0 \cdot 10^{-2}$	
$K_{\dot{p}} = -1.0 \cdot 10^{-3}$	$K_{\dot{r}} = -3.4 \cdot 10^{-5}$	$K_{pq} = -6.9 \cdot 10^{-5}$	$K_{qr} = 1.7 \cdot 10^{-2}$
$K_{\dot{v}} = 1.3 \cdot 10^{-4}$	$K_{p} = -1.1 \cdot 10^{-2}$	$K_{r} = -8.4 \cdot 10^{-4}$	$K_{vq} = -5.1 \cdot 10^{-3}$
$K_{wp} = -1.3 \cdot 10^{-4}$	$K_{wr} = 1.4 \cdot 10^{-2}$	$K_{v} = 3.1 \cdot 10^{-3}$	$K_{vw} = -1.9 \cdot 10^{-1}$
$K_{\delta_b/2} = 0.0$	$K_{pn} = -5.7 \cdot 10^{-4}$	$K_{prop} = 0.0$	
$M_{\dot{q}} = -1.7 \cdot 10^{-2}$	$M_{pp} = 5.3 \cdot 10^{-5}$	$M_{pr} = 5.0 \cdot 10^{-3}$	$M_{rr} = 2.9 \cdot 10^{-3}$
$M_{\dot{w}} = -6.8 \cdot 10^{-2}$	$M_{uq} = -6.8 \cdot 10^{-2}$	$M_{vp} = 1.2 \cdot 10^{-3}$	$M_{vr} = 1.7 \cdot 10^{-2}$
$M_{uw} = 1.0 \cdot 10^{-1}$	$M_{vv} = -2.6 \cdot 10^{-2}$	$M_{\delta_s} = -4.1 \cdot 10^{-2}$	$M_{\delta_b/2} = 3.5 \cdot 10^{-3}$
$M_{qn} = -1.6 \cdot 10^{-3}$	$M_{wn} = -2.9 \cdot 10^{-3}$	$M_{\delta_s n} = -5.2 \cdot 10^{-3}$	
$N_{\dot{p}} = -3.4 \cdot 10^{-5}$	$N_{\dot{r}} = -3.4 \cdot 10^{-3}$	$N_{pq} = -2.1 \cdot 10^{-2}$	$N_{qr} = 2.7 \cdot 10^{-3}$
$N_{\dot{v}} = 1.2 \cdot 10^{-3}$	$N_{p} = -8.4 \cdot 10^{-4}$	$N_{r} = -1.6 \cdot 10^{-2}$	$N_{vq} = -1.0 \cdot 10^{-2}$
$N_{wp} = -1.7 \cdot 10^{-2}$	$N_{wr} = 7.4 \cdot 10^{-3}$	$N_{v} = -7.4 \cdot 10^{-3}$	$N_{vw} = -2.7 \cdot 10^{-2}$
$N_{\delta_r} = -1.3 \cdot 10^{-2}$	$N_{prop} = 0.0$		

It is apparent from the above that the propeller rate of revolution command comprises a term that accelerates the vehicle in the desired measure ($\dot{u}_c(t)$), overcomes the linear drag ($u(t)|u(t)|$), and attenuates the perturbations due to disturbances and process noise ($\dot{\sigma} u(t)$).

4.2.2 Heading

The sliding surface for the subset of states governing the vehicle's heading is given below, in (43). The resulting sliding mode controller is contained in (44).

$$\sigma_r = -0.074\tilde{v}(t) + 0.816\tilde{r}(t) + 0.573\tilde{\varphi}(t) \tag{43}$$

$$\delta_r = 0.033v(t) + 0.1112r(t) + 2.58 \tanh \frac{0.074\tilde{v}(r) + 0.816\tilde{r}(t) + 0.573\tilde{\varphi}(t)}{0.1}$$

It should be noted that $\tilde{v}(r)$ seems to imply the possibility of defining some $v_c(t)$ for the vehicle to track. This is impractical. The *Aries*-precursor's thrust allocation and kinematics, nonholonomic in sway, would lead to severe degradation of this sliding mode controller's performance in its main objective – tracking the heading. Lienard (1990) provides a further detailed discussion of this and similar sliding mode controllers.

4.2.3 Pitch and depth

The main objective of the third of the three controllers onboard the *Aries*-precursor HIL simulator, that for the combination of pitch and depth, is to control *depth.* For a vehicle with the holonomic constraints and kinematics of the model used here, this is only possible by using the stern elevators δ_s to pitch the vehicle down and dive. Accordingly, the sliding surface is designed in (44), and the controller in (45).

$$\sigma_z(t) = \tilde{q}(t) + 0.520\tilde{\vartheta}(t) - 0.011\tilde{z}(t) \tag{44}$$

$$\begin{aligned} \delta_s(t) &= -5.143q(t) + 1.070\vartheta(t) + 4\tanh\frac{\sigma_z(t)}{0.4} \\ &= -5.143q(t) + 1.070\vartheta(t) \\ &\quad +4\tanh\frac{\tilde{q}(t) + 0.520\tilde{\vartheta}(t) - 0.011\tilde{z}(t)}{0.4} \end{aligned} \tag{45}$$

5. OBSTACLE CLASSIfiCATION, STATE ESTIMATION AND CONDITIONING THE CONTROL SIGNALS

In this section, the issues of obstacle classification will be addressed, giving the expressions for (d_i, n_i) of every type of obstacle considered, which are functions prerequisite to obtaining Pi-s through composition with one of (6, 8, 11). Also, full-state estimation of the AUV (modeled after the NPS *Aries*-precursor vehicle described in the preceding section), $\hat{\mathbf{x}} = [\hat{u}\,\hat{v}\,\hat{w}\,\hat{p}\,\hat{q}\,\hat{r}\,\hat{x}\,\hat{y}\,\hat{z}\,\hat{\varphi}\,\hat{\vartheta}\,\hat{\psi}]^T$ will be explored. Realistic plant and measurement noise $(\tilde{\mathbf{n}}, \tilde{\mathbf{y}})$, which can be expected when transposing this control system from HILS to a real application will be discussed and a scheme for the generation of non-stationary stochastic noise given. Finally, the section will address a scheme for conditioning / clamping the low-level control signals to values and dynamic ranges realizable by the AUV with the *Aries* body-plan. The conditioning adjusts the values in the low-level command vector $c = [a_c\ u_c\ r_c\ \phi_c]^T$ to prevent unfeasible commands which can cause saturation in the actuators and temporary break-down of feedback.

5.1 Obstacle classification

The problem of classification in a 2D waterspace represented by $\mathbb{R}^2$ is a well studied topic. We have adopted an approach based on modeling real-world features after a sparse set of geometrical primitives – circles, rectangles and ellipses.

In the ensuing expressions, $\{\mathbf{x}^{int}\}$ will be used for the closed, connected set comprising the interior of the obstacle being described. $\mathbf{T}_i$ shall be a homogeneous, isomorphic coordinate transform from the global reference coordinate system to the coordinate system attached to the obstacle, affixed to the *centroid* of the respective obstacle with a possible rotation by some ψ_i if applicable.

5.1.1 Circles

Circles are the simplest convex obstacles to formulate mathematically. The distance and normal vector to a circle defined by ($\boldsymbol{x}_i \in \mathbb{R}^2$, $r_i \in \mathbb{R}^+$), its center and radius respectively, are given below:

$$d_i : \quad \mathbb{R}^2 \setminus \left\{ x^{int} : \left\| x^{int} - x_i \right\| < r_i \right\} \to \mathbb{R}^+$$

$$d_i(x) = \|x - x_i\| - r_i \in \mathbb{R}^+ \tag{46}$$

$$n_i : \quad \mathbb{R}^2 \setminus \left\{ x^{int} : \left\| x^{int} - x_i \right\| < r_i \right\} \to \mathrm{SO}^2$$

$$n_i(x) = \frac{x - x_i}{\|x - x_i\|} \tag{47}$$

Robust and fast techniques of classifying 2D point-clouds as circular features are very well understood both in theory and control engineering practice. It is easy to find solid algorithms applicable to hard-real time implementation. Good coverage of the theoretic and practical aspects of the classification problem, solved by making use of the circular Hough transform is given in (Haule & Malowany, 1989; Illingworth & Kittler, 1987; Maitre, 1986; Rizon et al., 2007).

5.1.2 Rectangles

The functions for the distance and normal vector ($d_i(\mathbf{x}), \boldsymbol{n}_i(\mathbf{x})$), of a point with respect to a rectangle in Euclidean 2-space defined by $(x_i \in \mathbb{R}^2, a_i, b_i \in \mathbb{R}^+, \psi_i \in [-\pi, \pi))$, the center of the rectangle, the half-length and half-breadth and the angle of rotation of the rectangle's long side w.r.t. the global coordinate system, respectively, are given below:

$$d_i : \quad \mathbb{R}^2 \setminus \left\{ x^{int} : \left\| \begin{bmatrix} a_i & 0 \\ 0 & b_i \end{bmatrix}^{-1} \cdot T_i(x^{int}) \right\|_\infty < 1 \right\} \to \mathbb{R}^+$$

$$d_i(x) = \begin{cases} |\hat{\imath} \cdot T_i(x)| < a_i : & |\hat{\jmath} \cdot T_i(x)| - b_i \\ |\hat{\jmath} \cdot T_i(x)| < b_i : & |\hat{\imath} \cdot T_i(x)| - a_i \\ \text{otherwise :} & \left\| |T_i(x)| - \begin{bmatrix} \frac{a_i}{2} & \frac{b_i}{2} \end{bmatrix}^\mathrm{T} \right\| \end{cases} \tag{48}$$

$$n_i : \quad \mathbb{R}^2 \setminus \left\{ x^{int} : \left\| \begin{bmatrix} a_i & 0 \\ 0 & b_i \end{bmatrix}^{-1} \cdot T_i(x^{int}) \right\|_\infty < 1 \right\} \to \mathrm{SO}^2$$

$$n_i(x) = \begin{cases} |\hat{\imath} \cdot \mathcal{T}_i(x)| < \frac{a_i}{2} : & \mathcal{T}_i^{-1}\{\text{sign}[\hat{\jmath} \cdot \mathcal{T}_i(x)]\,\hat{\jmath}\} \\ |\hat{\jmath} \cdot \mathcal{T}_i(x)| < \frac{b_i}{2} : & \mathcal{T}_i^{-1}\{\text{sign}[\hat{\imath} \cdot \mathcal{T}_i(x)]\,\hat{\imath}\} \\ \text{otherwise} : & \mathcal{T}_i^{-1}\left\{\mathcal{T}_i(x) - \left[a_i\,\text{sign}[\hat{\imath} \cdot \mathcal{T}_i(x)]\ b_i\,\text{sign}[\hat{\jmath} \cdot \mathcal{T}_i(x)]\right]^{\mathrm{T}}\right\} \end{cases} \tag{49}$$

There is a large amount of published work dedicated to the extraction of the features of rectangles from sensed 2D point-clouds. Most of these rely on Hough space techniques (Hough & Powell, 1960) and (Duda & Hart, 1972) to extract the features of distinct lines in an image and determine whether intersections of detected lines are present in the image (He & Li, 2008; Jung & Schramm, 2004; Nguyen et al., 2009).

5.1.3 Ellipses

The method of solving for a distance of a point to an ellipse involves finding the roots of the quartic (57). Therefore, it is challenging to find *explicit analytical solutions,* although some options include Ferrari's method (Stewart, 2003) or algebraic geometry (Faucette, 1996). A computer-based control system can, however, employ a good, numerically stable algorithm to obtain a precise enough solution. The rudimentary part of analytic geometry that formulates the quartic to be solved is given below in (50 - 57).

The equation of the ellipse with the center in the origin and axes aligned with the axes of the coordinate system is:

$$\left(\frac{x}{a}\right)^2 + \left(\frac{y}{b}\right)^2 = 1 \tag{50}$$

The locus of its solutions is the ellipse, $\{\mathbf{x}_e = [x_e\ y_e]^{\mathrm{T}}\}$. The analysis proceeds by considering those $\mathbf{x} = [x\ y]^{\mathrm{T}} \in \mathbb{R}^2$ for which $\mathbf{x} - \mathbf{x}_e$ is normal to the ellipse. The equation of such a normal is:

$$x_n(\tau) = k\tau + x_e \tag{51}$$

Where $\tau \in \mathbb{R}$ is an independent parameter, the degree of freedom along the line and $\boldsymbol{k}$ is the direction vector of the line, given below:

$$k = \nabla\left\{\left(\frac{x_e}{a}\right)^2 + \left(\frac{y_e}{b}\right)^2 - 1\right\} = \left[\frac{x_e}{a^2}\ \frac{y_e}{b^2}\right]^{\mathrm{T}} \tag{52}$$

It follows that if $\tau = t = \arg x$, i.e. $x_n(t) \overset{\text{id}}{=} x$. Then, the following manipulation can be made:

$$[x - x_e\ y - y_e]^{\mathrm{T}} = \left[\frac{tx_e}{a^2}\ \frac{ty_e}{b^2}\right]^{1} \tag{53}$$

$$[x_e\ y_e]^{\mathrm{T}} = \left[\frac{a^2 x}{t+a^2}\ \frac{b^2 y}{t+b^2}\right]^{\mathrm{T}} \tag{54}$$

Substituting the right-hand side of (54) into (50), the quartic discussed is obtained as:

$$\left(\frac{ax}{t+a^2}\right)^2 + \left(\frac{by}{t+b^2}\right)^2 = 1 \tag{55}$$

$$(t+b^2)^2a^2x^2 + (t+a^2)^2b^2y^2 = (t+a^2)^2(t+b^2)^2 \tag{56}$$

$$(t+a^2)^2(t+b^2)^2 - (t+b^2)^2a^2x^2 - (t+a^2)^2b^2y^2 = 0 \tag{57}$$

The greatest root of (57), $\bar{t}$, allows for the calculation of ($d_i(\boldsymbol{x})$, $\boldsymbol{n}_i(\boldsymbol{x})$) in (51, 54), as given below:

$$d_i : \quad \mathbb{R}^2 \setminus \left\{ x^{int} : x^{\mathrm{T}} T_i \left\{ \begin{bmatrix} a & 0 \\ b & 0 \end{bmatrix} T_i^{-1}(x) \right\} < 1 \right\} \to \mathbb{R}^+$$

$$d_i(\boldsymbol{x}) = \|\boldsymbol{x} - \boldsymbol{x}_e\| = \left\|\boldsymbol{k}\bar{t}\right\| = \bar{t} \left\| \begin{bmatrix} \frac{a^2 x_e}{\bar{t}+a^2} & \frac{b^2 y_e}{\bar{t}+b^2} \end{bmatrix}^{\mathrm{T}} \right\|$$

$$= \bar{t}\sqrt{\frac{(\bar{t}+b^2)^2a^4x_e^2 + (\bar{t}+a^2)^2b^4y_e^2}{(\bar{t}+a^2)^2(\bar{t}+b^2)^2}} \tag{58}$$

$$= \bar{t}\sqrt{\frac{(\bar{t}+b^2)^2a^4\left[\hat{\imath}T_i(x)\right]^2 + (\bar{t}+a^2)^2b^4\left[\hat{\jmath}T_i(x))\right]^2}{(\bar{t}+a^2)^2(\bar{t}+b^2)^2}} \tag{59}$$

$$\boldsymbol{n}_i : \quad \mathbb{R}^2 \setminus \left\{ x^{int} : x^{\mathrm{T}} T_i \left\{ \begin{bmatrix} a & 0 \\ b & 0 \end{bmatrix} T_i^{-1}(x) \right\} < 1 \right\} \to \mathrm{SO}^2$$

$$T_i\boldsymbol{n}_i^{(i)}(x) = \frac{\boldsymbol{k}}{\|\boldsymbol{k}\|} = \frac{\begin{bmatrix} \frac{x_e}{\bar{t}+a^2} & \frac{y_e}{\bar{t}+b^2} \end{bmatrix}^{\mathrm{T}}}{\left\| \begin{bmatrix} \frac{x_e}{\bar{t}+a^2} & \frac{y_e}{\bar{t}+b^2} \end{bmatrix}^{\mathrm{T}} \right\|} \tag{60}$$

$$T_i\boldsymbol{n}_i(x) = \frac{\begin{bmatrix} \frac{\hat{\imath}T_i(x)}{\bar{t}+a^2} & \frac{\hat{\jmath}T_i(x)}{\bar{t}+b^2} \end{bmatrix}^{\mathrm{T}}}{\left\| \begin{bmatrix} \frac{\hat{\imath}T_i(x)}{\bar{t}+a^2} & \frac{\hat{\jmath}T_i(x)}{\bar{t}+b^2} \end{bmatrix}^{\mathrm{T}} \right\|} \tag{61}$$

$$n_i(x) = \mathcal{T}_i^{-1} \left\{ \frac{\left[\frac{\hat{\imath}\mathcal{T}_i(x)}{\bar{t}+a^2} \; \frac{\hat{\jmath}\mathcal{T}_i(x)}{\bar{t}+b^2} \right]^{\mathrm{T}}}{\left\| \left[\frac{\hat{\imath}\mathcal{T}_i(x)}{\bar{t}+a^2} \; \frac{\hat{\jmath}\mathcal{T}_i(x)}{\bar{t}+b^2} \right]^{\mathrm{T}} \right\|} \right\} \tag{62}$$

With the advent of cheap solid-state perception sensors in service robotics and aerial photography in the last decade, publication on fast and robust ellipse-fitting of 2D point clouds has intensified – (Ahn et al., 1999; Jiang et al., 2007; Pilu et al., 1996).

5.2 State estimation

The full state $\boldsymbol{x} = [\boldsymbol{v}^{\mathrm{T}} \; \omega^{\mathrm{T}} \; \boldsymbol{x}^{\mathrm{T}} \; \Theta^{\mathrm{T}}]^{\mathrm{T}} = [u \; v \; w \,|\, p q \; r \,|\, xy \; z \,|\, \varphi \vartheta \; \psi]^{\mathrm{T}}$ of the AUV will be estimated using the Scaled Unscented Transform Sigma-Point Kalman Filter (SP-UKF) introduced by van der Merwe (2004).

The Extended Kalman Filter formulations that feature more prominently in marine control engineering state-of-the-art are capable of estimating the states of nonlinear model dynamics by taking into account only first-order statistics of the states (with possible addition of plant / process noise. EKFs use Jacobians of the nonlinear operator(s) evaluated at the current state estimate. The Unscented Kalman Filters (UKF), in contrast, use the original non-linear model dynamics to propagate *samples* – called *sigma-points,* which are characteristic of the *current estimate of the statistical distribution of the states,* influenced by process and measurement noise. The Kalman gain is evaluated based on the covariance of state hypotheses thus propagated vs. the covariance of the samples characteristic of the current estimate of the statistical distribution of *the measurements.* The Kalman gain will award a higher gain to those measurements for which a significant correlation is discovered between the state and measurement hypotheses and for which the covariance of the measurement hypotheses themselves is relatively small.

5.3 Measurement and process noise

The AUV is assumed to carry a 4-beam DVL[4] which it can use to record the true 3D speed-over-ground measurement $\boldsymbol{v} = [u \; v \; w]^{\mathrm{T}}$. Furthermore, the AUV carries a 3-axial rate gyro package capable of measuring the body-fixed angular velocities $\omega = [p \; q \; r]^{\mathrm{T}}$. A low-grade commercial USBL[5] system is assumed to provide estimates of $[x \, y]^{\mathrm{T}}$. A fusion of the USBL estimate and the pressure gauge prior to the SP-UKF entry point is assumed to provide a relatively good-quality depth reading of z. A 3-axial middle-market strap-down AHRS6 is assumed to provide the Tait-Bryan angle readings, $\Theta = [\phi \; \vartheta \; \psi]^{\mathrm{T}}$.

0. Parameterization

0.1. Set α, the scaling parameter for the Scaled Unscented Transform

0.2 Set β, the parameter of accentuation of the central estimate

0.3 Set κ, the scaling parameter of the set of sigma-point drawn from the underlying distribution

0.4 $L = 12$, the number of states

0.5. $\lambda = \alpha^2(L + \kappa) - L$

0.6 $\mathbf{w}_c = \left\{ w_c^{(0)}\ w_c^{(1)}\ \cdots\ w_c^{(2L)} \right\},$
$w_c^{(0)} = \frac{\lambda}{L+\lambda} + (1 - \alpha^2 + \beta),$
$w_c^{(1...2L)} = w_m^{(1...2L)} = \frac{1}{2(L+\lambda)}$

0.7 $\mathbf{w}_m = \left\{ w_m^{(0)}\ w_m^{(1)}\ \cdots\ w_m^{(2L)} \right\},$
$w_m^{(0)} = \frac{\lambda}{L+\lambda}$

1. Initialization

1.1 Set $\hat{\mathbf{x}}(0|1) = \bar{\mathbf{x}}$ the initial a priori estimate

1.2 Set $\mathbf{P}_\mathbf{x}(k) = \mathrm{E}\left[(\mathbf{x} - \bar{\mathbf{x}})(\mathbf{x} - \bar{\mathbf{x}})^\mathrm{T}\right]$

1.3 Set $\mathbf{R}_f$ - the process noise covariance

Set $\mathbf{R}_n$ - the measurement noise covariance

2. Iteration for $k = 1 \ldots \infty$

2.1 Sigma-points and hypotheses of the states

2.1.1 $\boldsymbol{\mathcal{X}}^-(k) = \{\mathbf{x}^-\} = \left\{\mathbf{x}(k|k-1)\ \mathbf{x}(k|k-1) + \gamma\sqrt{\mathbf{P}_\mathbf{x}(k)}\ \mathbf{x}(k|k-1) - \gamma\sqrt{\mathbf{P}_\mathbf{x}(k)}\right\}$

2.2 Time-update

2.2.1 $\boldsymbol{\mathcal{X}}^{-*}(k) = \{\mathbf{x}^{-*}\} = \{\mathbf{F}(\boldsymbol{\mathcal{X}}^-(k))\}$

2.2.2 $\hat{\mathbf{x}}^-(k|k-1) = \sum_{i=0}^{2L} w_m^{(i)} \mathbf{x}^{-*(i)}$

2.2.3 $\mathbf{P}_\mathbf{x}^- = \sum_i^{2L} w_c^{(i)} \left(\boldsymbol{\mathcal{X}}^{-*(i)} - \hat{\mathbf{x}}^-(k|k-1)\right)\left(\boldsymbol{\mathcal{X}}^{-*(i)} - \hat{\mathbf{x}}^-(k|k-1)\right)^\mathrm{T}$

2.2.4. Re-draw the hypotheses taking into account the process noise covariance

$\mathbf{X}(k|k-1) = \left\{\hat{\mathbf{x}}^-\ \hat{\mathbf{x}}^- + \gamma\sqrt{\mathbf{R}_\mathbf{v}}\ \hat{\mathbf{x}}^- - \gamma\sqrt{\mathbf{R}_\mathbf{v}}\right\}$

2.2.5. $\mathbf{Y}(k|k-1) = \mathbf{H}(\mathbf{X})$ ·

2.2.6. $\hat{\mathbf{y}}^- = \sum_{i=0}^{2L} w_m^{(i)} \mathbf{Y}^{(i)}$

2.3 Measurement update

2.3.1 $\mathbf{P}_{\mathbf{y}} = \sum_{i=0}^{2L} w_c^{(i)} \left(\mathbf{Y}^{(i)} - \hat{\mathbf{y}}^{-}\right)\left(\mathbf{Y}^{(i)} - \hat{\mathbf{y}}^{-}\right)^{\mathrm{T}}$

2.3.2 $\mathbf{P}_{\mathbf{xy}} = \sum_{i=0}^{2L} w_c^{(i)} \left(\mathbf{X}^{(i)} - \hat{\mathbf{x}}^{-}\right)\left(\mathbf{Y}^{(i)} - \hat{\mathbf{y}}^{-}\right)^{\mathrm{T}}$

2.3.3 $\mathbf{K}(k) = \mathbf{P}_{\mathbf{xy}}\mathbf{Py}^{-1}$

2.3.4. $\hat{\mathbf{x}}(k|k) = \hat{\mathbf{x}}^{-}(k|k-1) + \mathbf{K}(k)\left(\mathbf{y} - \hat{\mathbf{y}}^{-}\right)$

2.3.5. $\mathbf{P}_{\mathbf{x}}(k) = \mathbf{P}^{-}\mathbf{x} - \mathbf{K}(k)\mathbf{P}_{\mathbf{y}}\,\mathbf{K}(k)^{\mathrm{T}}$

In the proposed HILS framework, measurement noises should mimic the actual experience during AUV fieldwork operations. Therefore, a noise generator which can produce non-stationary, varying noises is required. It is intended that these noises include errors in the sensor readings whose sources cannot be simply identified by recourse to first-order statistics, and which therefore cannot be easily calibrated (de-biased) for. Additionally, we wish to be able to generate *sporadic irrecoverable faults* i.e. events during which a sensor reading cannot be relied on in any meaningful way.

For this reason, we propose the use of a bank of *Gaussian Markov models* GMMs, for generating the additive measurement noise. Markov models are stochastic state-machines whose state-switching is governed by random number generators. GMMs ultimately output a normally distributed random number with the statistics dependent on the current state. Means and standard deviations (μ_i, σ_i) of each state i are designed into the GMM. In this chapter, a bank of $12n_{AUV}$ Gaussian Markov models, one for each state of each of the n_{AUV} AUVs is used. All of the GMM states contain separate *univariate rate-limited white noise generators* parameterized by (μ_i, σ_i, n_i), where n_i is the rate limit of the additive measurement noise in the *i*-th channel.

Relying on the MATLAB normally distributed random number generator invoked by the randn command, each state generates a number according to:

$$\tilde{y}_i^{-}(k) = \mu_i + \sigma_i \cdot \mathtt{randn} \tag{63}$$

$$\tilde{y}_i(k) = \mathrm{sign}(\tilde{y}^{-}(k) - \tilde{y}(k-1)) \cdot \min\left[\left|\tilde{y}^{-}(k) - \tilde{y}(k-1)\right|, \frac{n_i}{T}\right] \tag{64}$$

Where T is the sampling time.

To optimize between a realistic nature of the measurement noises and HILS complexity, each of the Markov models in the employed bank contains 6 states, {*nominal*, + *0reliable*, *reliable* + unreliable, unreliable, fault} . The 6-state Gaussian Markov models are initialized by a 6-tuple $\mathcal{M} = ((\mu_1, \sigma_1, n_1), \ldots, (\mu_6, \sigma_6, n_6))$ and a 6×6 *transition matrix* $\mathbf{T} = [t_{ij}]$ with t_{ij} being a priori probabilities of switching from state i to state j. The actual parameters used in the HILS simulation are presented in (65 – 74). Before adding them to idealized state measurements, the noise channels are mixed together (correlated) as $\mathbf{y} \rightarrow \mathbf{S}_y\mathbf{y}$ using the matrix $\mathbf{S}_y$ in (75), to mimic the physics of the relevant sensors' interdependence of measurements. Notice that Sy has a pronounced block-diagonal structure, indicative of the fact that the mentioned sensors (DVL, USBL, AHRS, gyro-compass and rate gyros) output several state measurements each. The correlation between the states measured by a single instrument is more pronounced than the one between measurements of mutually dislocated sensors operating along different physical principles.

$$\mathcal{M}_{\mathbf{y_v}} = \mathcal{M}_{\mathbf{y}_u} = \mathcal{M}_{\mathbf{y}_v} = \mathcal{M}_{\mathbf{y}_w} = \begin{cases} \text{State } \textit{nominal}: & (\mu_1, \sigma_1, n_1) = (0, 0.06, 0.03815) \\ \text{State } \pm \textit{reliable}: & (\mu_1, \sigma_1, n_1) = (\pm 0.09, 0.11, 0.05) \\ \text{State } \pm \textit{unreliable}: & (\mu_1, \sigma_1, n_1) - (\pm 0.2981, 0.24, 0.09) \\ \text{State } \textit{faulty}: & (\mu_1, \sigma_1, n_1) = (\text{NaN}, \text{NaN}, \text{NaN}) \end{cases} \tag{65}$$

$$\mathbf{T}_{\mathbf{y_v}} = \mathbf{T}_{\mathbf{y}_u} = \mathbf{T}_{\mathbf{y}_v} = \mathbf{T}_{\mathbf{y}_w} = \begin{bmatrix} 0.7542 & 0.1000 & 0.1000 & 0.0208 & 0.0208 & 0.0042 \\ 0.4739 & 0.3365 & 0.1043 & 0.0379 & 0.0284 & 0.0190 \\ 0.4739 & 0.1043 & 0.3365 & 0.0284 & 0.0379 & 0.0190 \\ 0.3825 & 0.1749 & 0.1749 & 0.0984 & 0.0984 & 0.0710 \\ 0.3825 & 0.1749 & 0.1749 & 0.0984 & 0.0984 & 0.0710 \\ 0.0270 & 0.1622 & 0.1622 & 0.2703 & 0.2703 & 0.1081 \end{bmatrix} \tag{66}$$

$$\mathcal{M}_{\mathbf{y}_\omega} = \mathcal{M}_{\mathbf{y}_p} = \mathcal{M}_{\mathbf{y}_q} = \mathcal{M}_{\mathbf{y}_r} = \begin{cases} \text{State } \textit{nominal}: & (\mu_1, \sigma_1, n_1) = (0, \frac{\pi}{85}, \frac{\pi}{227.608}) \\ \text{State } \pm \textit{reliable}: & (\mu_1, \sigma_1, n_1) = (\pm \frac{\pi}{72}, \frac{\pi}{60}, \frac{\pi}{144.201}) \\ \text{State } \pm \textit{unreliable}: & (\mu_1, \sigma_1, n_1) = (\pm \frac{\pi}{21.5}, \frac{\pi}{18.8}, \frac{\pi}{64.454}) \\ \text{State } \textit{faulty}: & (\mu_1, \sigma_1, n_1) = (\text{NaN}, \text{NaN}, \text{NaN}) \end{cases} \tag{67}$$

$$\mathbf{T}_{\mathbf{y}_\omega} = \mathbf{T}_{\mathbf{y}_p} = \mathbf{T}_{\mathbf{y}_q} = \mathbf{T}_{\mathbf{y}_r} = \begin{bmatrix} 0.5928 & 0.1596 & 0.1596 & 0.0423 & 0.0423 & 0.0033 \\ 0.4978 & 0.2489 & 0.1511 & 0.0356 & 0.0533 & 0.0133 \\ 0.4978 & 0.1511 & 0.2489 & 0.0533 & 0.0356 & 0.0133 \\ 0.5234 & 0.1963 & 0.0561 & 0.0935 & 0.0748 & 0.0561 \\ 0.5234 & 0.0561 & 0.1963 & 0.0748 & 0.0935 & 0.0561 \\ 0.0588 & 0.2941 & 0.2941 & 0.1176 & 0.1176 & 0.1176 \end{bmatrix} \tag{68}$$

$$\mathcal{M}_{\mathbf{y}_{xy}} = \mathcal{M}_{\mathbf{y}_x} = \mathcal{M}_{\mathbf{y}_y} = \begin{cases} \text{State } \textit{nominal}: & (\mu_1, \sigma_1, n_1) = (0, 1.0, 0.012) \\ \text{State } \pm \textit{reliable}: & (\mu_1, \sigma_1, n_1) = (\pm 1.3, 1.5, 0.06) \\ \text{State } \pm \textit{unreliable}: & (\mu_1, \sigma_1, n_1) = (\pm 3.85, 4.0, 1.28) \\ \text{State } \textit{faulty}: & (\mu_1, \sigma_1, n_1) = (\text{NaN}, \text{NaN}, \text{NaN}) \end{cases} \tag{69}$$

$$\mathbf{T}_{\mathbf{y}_{xy}} = \mathbf{T}_{\mathbf{y}_x} = \mathbf{T}_{\mathbf{y}_y} = \begin{bmatrix} 0.4809 & 0.1967 & 0.1967 & 0.0601 & 0.0601 & 0.0055 \\ 0.4160 & 0.3200 & 0.1440 & 0.0800 & 0.0320 & 0.0080 \\ 0.4160 & 0.1440 & 0.3200 & 0.0320 & 0.0800 & 0.0080 \\ 0.3689 & 0.2136 & 0.1359 & 0.1942 & 0.0777 & 0.0097 \\ 0.3689 & 0.1359 & 0.2136 & 0.0777 & 0.1942 & 0.0097 \\ 0.0102 & 0.2959 & 0.2959 & 0.1837 & 0.1837 & 0.0306 \end{bmatrix} \tag{70}$$

$$\mathcal{M}_{\mathbf{y}_z} = \begin{cases} \text{State } \textit{nominal}: & (\mu_1, \sigma_1, n_1) = (0, 0.08, 0.012) \\ \text{State } \pm \textit{reliable}: & (\mu_1, \sigma_1, n_1) = (\pm 0.11, 0.1208, 0.06) \\ \text{State } \pm \textit{unreliable}: & (\mu_1, \sigma_1, n_1) = (\pm 0.55, 0.71, 1.28) \\ \text{State } \textit{faulty}: & (\mu_1, \sigma_1, n_1) = (\text{NaN}, \text{NaN}, \text{NaN}) \end{cases} \tag{71}$$

$$\mathbf{T}_{\mathbf{y}_z} = \begin{bmatrix} 0.5198 & 0.1762 & 0.1762 & 0.0617 & 0.0617 & 0.0044 \\ 0.4020 & 0.4020 & 0.1106 & 0.0503 & 0.0302 & 0.0050 \\ 0.4020 & 0.1106 & 0.4020 & 0.0302 & 0.0503 & 0.0050 \\ 0.3704 & 0.2667 & 0.1481 & 0.1481 & 0.0593 & 0.0074 \\ 0.3704 & 0.1481 & 0.2667 & 0.0593 & 0.1481 & 0.0074 \\ 0.0667 & 0.2000 & 0.2000 & 0.2000 & 0.2000 & 0.1333 \end{bmatrix} \tag{72}$$

$$\mathcal{M}_{\mathbf{y}_\Theta} = \mathcal{M}_{\mathbf{y}_\varphi} = \mathcal{M}_{\mathbf{y}_\theta} = \mathcal{M}_{\mathbf{y}_\psi} = \begin{cases} \text{State } \textit{nominal}: & (\mu_1, \sigma_1, n_1) = \left(0, \frac{\pi}{220}, \frac{\pi}{98.05}\right) \\ \text{State } \pm \textit{reliable}: & (\mu_1, \sigma_1, n_1) = \left(\pm\frac{\pi}{192}, \frac{\pi}{176}, \frac{\pi}{42.60}\right) \\ \text{State } \pm \textit{unreliable}: & (\mu_1, \sigma_1, n_1) = \left(\pm\frac{\pi}{60}, \frac{\pi}{42}, \frac{\pi}{10}\right) \\ \text{State } \textit{faulty}: & (\mu_1, \sigma_1, n_1) = (\mathtt{NaN}, \mathtt{NaN}, \mathtt{NaN}) \end{cases} \tag{73}$$

$$\mathbf{T}_{\mathbf{y}_\Theta} = \mathbf{T}_{\mathbf{y}_\varphi} = \mathbf{T}_{\mathbf{y}_\theta} = \mathbf{T}_{\mathbf{y}_\psi} = \begin{bmatrix} 0.4686 & 0.2301 & 0.2301 & 0.0335 & 0.0335 & 0.0042 \\ 0.4014 & 0.2721 & 0.1769 & 0.1020 & 0.0340 & 0.0136 \\ 0.4014 & 0.1769 & 0.2721 & 0.0340 & 0.1020 & 0.0136 \\ 0.3982 & 0.1403 & 0.0995 & 0.1719 & 0.0995 & 0.0905 \\ 0.3982 & 0.0995 & 0.1403 & 0.0995 & 0.1719 & 0.0905 \\ 0.0526 & 0.1579 & 0.1579 & 0.2105 & 0.2105 & 0.2105 \end{bmatrix} \tag{74}$$

$$\mathbf{S}_y = \begin{bmatrix} 1 & 1.0\cdot10^{-4} & 1.0\cdot10^{-4} & 3.0\cdot10^{-5} & 3.0\cdot10^{-5} & 3.0\cdot10^{-5} & 0 & 0 & 0 & 5.0\cdot10^{-7} & 5.0\cdot10^{-7} & 5.0\cdot10^{-7} \\ 1.0\cdot10^{-4} & 1 & 1.0\cdot10^{-4} & 3.0\cdot10^{-5} & 3.0\cdot10^{-5} & 3.0\cdot10^{-5} & 0 & 0 & 0 & 5.0\cdot10^{-7} & 5.0\cdot10^{-7} & 5.0\cdot10^{-7} \\ 1.0\cdot10^{-4} & 1.0\cdot10^{-4} & 1 & 3.0\cdot10^{-5} & 3.0\cdot10^{-5} & 3.0\cdot10^{-5} & 0 & 0 & 0 & 5.0\cdot10^{-7} & 5.0\cdot10^{-7} & 5.0\cdot10^{-7} \\ 3.0\cdot10^{-5} & 3.0\cdot10^{-5} & 3.0\cdot10^{-5} & 1 & 6.0\cdot10^{-4} & 6.0\cdot10^{-4} & 0 & 0 & 0 & 0 & 0 & 0 \\ 3.0\cdot10^{-5} & 3.0\cdot10^{-5} & 3.0\cdot10^{-5} & 6.0\cdot10^{-4} & 1 & 6.0\cdot10^{-4} & 0 & 0 & 0 & 0 & 0 & 0 \\ 3.0\cdot10^{-5} & 3.0\cdot10^{-5} & 3.0\cdot10^{-5} & 6.0\cdot10^{-4} & 6.0\cdot10^{-4} & 1 & 0 & 0 & 0 & 0 & 0 & 0 \\ 0 & 0 & 0 & 0 & 0 & 0 & 1 & 3.0\cdot10^{-4} & 4.0\cdot10^{-6} & 2.0\cdot10^{-5} & 2.0\cdot10^{-5} & 2.0\cdot10^{-5} \\ 0 & 0 & 0 & 0 & 0 & 0 & 3.0\cdot10^{-4} & 1 & 4.0\cdot10^{-6} & 2.0\cdot10^{-5} & 2.0\cdot10^{-5} & 2.0\cdot10^{-5} \\ 0 & 0 & 0 & 0 & 0 & 0 & 4.0\cdot10^{-6} & 2.0\cdot10^{-5} & 1 & 0 & 3.0\cdot10^{-5} & 3.0\cdot10^{-5} \\ 5.0\cdot10^{-7} & 5.0\cdot10^{-7} & 5.0\cdot10^{-7} & 0 & 0 & 0 & 2.0\cdot10^{-5} & 4.0\cdot10^{-6} & 1 & 0 & 0 & 0 \\ 5.0\cdot10^{-7} & 5.0\cdot10^{-7} & 5.0\cdot10^{-7} & 0 & 0 & 0 & 2.0\cdot10^{-5} & 2.0\cdot10^{-5} & 0 & 3.0\cdot10^{-5} & 1 & 3.0\cdot10^{-5} \\ 5.0\cdot10^{-7} & 5.0\cdot10^{-7} & 5.0\cdot10^{-7} & 0 & 0 & 0 & 2.0\cdot10^{-5} & 2.0\cdot10^{-5} & 0 & 3.0\cdot10^{-5} & 3.0\cdot10^{-5} & 1 \end{bmatrix} \tag{75}$$

5.3.1 Sensor fault simulation

The last mentioned state, "*fault*" doesn't generate random additive measurement noises, but outputs NaN[7] values, which are ignored by the SP-UKF. On detecting the NaN value in a measurement channel, the SP-UKF sets the corresponding, i-th row of the column vector of measurements y_i to the value of $\hat{y}_i$. This results in the corresponding rows and columns of the $\mathbf{P}_\mathbf{y}$ covariance matrix being zero, an update to the corresponding row of the estimate column vector $\hat{\mathbf{x}}(k|k)$ being omitted, and the corresponding rows and columns of $\mathbf{P}_\mathbf{x}(k)$ growing rather than falling. The latter signifies a decrease in the trustworthiness of the estimate, and *compromises the stability of the SP-UKF* should a string of faulty readings continue overlong.

5.3.2 Outlier rejection

In order for the SP-UKF to remain stable and deliver trustworthy state estimates to the feedback of the relevant controllers, outlier measurements are rejected. Rejection of an outlier is dealt with identically to a faulty measurement, the appropriate row of the measurement column vector $\mathbf{y}$ being over-written with NaN as if there were a sensor fault occurring in the outlier measurement channel.

Outlier rows of y are considered those for which any values, inspected column-wise, fulfill:

$$\mathbf{R}_y(k) = \mathbf{y}(k)^{\mathrm{T}} \cdot \mathbf{y}(k) = \left[r_y^{(i,j)}\right] \tag{76}$$

$$i_{out} = \left\{\arg_i\left(\exists j, r_y^{(i,j)} > 16 \cdot \mathbf{P}_{\mathbf{x}}(k)\right)\right\} \tag{77}$$

$$\mathbf{y}\left[\{i_{out}\}\right] \stackrel{\text{redef}}{=} \mathtt{NaN} \tag{78}$$

Where $\mathbf{P}_{\mathbf{x}}(k)$ is the covariance matrix of the estimate of the full-state vector rendered by the SP-UKF.

5.3.3 Process noise

The additive process noise is assumed to be multivariate rate-limited white noise without bias. The used covariance matrix is given in (79) and the rate limits in a vector in (80).

$$\mathbf{R}_v^{(true)} = \begin{bmatrix} 0.2 & 0.01 & 0.01 & 0 & 0.005 & 0.005 \\ 0.01 & 0.1 & 0.0275 & 0.01 & 0.003 & 0.035 \\ 0.01 & 0.0275 & 0.1 & 0.01 & 0.003 & 0 \\ 0 & 0.01 & 0.01 & 0.0011 & 0.0001 & 0.00015 \\ 0.005 & 0.003 & 0.03 & 0.0001 & 0.002 & 0 \\ 0.005 & 0.035 & 0 & 0.00015 & 0 & 0.0022 \end{bmatrix} \tag{79}$$

$$\mathbf{n}_v^{(true)} = \left[0.2\ 0.2\ 0.2\ \tfrac{\pi}{36}\ \tfrac{\pi}{36}\ \tfrac{\pi}{36}\right]^{\mathrm{T}} \tag{80}$$

5.4 Control signals conditioning

The solution of $f_i(k) \leftarrow \mathrm{E}(\hat{x}_{AUV}(k))$, obtained by evaluating (2), or more precisely (29), at the estimate $\hat{x}_{AUV}(k)$ rendered by the SP-UKF, will be used for the formation of commands accepted by the forward speed and heading controllers in (41, 44), $\mathbf{c} = [\dot{u}_c\ u_c\ \dot{\psi}_c\ \psi_c]$.

The task of the low-level control system of the AUV is to try to recreate a motion that would result from applying $\boldsymbol{f}(k)$ to an unconstrained point unit mass, i.e. a holonomic 2D double integrator model, up to the thrust allocation and kinematic and dynamic constraints of the actual vehicle. In the following equations, the sampling with time T of the integration over the $\mathbb{R}^2$ space is assumed in the form of *the Euler backwards formula.* So naively:

$$u(k) = \sqrt{u(k-1)^2 + T^2 f^2 + 2Tuf_{||}} \tag{81}$$

$$\dot{u}(k) = \frac{1}{T}(u(k) - u(k-1))$$

$$= \frac{1}{T}\sqrt{u(k-1)^2 + T^2 f^2 + 2Tf_{\|}u(k-1)} - \frac{u(k-1)}{T} \tag{82}$$

$$\dot{\psi}(k) = \frac{1}{T}\arctan\left(\frac{Tf_{\perp}}{Tf_{\|} + u(k-1)}\right) \tag{83}$$

$$\ddot{\psi}(k) = \frac{1}{T}\left[\frac{1}{T}\arctan\left(\frac{Tf_{\perp}}{Tf_{\|} + u(k-1)}\right) - \dot{\psi}(k-1)\right] \tag{84}$$

Where:

$f \overset{\text{id}}{=} \|f\|$ is the norm of the total controlling force, admitting decomposition into $[f_{\|}\, f_{\perp}]^{\mathrm{T}}$, the components parallel and perpendicular to the direction of heading of the AUV (given by $\psi(k - 1)$, notwithstanding possible *sideslip* resulting from $\mathbf{x}_2 = \mathbf{v}_2 = v \neq 0$).

At this point, it is assumed that an AUV has a specified *performance envelope* of $(\overline{u}, \overline{\dot{u}}, \overline{\dot{\psi}}, \overline{\ddot{\psi}})$.

Given these independent parameters, the manipulation of (81 - 84) establishes restrictions that define the permissible ranges to which f must be restricted to prevent exceeding the typical operating range of the AUV's low-level controllers.

5.4.1 The constraint inequalities

The locus of solutions of (81) for ***f*** in easier to visualize in an AUV-fixed coordinate system (one with the origin in x_i, with the x-axis aligned with $u(k - 1)\angle\psi(k - 1)$). In it, the *admissible solution locus* is a disc offset along the x-axis by $u(k - 1)/T$, given by the implicit expression:

$$\left\|\left[f_{\|} + \tfrac{u(k-1)}{T} \quad f_{\perp}\right]^{\mathrm{T}}\right\| \leq \frac{\overline{u}}{T} \tag{85}$$

The locus of solutions of (82) is, similarly to the preceding case, an offset annulus concentric to the disc described by (85):

$$\frac{u(k-1)}{T} - \overline{\dot{u}} \leq \left\|\left[f_{\|} + \tfrac{u(k-1)}{T} \quad f_{\perp}\right]^{\mathrm{T}}\right\| \leq \frac{u(k-1)}{T} + \overline{\dot{u}} \tag{86}$$

The locus of solutions of (83) is an angular sector of an infinite disk (a 1-cone) concentric to the previous two loci, expressed in terms of:

$$\left|\arctan\left(\frac{a_{\perp}}{a_{\|} + \frac{u(k-1)}{T}}\right)\right| \leq T\overline{\dot{\psi}} \tag{87}$$

The locus of solutions of (84) is likewise an *angular sector of an infinite disk* (a 1-cone) concentric to all the other loci, which satisfies the inequality:

$$T\dot{\psi}(k-1) - \frac{T^2}{2}\overline{\ddot{\psi}} \leq \arctan\left(\frac{f_\perp}{f_\parallel + \frac{u(k-1)}{T}}\right)$$
$$\leq T\dot{\psi}(k-1) + \frac{T^2}{2}\overline{\ddot{\psi}} \qquad (88)$$

A solution for ***f*** (*k*) is legal if it meets all of the criteria stated in (85–88), i.e. if it belongs to a subset of $\mathbb{R}^2$ shaped as an annular sector.

5.4.2 Clamping the total controlling force

If (85 – 88) are not fulfilled, a non-linear procedure for the clamping ***f*** to an admissible range is employed. Consequently, the low-level controllers' operating point(s) remain within a quasi-linear vicinity of the sliding surfaces. The procedure is pseudo-coded in Algorithm After clamping by the presented algorithm, (81 – 84) are used to form the commands $\mathbf{c} = [a_c\ u_c\ r_c\ \psi_c]^T$ for the low-level controllers 41, 44).

6. SIMULATION RESULTS

Combining the presented virtual potentials framework and the HILS presented in the previous chapters, a full simulation is presented for a group of 4 simulated *Aries*-precursor AUVs cruising in formation.

6.1 Simulation 1

The first simulation presents a cruise in formation down an unobstructed channel in between two obstacles towards the way-point. Figure 10 presents the actual paths traveled by the AUVs. Figure 11 presents the speeds of all four vehicles. The initial dips in the path occur due to the non-holonomic nature of the vehicles' kinematics, due to which they cannot initialize the manoeuvres from zero starting speed that would preserve the initial formation perfectly and still commence navigation to the way-point. This is especially exacerbated by the fact that at near-zero speeds, the control surfaces (δ_r, δ_s) are terribly ineffective. The final dips in the area of the paths around the way-point occur after the AUVs have parked in the stable formation configuration. Near the waypoint and at low speeds, the drift in the state estimates is accentuated by a lack of passive stability provided by AUVs' streamlining at higher speeds. This, in hand with non-holonomic kinematics, causes the vehicles to momentarily break formation. It is only after accumulating enough speed that vehicles can turn within a small enough radius to re-establish the formation. Dips in the path correspond to the dips in the speed graphs for the vehicles, as their commanded speed shoots up again in order to re-establish the formation.

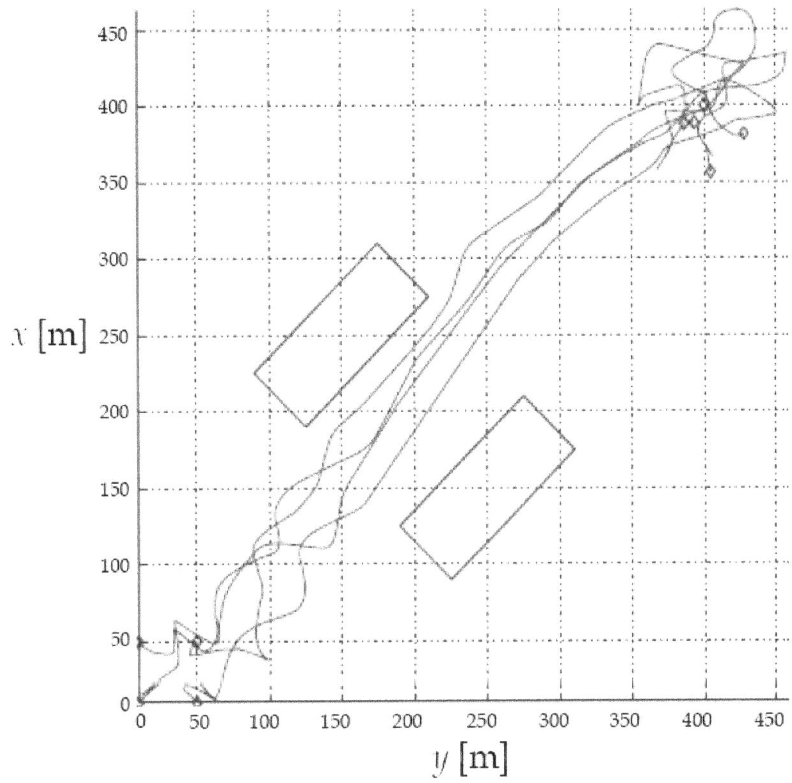

Fig. 10. Paths of the 4 HILS models of AUVs based on the precursor to the NPS Aries vehicle cruising in an uncluttered environment.

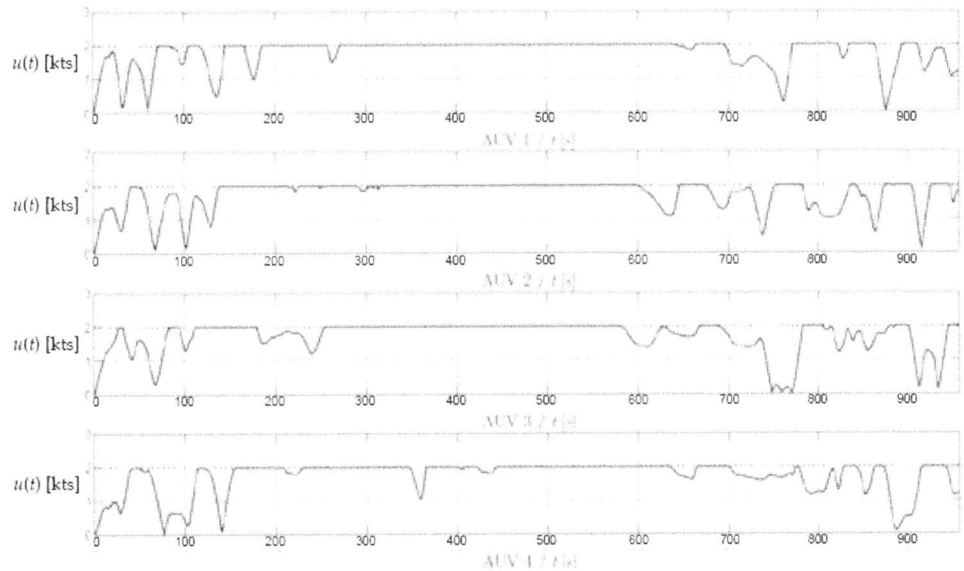

Fig. 11. Speeds of the 4 HILS models of AUVs based on the precursor to the NPS *Aries* vehicle cruising in an uncluttered environment.

if not (85) or (86)

if $f||$ > 0

— *this indicates that the required dominant behaviour produced by the virtual potentials framework is to accelerate the AUV further in the forward direction; Trying to follow through on this behaviour, having in mind the operational wisdom concerning real-world missions and experiences, it is much safer to give priority to the component that manoeuvres the AUV to the side, in order for the AUV to be able to circumnavigate the obstacles it might be heading towards.*

$$f_{\perp} = \operatorname{sign}(f_{\perp}) \min\{|f_{\perp}|, \overline{u}/T, u(k-1)/T + \overline{\dot{u}}\}$$
$$f_{\|} = \sqrt{\min\{\overline{u}/T, u(k-1)/T + \overline{\dot{u}}\}^2 - f_{\perp}^2} - u(k-1)/T$$

— *therefore, require the* $f\perp$ component to be *as large as admissible, and adjust the* $f||$ accordingly

else

— otherwise the required dominant behaviour produced by the virtual potentials framework is to decelerate or break the AUV; Trying to follow through on this behaviour, it is much safer to give priority to the component that decreases the forward speed of the AUV, in order for the AUV to avoid colliding with possible obstacles it is heading towards.

$$f_{\|} = \max\{f_{\|}, -u(k-1)/T, -\overline{\dot{u}}\}$$
$$f_{\perp} = \operatorname{sign}(f_{\perp}) \min\{|f_{\perp}|, \sqrt{(\overline{\dot{u}}/T)^2 - (f_{\|} + u(k-1)/T)^2}\}$$

— *therefore, require that* $f||$ *component to be as large as admissible, and adjust the* $f\perp$ *accordingly*

end if

end if

$\psi_{crit} \stackrel{\text{def}}{=}$ minimum right-hand side of (87) and (88)

if either of the left-hand sides of (87) or (88) > $\psi crit$

— *if, after preceding adjustments,* $\angle$ $(\boldsymbol{f} - \boldsymbol{f} \cdot \mathbf{u}(k)/||\mathbf{u}(k)||)$ *is inadmissible*

if $f||$ > 0

$f|| = f|| \cot \psi_{crit} + f\perp \cos^2 \psi_{crit} - (u(k-1)/\mathrm{T}) \sin^2 \psi_{crit}$

— *again, pay attention to the intention of the manoeuvre, and if the dominant behavior is acceleration, decrease the accelerating component of* f, $f||$, *further*

end if

$f\perp = \operatorname{sign}(f\perp) . (f|| + u(k-1)/T) \tan \psi_{crit}$

— *clamp the* $f\perp$ *component to an admissible value*

end if

6.2 Simulation 2

The second simulation presents a cruise in formation down a heavily cluttered corridor defined by two larger rectangular obstacles. Figure 12 presents the actual paths traveled. Figure 13 presents the speeds of the four vehicles.

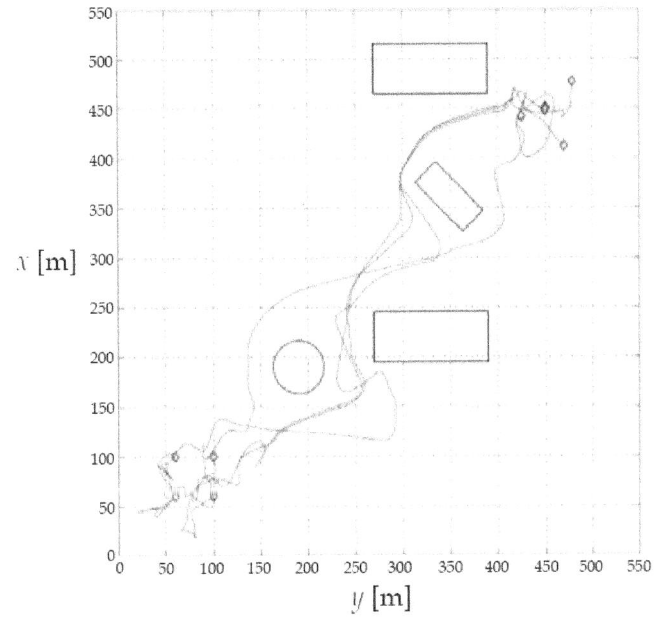

Fig. 12. Paths of the 4 HILS models of AUVs based on the precursor to the NPS Aries vehicle cruising in a cluttered environment.

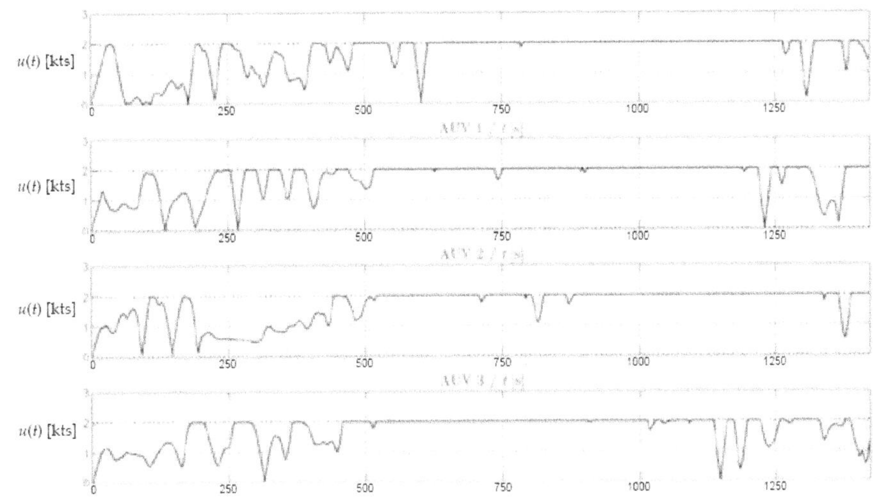

Fig. 13. Speeds of the 4 HILS models of AUVs based on the precursor to the NPS *Aries* vehicle cruising in a cluttered environment.

6.2.1 Cruise phase 1

A cruise is started in a formation. However, very soon the formation encounters the first obstacle. As the leading formation-members are momentarily slowed down before they circumnavigate to either side of the obstacle, the trailing members "pile up" in front of this artificial potential barrier (especially the AUV closest to the origin). This is evident in the dips and temporary confusion before the first, circular obstacle. However, the operational safety approach that is implicitly encapsulated by the *cross-layer design* is preserved. The vehicles break formation, so that one of the vehicle circumnavigates the first obstacle on the left, and the others on the right. Since this produces a significantly different trajectory from the rest of the group, vehicle 1 isn't able to rejoin the formation until much later.

6.2.2 Cruise phase 2

The other 3 vehicles (2, 3 and 4), before being able to restore a formation encounter the first of the two large rectangular obstacles. Note that vehicles 2 and 3 remain in the leader-follower arrangement as evidenced by their closely matching trajectories in this phase of the cruise. The "outrigger" vehicle 4, trying to keep in formation with 2 and 3, encounters the large rectangular obstacle at a bearing much closer to head-on. Therefore, it executes a significant course change manoeuvre, during which it cannot satisfactorily compromise between safe avoidance of the obstacle and staying in formation with 2 and 3. As vehicles 2 and 3 navigate in formation through the strait in between the circular and the first large rectangular obstacle, the leader vehicle 3 starts to manoeuvre to starboard towards the way-point. This manoeuvre causes the formation cell vertex trailing behind vehicle 3 that represents the dominant navigation goal for vehicle 2 to start accumulating speed in excess of what 2 is able to match. This is due to the fact that as 3 swings to starboard, the formation cell vertex "sweeps" through water with a velocity that consists of the sum of linear velocity of vehicle 3 and the tangential velocity contributed by the "arm" of the formation cell *f*. Therefore, the formation is temporarily completely broken.

6.2.3 Cruise phase 3

However, the breaking of formation between 2 and 3 occurs at such a time that 1 catches up with 2 before 2 gets much farther afield, presenting its trailing cell vertex as a local navigation goal to 2. That is why 2 exhibits a hard break to starboard, trying to form itself up as a follower of vehicle 1. However, just as 2 is completing its formation, vehicle 1, manoeuvres around the final obstacle – the small diagonally presented rectangle. As the trailing cell vertex of 1 is, from 2's viewpoint, shadowed by the obstacle's repulsion, it reorients towards what until then is a secondary navigation goal in its vicinity – the cell vertex of the "latecomer" of Phase 2, vehicle 4. This reorientation is what contributes to 2's "decision" to circumnavigate the diagonal rectangle to starboard, rather than to port, as would bc optimal if no formation influences were present. Phase 3 finishes as vehicle 2 is trying to pursue vehicle 4, and vehicle 4 corners the diagonal rectangle, getting away from vehicle 2.

6.2.4 Cruise phase 4

Phase 4 is entered into without formations. This phase is characterized by converging on the way-point, which all the vehicles reach independently, followed by re-establishing the formation. However, an ideal formation is impossible due to operational safety, as no vehicle is "willing" to approach the second large rectangle (towards the top of the figure). This is exacerbated by reduced manoeuvring capabilities, as all vehicles reduce speed in the vicinity of the way-point.

7. CONCISE SUMMARY

The chapter has introduced a decentralized formation guidance framework based on virtual potentials that functions in two dimensions. The framework ensures trajectory stability, convergence to the way-point as the global navigation objective, and evasion of prominent, dangerous obstructions. The framework provides a cross-layer methodology to integrate two conflicting behaviors that Autonomous Underwater Vehicles (AUVs) in a formation guidance system must reconcile: the prioritization of formation maintenance versus the imperative of operational safety in navigating obstacles within a cluttered environment. In addition to the theoretical contribution, a comprehensive functional hardware-in-the-loop system (HILS) for realistic simulation analysis was offered.

- The system incorporates multiple levels of realistic dynamic behavior.
- A full-state coupled model dynamics of a seaworthy, long-autonomy AUV model based on rigid-body physics and hydrodynamics of viscous fluids like water,
- An unbiased rate-limited white noise model of the process noise,
- A non-stationary generator of measurement noise based on Gaussian Markov models with an explicitly included fault-mode,
- An outlier-elimination scheme based on the evaluation of the state estimate covariance returned by the employed estimator,
- A Scaled Unscented Transform Sigma-Point Kalman Filter (SP-UKF) that can work either in the filtering mode, or a combination of filtering and pure-prediction mode when faulty measurements are present, utilizing a full-state non-linear coupled AUV model dynamics,

- A command signal adaptation mechanism that accents operational safety concerns by prioritizing turning manoeuvres while accelerating, and "pure" braking / shedding forward speed when decelerating.

7.1 Subsequent tasks

Multiple specific study domains, grounded on the established HILS framework, are yet to evaluate the efficacy of the proposed virtual potential-based decentralized cooperative framework. These are essential to provide the framework for application in expensive and logistically complex operations in the actual oceanic environment.

- Accurately simulate the local representation of knowledge regarding other AUVs on each AUV.

1. This can be approached on several fronts:
   - Exploring the realistic statistics of the sensing process when applied to sensing other AUVs as opposed to salient obstacles in the waterspace. Exploring and modeling the beam-forming issues arising with mechanically scanning sonars vs. more complex and costlier multi-beam imaging sonars,
   - Exploring the increases in complexity (and computer resource management), numerical robustness and stability issues of AUV-local estimation of other AUVs in the formation,
   - Dealing with the issues of the instability of the "foreign" AUVs' state estimates covariance matrix by one of three ways: (*i*) using synchronous, pre-scheduled hydroacoustic communication. Communication would entail improved estimates coming from on-board the AUVs, where the estimates are corrected by collocated measurement; (*ii*) exploring an on-demand handshake-based communication scheme. Handshaking would be initiated by an AUV polling a team-member for a correction to the local estimate featuring unacceptably large covariance; (*iii*) exploring a predictive communication scheme where the AUVs themselves determine to broadcast their measurements without being polled. This last option needs to involve each AUV continually predicting how well other AUVs are keeping track of its own state estimates.
2. Explore the applicability of the framework to non-conservative, energetic manoeuvring in 3D, i.e. use the same framework to generate commands for the depth / pitch low-level controllers. Explore the behaviour of 3D-formations based on the *honeycombs* (3D tesselations) of the vector space of reals.

REFERENCES

Abkowitz, M. (1969). *Stability and Motion Control of Ocean Vehicles,* MIT Press Cambridge MA USA.

Ahn, S., Rauh, W. & Recknagel, M. (1999). Ellipse fitting and parameter assessment of circular object targets for robot vision, *Intelligent Robots and Systems 1999. IROS '99 Proceedings. 1999 IEEE/RSJ International Conference on,* Vol. 1, pp. 525 –530.

Allen, B., Stokey, R., Austin, T., Forrester, N., Goldborough, R., Purcell, M. & von Alt, C. (1997). REMUS A Small Low Cost AUV system description field trials and performance results, *OCEANS 1997 Conference Record (IEEE),* Vol. 2, pp. 994–100.

Allen, T., Buss, A. & Sanchez, S. (2004). Assessing obstacle location accuracy in the REMUS unmanned underwater vehicle, *Proceedings - Winter Simulation Conference,* Vol. 1, pp. 940–948.

An, P., Healey, A., Park, J. & Smith, S. (1997). Asynchronous data fusion for auv navigation via heuristic fuzzy filtering techniques, *OCEANS '97. MTS/IEEE Conference Proceedings,* Vol. 1, pp. 397 –402 vol.1.

Barisic, M., Vukic, Z. & Miskovic, N. (2007a). Kinematic simulative analysis of virtual potential field method for AUV trajectory planning, *in* K. Valavanis & Z. Kovacic (eds), *15th Mediterranean Conference on Control and Automation, 2007. Proceedings of the,* p. on CD.

Barisic, M., Vukic, Z. & Miskovic, N. (2007b). A kinematic virtual potentials trajectory planner for AUV-s, *in* M. Devy (ed.), *6th IFAC Symposium on Intelligent Autonomous Vehicles, 2007. Proceedings of the,* p. on CD.

Bildberg, R. (2009). Editor's foreword, *in* B. R. (ed.), *Proceedings of the 16th International Symposium on Unmanned Untethered Submersibles Technology,* Autonomous Undersea Systems Institute. on CD.

Boncal, R. (1987). *A Study of Model Based Maneuvering Controls for Autonomous Underwater Vehicles,* Master's thesis, Naval Postgraduate School Monterey CA USA.

Carder, K., Costello, D., Warrior, H., Langebrake, L., Hou, W., Patten, J. & Kaltenbacher, E. (2001). Ocean-science mission needs: Real-time AUV data for command control and model inputs, *IEEE Journal of Oceanic Engineering* 26(4): 742–751.

Clegg, D. & Peterson, M. (2003). User operational evaluation system of unmanned underwater vehicles for very shallow water mine countermeasures, *OCEANS 2003 Conference Record (IEEE),* Vol. 3, pp. 1417–1423.

Curtin, T., Bellingham, J., Catipovic, J. & Webb, D. (1993). Autonomous oceanographic sampling networks, *Oceanography* 6(3): 86–94.

Duda, R. O. & Hart, P. E. (1972). Use of the Hough transformation to detect lines and curves in pictures, *Communications of the ACM* 15(1): 11–15.

Eisman, D. (2003). Navy ships seize boats carrying mines in Iraqi port, *The Virginian-pilot.*

Farrell, J., S., P. & Li, W. (2005). Chemical plume tracing via an autonomous underwater vehicle, *IEEE Journal of Oceanic Engineering* 30(2): 428–442.

Faucette, W. M. (1996). A geometric interpretation of the solution of the general quartic polynomial, *The American Mathematical Monthly* 103(1): 51–57.

Gertler, M. & Hagen, G. (1967). Standard Equations of Motion for Submarine Simulation, *Technical report,* David W. Taylor Naval Ship Research and Development Center Bethesda MD USA.

Haule, D. & Malowany, A. (1989). Object recognition using fast adaptive Hough transform, *Communications Computers and Signal Processing 1989. Conference Proceedings IEEE Pacific Rim Conference on,* pp. 91 –94.

He, Y. & Li, Z. (2008). An effective approach for multi-rectangle detection, Young Computer Scientists 2008. ICYCS. The 9th *International Conference* for, pp. 862 –867.

Healey, A. & Lienard, D. (1993). Multivariable sliding mode control for autonomous diving and steering of unmanned underwater vehicles, *Oceanic Engineering IEEE Journal of* 18(3): 327 –339.

Hough, P. & Powell, B. (1960). A method for faster analysis of bubble chamber photographs, Vol. 18, Italian Physical Society, pp. 1184 –1191.

Illingworth, J. & Kittler, J. (1987). The adaptive Hough transform, *Pattern Analysis and Machine Intelligence IEEE Transactions on* PAMI-9(5): 690 –698.

Jiang, X., Huang, X., Jie, M. & Yin, H. (2007). Rock detection based on 2D maximum entropy thresholding segmentation and ellipse fitting, *Robotics and Bioimetics* 2007. ROBIO 2007. *IEEE International Conference on,* pp. 1143 –1147.

Jung, C. & Schramm, R. (2004). Rectangle detection based on a windowed Hough transform, *Computer Graphics and Image Processing 2004. Proceedings, 17th Brazilian Symposium on,* pp. 113 – 120.

Lienard, D. (1990). *Sliding Mode Control for Multivariable AUV Autopilots,* Master's thesis, Naval Postgraduate School Monterey CA USA.

Maitre, H. (1986). Contribution to the prediction of performances of the Hough transform, *Pattern Analysis and Machine Intelligence IEEE Transactions on* PAMI-8(5): 669 –674.

Marco, D. & Healey, A. (2000). Current developments in underwater vehicle control and navigation: the NPS Aries AUV, OCEANS *Conference record (IEEE)*, Vol. 2, pp. 1011–1016.

Marco, D. & Healey, A. (2001). Command control and navigation experimental results with the NPS Aries AUV, *IEEE Journal of Oceanic Engineering* 26(4): 466–476.

Nguyen, T., Pham, X. & Jeon, J. (2009). Rectangular object tracking based on standard Hough transform, *Robotics and Biomimetics 2008. ROBIO IEEE International Conference on*, pp. 2098 –2103.

Pang, S. (2006). Development of a guidance system for AUV chemical plume tracing, *OCEANS 2006 Conference Record (IEEE)*, pp. 1–6.

Pang, S., Arrieta, R., Farell, J. & Li, W. (2003). AUV reactive planning: Deepest Point, *OCEANS 2009 Conference Record (IEEE)*, Vol. 4, pp. 2222–2226.

Papoulias, F., Cristi, R., Marco, D. & Healey, A. (1989). Modeling, sliding mode control design, and visual simulation of auv dive plane dynamic response, *Unmanned Untethered Submersible Technology, 1989. Proceedings of the 6th International Symposium on*, pp. 536 –547.

Pilu, M., Fitzgibbon, A. & Fisher, R. (1996). Ellipse-specific direct least-square fitting, *Image Processing 1996. Proceedings. International Conference on*, Vol. 3, pp. 599 –602.

Plueddemann, A., Packard, G., Lord, J. & Whelan, S. (2008). Observing Arctic Coastal Hydrography Using the REMUS AUV, *Autonomous Underwater Vehicles 2008. AUV 2008. IEEE/OES*, pp. 1–4.

Rizon, M., Yazid, H. & Saad, P. (2007). A comparison of circular object detection using Hough transform and chord intersection, *Geometric Modeling and Imaging 2007. GMAI '07*, pp. 115 –120.

Stewart, I. (2003). *Galois Theory*, Chapman & Hall / CRC Press, 6000 Broken Sound Pkwy. Suite 300 Boca Raton FL 33487 USA.

Tuohy, S. T. (1994). A simulation model for AUV navigation, *Proceedings of the 1994 Symposium on Autonomous Underwater Vehicle Technology*, pp. 470–478.

US Navy (2004). The US Navy Unmanned Underwater Vehicles Master Plan, http://www.navy.mil/navydata/technology/uuvmp.pdf.

van der Merwe, R. (2004). *Sigma-Point Kalman Filters for Probabilistic Inference in Dynamic State-Space Models*, PhD thesis, OGI School of Science & Engineering, Oregon Health & Science University, Portland, OR, USA.

6 Chapter

MODELING AND MOTION CONTROL STRATEGY FOR AUTONOMOUS UNDERWATER VEHICLES (AUV)

1. INTRODUCTION: AN OVERVIEW

The control systems for speed and position in Autonomous Underwater Vehicles (AUVs) have garnered heightened attention regarding performance and safety, attributed to their growing utilization in commercial and military applications, as well as research challenges over recent decades. These applications encompass underwater resource exploration, oceanographic mapping, undersea wreckage recovery, cable installation, geographical surveying, inspection of coastal and offshore structures, harbor security assessments, and mining operations, including countermeasures (Fossen, 2002). The advancement of an intelligent underwater work system would significantly increase various ocean operations, necessitating more stringent standards for the control systems of underwater vehicles. The control system must possess the intelligence to collect environmental data and formulate its own management methods autonomously, without human involvement (Yuh, 1990; Venugopal and Sudhakar, 1992). Underwater vehicle dynamics is significantly interconnected and extremely nonlinear due to the influence of increased hydrodynamic mass, as well as lift and drag forces operating on the vehicle. The engineering challenges posed by the high-density, non-uniform, and unstructured saltwater environment, along with the nonlinear responses of vehicles, complicate the attainment of a high level of autonomy. Consequently, six degrees of freedom vehicle modeling and simulation are essential and beneficial for the advancement of subsea vehicle control systems (Yuh, 1990; Fossen, 1991; Li et al., 2005). The autonomy of the AUV is crucial for operations in a highly dangerous and unfamiliar environment. The motion control architecture, a critical subsystem of underwater vehicles, is a framework that regulates both the sensory and actuator systems (Gan et al., 2006), thereby allowing the robot to execute a user-defined mission. This chapter derives a general mathematical model for nonlinear vehicle systems, which is sufficiently robust to be applicable to several underwater vehicles based on their physical attributes, hence facilitating model simplification. A simulation platform, "AUV-XX," is built to evaluate the motion characteristics of the vehicle based on this model. The motion control system, encompassing position, speed, and depth control, was examined for various vehicle duty assignments. An enhanced S-surface control utilizing a capacitor model was created, offering various gain options with distinct physical significance. The outcomes of motion control on the simulation platform "AUV-XX" are delineated.

2. MATHEMATICAL MODELING AND SIMULATION

Six degrees of freedomVehicle simulations are essential and beneficial in the advancement of subsea vehicle control systems.The simulation must model various processes, including vehicle hydrodynamics, rigid body dynamics, and actuator dynamics, among others.

2.1 AUV kinematics and dynamics

The mathematical models of marine vehicles consist of kinematic and dynamic part, where the kinematic model gives the relationship between speeds in a body-fixed frame and derivatives of positions and angles in an Earth-fixed frame, see Fig.1. The vector of positions and angles of an underwater vehicle $\eta = [\ x,\ y,\ z,\ \phi, \theta\ , \psi\ ]^T$ is defined in the Earth-fixed coordinate system(E) and vector of linear and angular v $v = [u, v, w, p,q,r]^T$ elocities is defined in a body-fixed(B) coordinate system, representing surge, sway, heave, roll, pitch and yaw velocity, respectively.

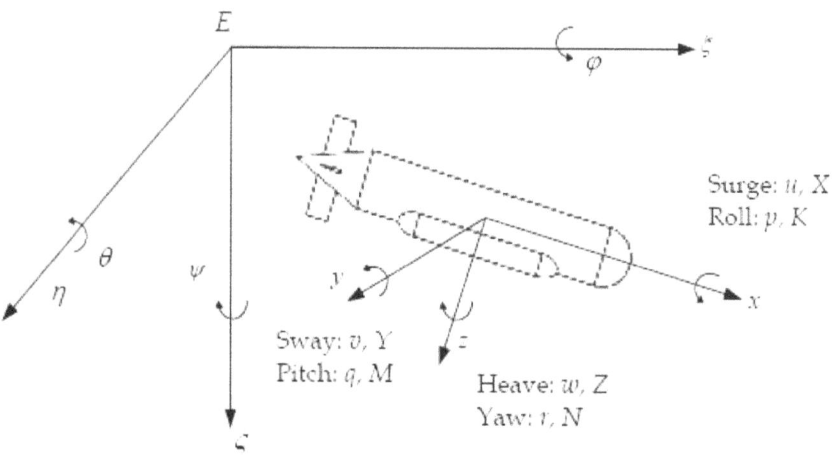

Fig. 1. Earth-fixed and body-fixed reference frames

According to the Newton-Euler formulation, the 6 DOF rigid-body equations of motion in the body-fixed coordinate frame can be expressed as:

$$\begin{cases} m\left[(\dot{u}_r - v_r r + w_r q) - x_G(q^2 + r^2) + y_G(pq - \dot{r}) + z_G(pr + \dot{q})\right] = X \\ m\left[(\dot{v}_r - w_r p + u_r r) - y_G(r^2 + p^2) + z_G(qr - \dot{p}) + x_G(qp + \dot{r})\right] = Y \\ m\left[(\dot{w}_r - u_r q + v_r p) - z_G(p^2 + q^2) + x_G(rp - \dot{q}) + y_G(rq + \dot{p})\right] = Z \\ I_x\dot{p} + (I_z - I_y)qr + m\left[y_G(\dot{w}_r + pv_r - qu_r) - z_G(\dot{v}_r + ru_r - pw_r)\right] = K \\ I_y\dot{q} + (I_x - I_z)rp + m\left[z_G(\dot{u}_r + w_r q - v_r r) - x_G(\dot{w}_r + pv_r - u_r q)\right] = M \\ I_z\dot{r} + (I_y - I_x)pq + m\left[x_G(\dot{v}_r + u_r r - pw_r) - y_G(\dot{u}_r + qw_r - v_r r)\right] = N \end{cases} \tag{1}$$

where m is the mass of the vehicle, I_x , I_y and I_z are the moments of inertia about the x_b , y_b and z_b -axes, x_g, y_g and z_g are the location of center of gravity, u_r , v_r , w_r are relative translational velocities associated with surge, sway and heave to ocean current in the body- fixed frame, here assuming the sea current to be constant with orientation in yaw only, which can be described by the vector $\boldsymbol{U}_c = [u_c\ ,\ v_c\ ,\ w_c\ ,\ 0,\ 0,\ \alpha_c]T$. The resultant forces X, Y, Z, K, M, N includes positive buoyant $B - W = \Delta P$ (since it is convenient to design underwater vehicles with positive buoyant such that the vehicle will surface automatically in the case of an emergency), hydrodynamic forces X_H , Y_H , Z_H , K_H , M_H , N_H and thruster forces.

2.2 Thrust hydrodynamics modeling

The modeling of thruster is usually done in terms of advance ratio J_0, thrust coefficients K_T and torque coefficient K_Q. By carrying out an open water test or a towing tank test, a unique curve where J_0 is plotted against K_T and K_Q can be obtained for each propeller to depict its performance. And the relationship of the measured thrust force versus propeller revolutions for different speeds of advance is usually least-squares fitting to a quadratic model.

Here we introduce a second experimental method to modeling thruster dynamics. Fig. 2 shows experimental results of thrusters from an open water test in the towing tank of the Key Lab of Autonomous Underwater Vehicles in Harbin Engineering University. The results are not presented in the conventional way with the thrust coefficient K_T plotted versus the open water advance coefficient J_0, for which the measured thrust is plotted as a function of different speeds of vehicle and voltages of the propellers.

The thrust force of the specified speed of vehicle under a certain voltage can be finally approximated by Atiken interpolation twice. In the first interpolation, for a certain voltage, the thrust forces with different speeds of the vehicle (e.g. 0m/s, 0.5m/s, 1.0m/s, 1.5m/s) can be interpolated from Fig.1, and plot it versus different speeds under a certain voltage. Then based on the results of the first interpolation, for the second Atiken interpolation we can find the thrust force for the specified speed of the vehicle.

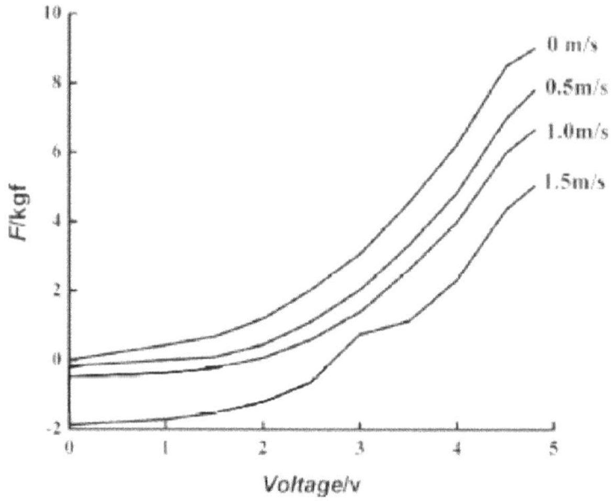

Fig. 2. Measured thrust force as a function of propeller driving voltage for different speeds of vehicle

Compared with conventional procedure to obtain thrust that is usually done firstly by linear approximating or least-squares fitting to K_T - J_0 plot (open water results), then using formulation $F_t = K_t n^2 D^4$ to compute the thrust F_t. The experimental results of open water can be directly used to calculate thrust force without using the formulation, which also can be applied to control surface of rudders or wings, *etc.*

2.3 General dynamic model

To provide a form that will be suitable for simulation and control purposes, some rearrangements of terms in Eq.(1) are required. First, all the non-inertial terms which have velocity components were combined with the fluid motion forces and moments into a fluid vector denoted by the subscript vis (viscous). Next, the mass matrix consisting all the coefficient of rigid body's inertial and added inertial terms with vehicle acceleration components $\dot{u}, \dot{v}, \dot{w}, \dot{p}, \dot{q}, \dot{r}$ was defined by matrix **E**, and all the remaining terms were combined into a vector denoted by the subscript else, to produce the final form of the model:

$$E\dot{X} = F_{\text{vis}} + F_{\text{else}} + F_t \tag{2}$$

where $\mathbf{X} = [u, v, w, p,q,r]^T$ is the velocity vector of vehicle with respect to the body-fixed frame.

Hence, the 6 DOF equations of motion for underwater vehicles yield the following general representation:

$$\begin{cases} \dot{X} = E^{-1}(F_{\text{vis}} + F_{\text{else}} + F_t) \\ \dot{\eta} = J(\eta)X \end{cases} \tag{3}$$

with

$$E = \begin{bmatrix} m-X_{\dot{u}} & 0 & 0 & 0 & mz_G & -my_G \\ 0 & m-Y_{\dot{v}} & 0 & -mz_G-Y_{\dot{p}} & 0 & mx_G-Y_{\dot{r}} \\ 0 & 0 & m-Z_{\dot{w}} & my_G & -mx_G-Z_{\dot{q}} & 0 \\ 0 & -mz_G-K_{\dot{v}} & 0 & I_x-K_{\dot{p}} & 0 & -K_{\dot{r}} \\ mz_G & 0 & -mx_G-M_{\dot{w}} & 0 & I_y-M_{\dot{q}} & 0 \\ 0 & mx_G-N_{\dot{v}} & 0 & -N_{\dot{p}} & 0 & I_z-N_{\dot{r}} \end{bmatrix} \tag{4}$$

where $J(\eta)$ is the transform matrix from body-fixed frame to earth-fixed frame, η is the vector of positions and attitudes of the vehicle in earth-fixed frame.

The general dynamic model is sufficiently robust to be applied to various types of underwater vehicles based on their physical characteristics, including body symmetry planes, available control degrees of freedom, and actuator configurations, thereby serving as an effective testing tool for vehicle control design.

3. MOTION CONTROL STRATEGY

This section delineates the design of the motion control system for AUV-XX. The control system can be delineated into two distinct designs: one for position and speed control in the horizontal plane, and another for the integrated heave and pitch control for descent in the vertical plane. An enhanced S-surface control algorithm founded on the capacitor plate model has been developed.

3.1 Control algorithm

As a nonlinear function method to construct the controller, S-surface control has been proven quite effective in sea trial for motion control of AUV in Harbin Engineering University (Li et al., 2002). The nonlinear function of S-surface is given as:

$$u = 2.0/(1.0 + \exp(-k_1 e - k_2 \dot{e})) - 1.0 \tag{5}$$

where e, $\dot{e}$ are control inputs, and they represent the normalized error and change rate of the error, respectively; u is the normalized output in each degree of freedom; k_1, k_2 are control parameters corresponding to control inputs e and e· respectively, and we only need to adjust them to meet different control requirements.

Based on the experiences of sea trials, the control parameters k_1, k_2 can be manually adjusted to meet the fundamental control requirements, however, whichever combination of k_1, k_2 we can adjust, it merely functions a global tuning which dose not change control structure. Here the improved S-surface control algorithm is developed based on the capacitor with each couple of plates putting restrictions on the control variables e , $\dot{e}$ respectively, which can provide flexible gain selection with proper physical meaning.

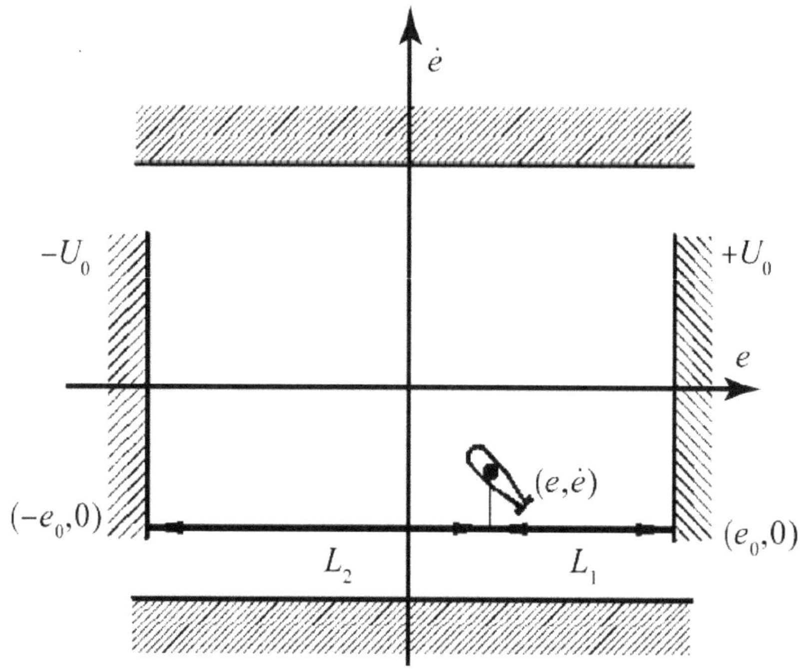

Fig. 3. Capacitor plate model

The capacitor plate model as shown in Fig.3 demonstrates the motion of a charged particle driven by electrical field in capacitor is coincident with the motion of a controlled vehicle from current point $(e,\dot{e})$ to the desired point, for which the capacitor plate with voltage serves as the controller, and the equilibrium point of electrical field is the desired position that the vehicle is supposed to reach.

Due to the restriction of two couples of capacitor plates put on control variables e and $\dot{e}$, the output of model can be obtained as

$$y = u^{+U_0} + u^{-U_0} = F(L_1, L_2)(+U_0) + F(L_2, L_1)(-U_0) \qquad (6)$$

where L_1 ,L_2 are horizontal distances from the current position of the vehicle to each capacitor plate, respectively, and the restriction function $F(*, *)$ is defined to be hyperbolic function of L_1 ,L_2 by Ren and Li (2005):

$$\begin{cases} F(L_1,L_2)=\dfrac{L_1^{-k}}{L_1^{-k}+L_2^{-k}} \\ F(L_2,L_1)=\dfrac{L_2^{-k}}{L_1^{-k}+L_2^{-k}} \end{cases} \tag{7}$$

The restriction function $F(*, *)$ reflects the closer the current position $(e,\dot{e})$ of vehicle moving to capacitor plate, the stronger the electrical field is. Choosing $U_0 = 1$, the output of capacitor plate model yields:

$$u=\frac{L_1^{-k}-L_2^{-k}}{L_1^{-k}+L_2^{-k}}U_0=\frac{(e_0+e)^k-(e_0-e)^k}{(e_0+e)^k+(e_0-e)^k} \tag{8}$$

where e_0 is the distance between the plate and field equilibrium point of capacitor. An improved S-surface controller based on the capacitor plate model is proposed, that is

$$\begin{cases} u_{ei}=[2.0\Big/\left(1.0+(\frac{e_0-e_i}{e_0+e_i})^{ki1}\right)-1.0] \\ u_{\dot{e}i}=[2.0\Big/\left(1.0+(\frac{e_0-\dot{e}_i}{e_0+\dot{e}_i})^{ki2}\right)-1.0] \\ f_i=K_{ei}\cdot u_{ei}+K_{\dot{e}i}\cdot u_{\dot{e}i} \end{cases} \tag{9}$$

where f_i is the outputted thrust force of controller for each DOF, and $K_{ei}=K_{\dot{e}i}=K_i$ is the maximal thrust force in i th DOF, therefore the control output can be reduced to

$$\begin{cases} u_i=u_{ei}+u_{\dot{e}i}=[2.0\Big/\left(1.0+(\frac{e_0-e_i}{e_0+e_i})^{ki1}\right)-1.0]+[2.0\Big/\left(1.0+(\frac{e_0-\dot{e}_i}{e_0+\dot{e}_i})^{ki2}\right)-1.0] \\ f_i=K_i\times u_i \end{cases} \tag{10}$$

The capacitor model's S-surface control can provide flexible gain selections with different forms of restriction function to L_1, L_2 to meet different control requirements for different phases of control procedure.

3.2 Velocity and spatial regulation in the horizontal plane

AUV-XX is outfitted with two transverse tunnel thrusters located at the front and rear of the vehicle, as well as two main thrusters positioned horizontally at the rear, one on the starboard side and one on the port side. These thrusters generate the requisite force in the x-direction for transit and in the y-direction for maneuvering. Both speed and position controls are built for the horizontal plane. Speed control involves monitoring the target surge velocity while maintaining a constant yaw angle and depth, commonly employed in the long-distance transit of underwater vehicles. Prior to executing certain underwater operations, the vehicle must undergo extensive journey to reach the destination. This chapter discusses a forward speed controller in surge, based on the previously mentioned control algorithm. Its purpose is to enable the vehicle to maintain a target velocity while ensuring stable attitudes, such as fixed yaw and depth.

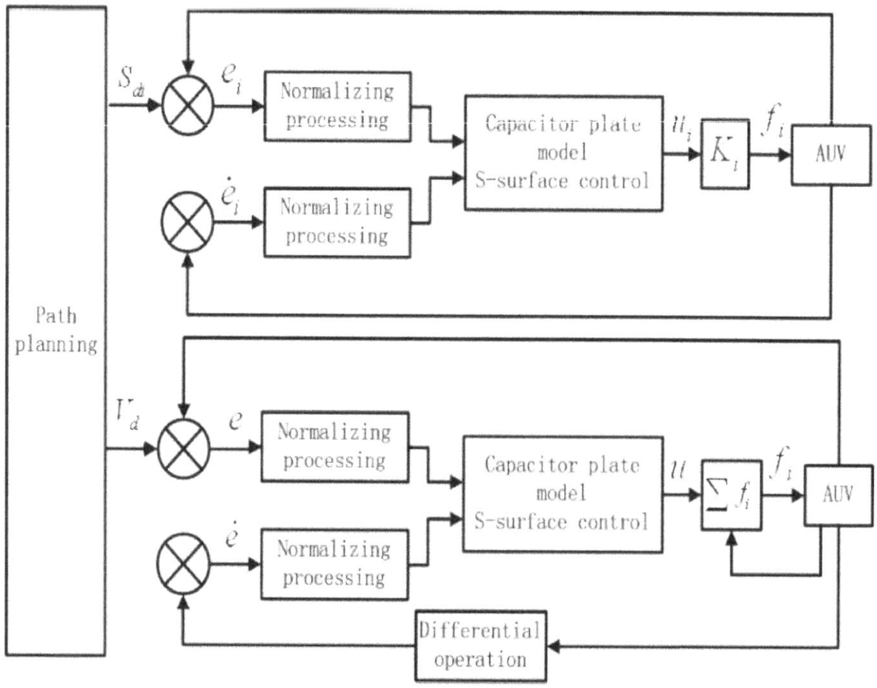

Fig. 4. Position and speed control loop

Position control enables the vehicle to perform various position-keeping functions, such as maintaining a steady position to perform a particular task, following a prescribed trajectory to search for missing or seek after objects. Accurate position control is highly desirable when the vehicle is performing underwater tasks such as cable laying, dam security inspection and mine clearing. To ensure AUV-XX to complete work assignments of obstacles avoiding, target recognition, and mine countermeasures, we design position controllers for surge, sway, yaw and depth respectively for equipping the vehicle with abilities of diving at fixed deepness, navigating at desired direction, sailing to given points and following the given track, *etc.*

As for the desired or target position or speed in the control system, it is the path planning system who decides when to adopt and switch control scheme between position and speed, the desired position that the vehicle is supposed to reach, and the velocity at which the vehicle should navigate with respect to the present tasks and motion states of the vehicle as well as operation environment.

Fig.4 shows both position and speed control procedures. It can be seen that it is easier to realize the control algorithm. For position control of *i* th DOF, the control inputs are the position error and the change rate of position error, that is the velocity obtained from motion sensors; while for speed control, the velocity error and acceleration are control inputs, since AUV-XX is not equipped with IGS to acquire the acceleration of the vehicle, acceleration is calculated differential the velocity in each control step.

3.3 Combined control of pitch and heave in vertical plane

As the vehicle attains high speeds, the thrust generated by tunnel thrusters significantly diminishes, rendering depth control only through tunnel thrusters challenging. Given that alterations in the vehicle's depth or height will consequently affect its pitch, we integrate pitch with heave control for diving while the vehicle operates at elevated speeds to offset the thrust loss of tunnel thrusters. The intended pitch angle can be formulated as a function of the vehicle's surge velocity:

$$\theta_T = \begin{cases} -\dfrac{k_\theta \cdot \Delta z}{\sqrt{u}} - \theta_0 & u \geq 0.8m/s \\ 0 & u < 0.8m/s \end{cases} \tag{11}$$

where θ_T is the target pitch angle, k_θ is a positive parameter to be adjusted, Δz is the depth deviation, u is the surge velocity of vehicle, θ_0 is the pitch angle when the vehicle is in static equilibrium. Since the change of pitch angle is usually associated with depth change and they affect each other, the target pitch is the output of the proportional control with respect to depth change with the proportional parameters k_θ, the velocity threshold of 0.8m/s is chosen based on the engineering experience of the sea trial and the capability of the thruster system of the vehicle.

When the vehicle is moving at a low speed($u < 0.8m / s$), tunnel thrusters can normally provide the needed force, so the command of target pitch will not be sent to motion control. According to control law (9), we can get the output of the heave controller. Since the vehicle is usually designed with positive buoyancy, the final output of control of heave can be obtained by

$$f_3 = K_3 \cdot u_3 + \Delta P \tag{12}$$

where ΔP is the positive buoyancy.

And as the surge velocity grows, the thrust of vertical tunnel will experience worse reduction and degradation, so the role it plays in the depth control will be greatly abridged, as a result, the output of pitch control of the main vertical thrusters aft of the vehicle should compensate for that, so the output of control will finally yields as

$$f_5 = \begin{cases} K_5 u_5 & u \le 0.8 \\ K_5(\varepsilon_1 \cdot u_3 + \varepsilon_2 \cdot u_5) & u > 0.8 \end{cases} \tag{13}$$

with

$$\begin{cases} \varepsilon_1 = \alpha_1 / \sqrt{\beta u} \\ \varepsilon_2 = \alpha_2 \cdot \exp(u \cdot u / 10) \end{cases} \tag{14}$$

where α_1, α_2, β can be manually adjusted based on experiences. The block diagram of combined control of pitch and depth in vertical plane can be in Fig.5.

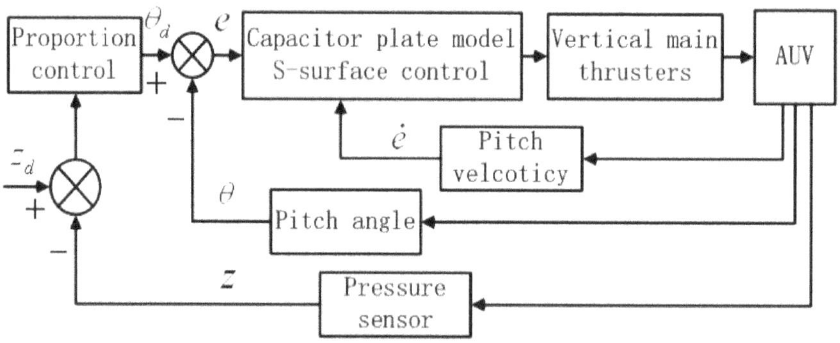

Fig. 5. Combined control of pitch and depth in vertical plane

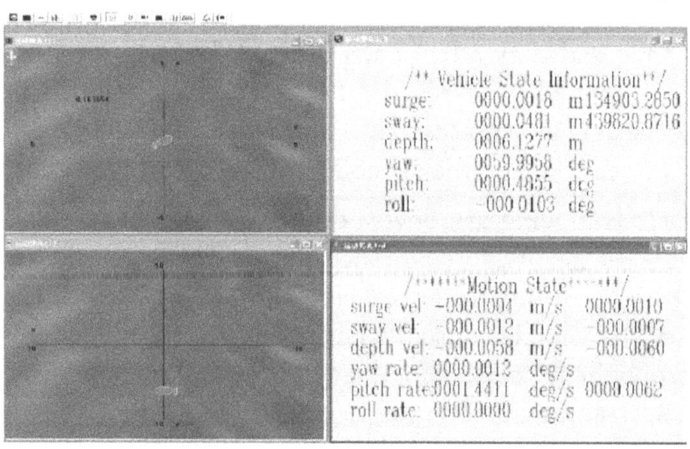

Fig. 6. AUV-XX simulation platform

4. SIMULATION RESULTS AND ANALYSIS

To verify the feasibility and effectiveness of the motion control system for the vehicle AUV- XX, simulations are carried out in the AUV-XX simulation platform. The vehicle researched in this chapter named by AUV-XX, AUV-XX's configuration is basically a long cylinder of 0.5m in diameter and 5m in length with crossed type wings near its rear end. On each edge of the wings, a thruster is mounted, which is used for both turn and dive. AUV-XX is also equipped with a couple of lateral tunnel thrusters for sway and a couple of vertical tunnel thrusters fore and aft of the vehicle, respectively. Based on the modeling method described in above section, we established the AUV-XX simulation platform to carry out fundamental tests on its motion characteristics, stability and controllability. The states of the vehicle including positions, attitudes and velocities are obtained at each instant by solving the mathematical model equation by integration using a time step of 0.5s. Fig. 6 shows the interface of AUV-XX simulation platform. Fig. 7 shows the data flow of the simulation platform connecting with motion control system.

Figs. 8–10 show the simulation results of capacitor plate model's S-surface control for separate position and speed control in surge, sway and yaw respectively, and also combined control of heave and pitch for diving of the vehicle. The roll is left uncontrolled. And all the vehicle states, including speed, are initialized to zero at the beginning of the each of the simulation. The solid lines denote the actual responses of the vehicle while the dashed denote the desired position or speed that the vehicle is commanded to achieve.

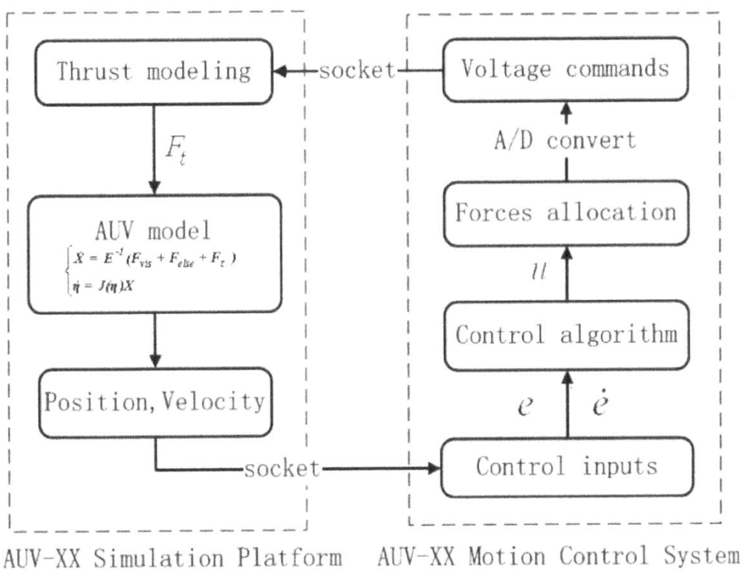

Fig. 7. Data flow of AUV-XX motion control in simulation platform

It can be seen from Fig. 8 that, the vehicle is commanded to move to some specified positions in surge, sway and yaw, respectively. For the case of surge as shown in Fig. 8(a) and Fig. 8(b), the larger position deviation from the target position produces faster response of surge velocity. Compared with responses of surge and yaw, the sway is slower with a rise time of 150s for the desired position is 16m, which may result from that the lateral resistance is much larger longitudinally and the thrust of transverse tunnel thrusters can provide is smaller than the main aft thrusters.

Fig.9 shows the speed control results of surge with constant yaw and depth keeping. The desired velocity is 1m/s and once the vehicle is moving stably with such speed then the vehicle is commanded to track the specified depth (5m) and yaw (45o) commands. It can be seen that there is no overshoot in surge speed and system response is fast with the yaw experiencing an accepted overshoot of ± 2o, and the depth is stably maintained, hence the vehicle is able to move at a desired speed with a desired fixed heading and depth. The speed control simulation results prove the feasibility of the proposed speed control strategy.

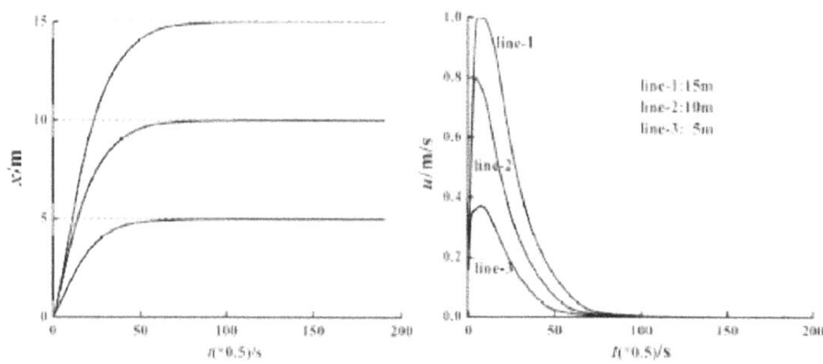

(a) Surge displacement (b) Surge velocity response

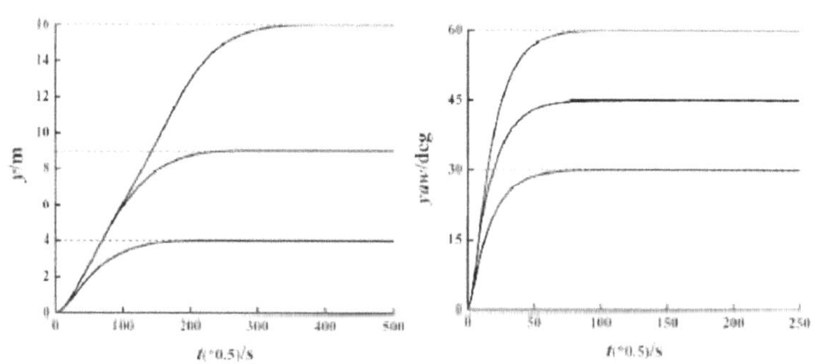

(c) Sway displacement (d) Yaw displacement

Fig. 8. Position control results of surge, sway and yaw

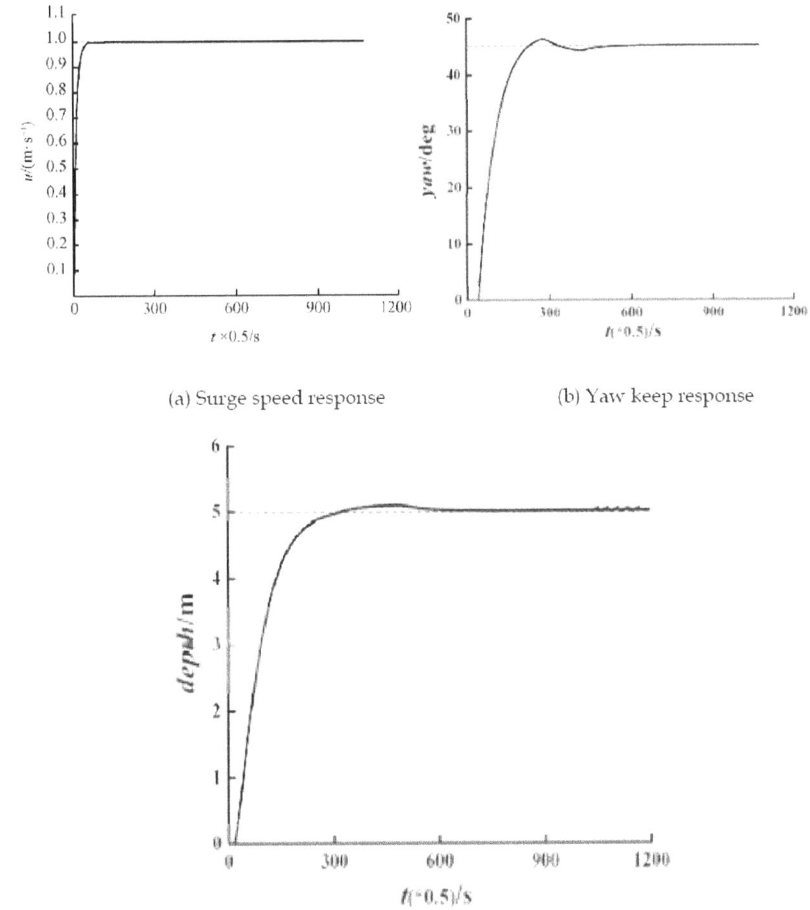

(c) Depth keep response

Fig. 9. Speed control results of surge with yaw and depth keeping

Fig. 10 shows the combined control of heave and pitch for diving in vertical plane. For this case, the velocity of surge that the vehicle is commanded to track is 1.5m/s, which is not very large so that the vertical tunnel thrusters will suffer thrust reduction to some extent but still can work to provide a portion of vertical thrust for diving. Hence when the vehicle is commanded to dive, the pitch will not experience a large change, which is reasonable design consideration in the case of large inertial vehicles.

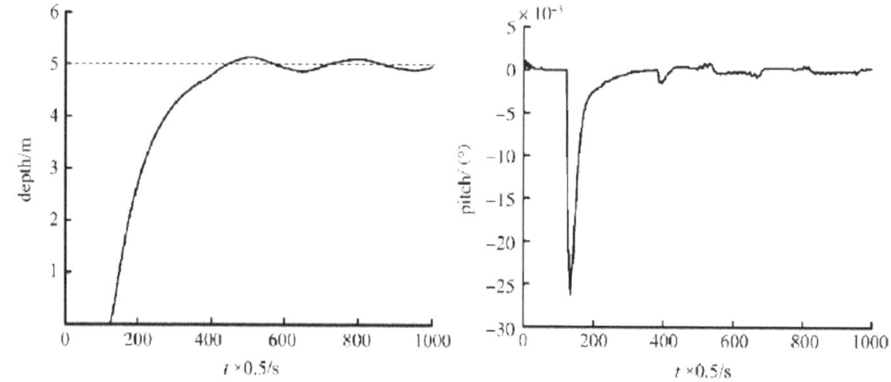

(c) Depth keep response (d) Pitch response

Fig. 10. Combined depth control of heave and pitch

5. CONCISE SUMMARY

This chapter delineates the design of a motion control system for Autonomous Underwater Vehicles, encompassing both position and velocity control in the horizontal plane, as well as integrated control of heave and pitch in the vertical plane. A comprehensive mathematical model of underwater vehicles with 6 degrees of freedom was developed for the construction of the control system, enabling its application to many types of underwater vehicles based on their specific physical characteristics. A simulation platform was developed to evaluate the motion characteristics, stability, and controllability of the vehicle based on the general mathematical model. Simulations conducted on the AUV-XX platform illustrate the efficacy of the developed controller, with the capacitor plate model S-surface control exhibiting commendable performance.

REFERENCES

Fossen T. I. (1991). *Nonlinear modeling and control of underwater vehicles.* Ph.D. thesis, Norwegian Institute of Technology-NTH, Trondheim, Norway

Fossen T. I. (2002). *Marine control system: guidance, navigation and control of ships, rigs and underwater vehicles.* Marine Cybernetics, Trondheim, 254-260

Gan Yong, Sun Yushan, Wan Lei, Pang Yongjie (2006). Motion control system architecture of underwater robot. *Proceedings of the 6th World Congress on Intelligent Control and Automation,* Dalian, China, 8876-8880

Gan Yong, Sun Yushan, Wan Lei, Pang Yongjie (2006). Motion control system of underwater robot without rudder and wing. Proceeding of the 2006 IEEE *International Conference on Intelligent Robotics and Systems,* Beijing, China, 3006-3011

Li Xuemin, Xu Yuru (2002). S-control of automatic underwater vehicles. *The Ocean Engineering,* 19(3), 81-84

Li Ye, Liu Jianchen, Shen Minxue (2005). Dynamics model of underwater robot motion control in 6 degrees of freedom. *Journey of Harbin Institute of Technology,* 12(4), 456- 459

Ren Yongping, Li Shengyi (2005). Design method of a kind of new controller. *Control and Decision,* 20(4), 471-474

Venugopal KP, Sudhakar R (1992). On-line learning control of autonomous underwater vehicles using feedforward neural networks. *IEEE Journal of Oceanic Engineering,* 17(4), 308-319

Yuh J (1990). Modeling and control of underwater robotic vehicles. *IEEE Transaction on Systems, Man, and Cybernetics,* 20(6), 1475-1483

7 Chapter

COMPREHENSIVE COUPLED SIX DEGREE-OF-FREEDOM CONTROL OF AN OVER-ACTUATED AUTONOMOUS UNDERWATER VEHICLE

1. INTRODUCTION: AN OVERVIEW

Unmanned underwater vehicles (UUVs) are progressively employed by civilian and military operators for increasingly intricate and perilous missions. This results from the inherent attributes of safety and cost efficiency relative to crewed vehicles. UUVs eliminate the necessity for human operators to endure the perilous circumstances of the underwater environment, hence significantly reducing or entirely mitigating the risk to human life. Cost efficiency, in terms of both time and finances, is derived from a significantly smaller vehicle that lacks the numerous subsystems necessary for underwater life support, as well as smaller, less powerful actuators that do not impose the same levels of stress and strain on the vehicle as a manned counterpart. This results in a significantly smaller team necessary for the routine maintenance required to maintain a vehicle's operability. Considering these two primary causes, the transition from human to unmanned vehicles is a rational advancement in the oceanographic sector. The category of UUVs encompasses both remotely operated vehicles (ROVs) and autonomous underwater vehicles (AUVs). Both vehicle types have been effectively utilized in industry, and their intrinsic distinctions dictate the suitability of each for specific missions. The primary distinction between the two is that a ROV necessitates a tether to a base station, while an AUV operates independently without such a connection. The tether links the ROV to a human operator, enabling observation of the vehicle's current status and facilitating control during its task execution. This tether, contingent upon the vehicle's setup, can also supply electricity to power the vehicle's actuators, sensors, and numerous internal electronic components. AUVs possess an advantage over ROVs by eliminating the need for a tether, resulting in two primary benefits. An AUV necessitates minimal or no human intervention throughout its operational tasks. The spacecraft is pre-configured with specified mission objectives and, upon launch, endeavors to accomplish these objectives autonomously, without people interaction from the base station. This mitigates the impact of human error during vehicle operation. The second advantage is the enhanced maneuverability achievable without a tether consistently connected to the AUV. This cable may become entangled with undersea structures, thus restricting the operational surroundings of a ROV and inducing drag on the vehicle's movement, thereby impacting its maneuverability. The vehicle's potential range from the base station is limited by the tether's length. The two primary advantages of AUVs compared to ROVs render autonomous vehicles preferable for survey operations in intricate, dynamic, and hazardous underwater settings; hence, AUVs are the focus of this chapter. Moreover, integrating the requisite performance attributes of AUVs with the previously described intricate operational environment necessitates that the controllers employed in AUVs be precise, accurate, and resilient to shocks and uncertainties. This chapter will concentrate on the accurate and reliable control of AUVs. The autonomy architecture of AUVs comprises three primary systems. The components are as follows: the guidance system, responsible for generating the vehicle's trajectory; the navigation system, which estimates the vehicle's current state; and the control system, which computes and applies the necessary forces to maneuver the vehicle (Fossen, 2002). This chapter will concentrate on the control system and its two primary subsystems: the control law and control allocation. The chapter will be segmented into three principal sections: the first will concentrate on the design and analysis of the control law, the second will examine control allocation, and the third will illustrate the integration of these two systems to constitute the overall control system. In the control law design and analysis section, the interactions among the various systems within an AUV will be examined, with a specific focus on their relationship to the control system. A comprehensive overview of the undersea environment will be provided, illustrating the intricacies of potential disturbances affecting a vehicle. This will be succeeded by an examination of the equations of motion, specifically the kinematic and kinetic equations that dictate the movement of a rigid body within a fluid. A description of the pertinent frames of reference utilized in the context of underwater vehicles will be provided. This section of the chapter will finish with a study of the control laws commonly employed in underwater vehicles. The second section will examine the function of control allocation in dispersing the requisite control forces across a vehicle's actuators. An examination of the primary categories of actuators now accessible to underwater vehicles will be performed, including their advantageous characteristics and inherent constraints. This section will finish with a summary of diverse strategies for executing control allocation, characterized by differing levels of computational complexity. The concluding section of this chapter will provide an example of a comprehensive control system for integration into AUV architecture. This example will illustrate the interaction between the control law and control allocation subsystems to achieve the necessary trajectory tracking performance utilizing the vehicle's numerous actuators.

2. DESIGN AND STUDY OF CONTROL LAWS

Prior to examining the regulations that dictate how a certain control system generates a corrective signal, it is essential to consider the requirements of the different systems inside an AUV. This will elucidate the role of the control system within the broader autonomous architecture of an AUV. It is essential to recognize the various sorts of disturbances to mitigate their effects, and the equations governing the vehicle's dynamic motion must be examined. Control law design and analysis can only proceed after evaluating these factors.

2.1 Requirements

As previously stated, the various components that make up the autonomy architecture of AUVs are the guidance system, navigation system, and control system. All three of these systems have their own individual tasks to complete, yet must also work cooperatively in order to reliably allow a vehicle to complete its objectives. Figure 1 shows a block diagram of how these various systems interact.

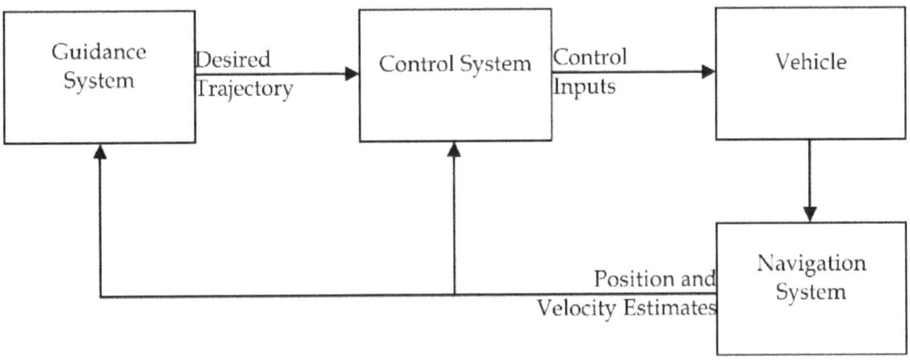

Fig. 1. Guidance, Navigation and Control Block Diagram.

2.1.1 Guidance

The guidance system generates the intended trajectory for the vehicle to adhere to. The task is accomplished by utilizing the predetermined waypoints established prior to the mission, while potentially including external environmental factors, to build a trajectory for the vehicle to navigate towards each subsequent waypoint (Fossen, 1994, 2002). Data concerning the vehicle's present state, including actuator configuration and potential malfunctions, can be employed to establish a plausible trajectory for the vehicle to adhere to. This trajectory establishes the vehicle's intended state, encompassing the desired position, orientation, velocity, and acceleration data.

2.1.2 Navigation

The navigation system performs the function of ascertaining the vehicle's current status. Global Positioning System (GPS) is readily accessible for surface, land, and airborne vehicles, frequently utilized to provide continuous and precise positioning data to navigation systems. Nevertheless, owing to the severely restricted transmission of these signals in water, GPS is predominantly inaccessible for underwater vehicles. The navigation system's objective is to calculate an optimal approximation of the vehicle's current state utilizing various readings from both proprioceptive and exteroceptive sensors, employing GPS solely when accessible. This is accomplished through sensor fusion techniques, such as Kalman filtering or Particle filtering (Lammas et al., 2010, 2008), to derive an optimal estimate of the current operating condition and facilitate the incorporation of a correction mechanism when GPS is accessible, such as when the vehicle is surfaced. The primary function of the navigation system is to deliver the most accurate assessment of the vehicle's current situation, irrespective of the available sensor data.

2.1.3 Control system

The control system is tasked with delivering corrective signals to ensure the vehicle adheres to a specified trajectory. This is accomplished by obtaining the vehicle's desired state from the guidance system and its current condition from the navigation system. The control system subsequently computes and implements a corrective force via the vehicle's actuators to reduce the disparity between the desired and actual states (Fossen, 1994, 2002). This enables the vehicle to follow a specified course despite unknown perturbations. Although each of the aforementioned systems is accountable for its own function, they must also collaborate to attain complete autonomy within an underwater vehicle context.

2.2 Environment

Underwater settings are exceedingly intricate and dynamic, rendering the handling of an AUV a formidable challenge. Currents and waves are constant disturbances that must be recognized for an AUV to navigate such an environment.

2.2.1 Ocean Currents

Ocean currents, the large scale movement of water, are caused by many sources. One component of the current present in the upper layer of the ocean is due to atmospheric wind conditions at the sea surface. Differing water densities, caused by combining the effect of variation of salinity levels with the exchange of heat that occurs at the ocean surface, cause additional currents known as thermohaline currents, to exist within the ocean. Coriolis forces, forces due to the rotation of the Earth about its axis, also induce ocean currents, while gravitational forces due to other planetary objects, such as the moon and the sun, produce yet another effect on ocean currents (Fossen, 1994, 2002). Combining all of these sources of water current, with the unique geographic topography that are present within isolated coastal regions, leads to highly dynamic and complex currents existing within the world's oceans.

2.2.2 Wind generated waves

There are many factors that lead to the formation of wind generated waves in the ocean. Wind speed, area over which the wind is blowing, duration of wind influencing the ocean surface, and water depth, are just some of the elements that lead to the formation of waves. Due to the oscillatory motion of these waves on the surface, any vehicle on the surface will experience this same oscillatory disturbance. Moreover, an underwater vehicle will experience both translational forces and rotational moments while at or near the surface due to this wave motion.

2.3 Dynamics

All matter that exists in our universe must adhere to certain differential equations determining its motion. By analysing the physical properties of an AUV, a set of equations can be derived that determine the motion of this vehicle through a fluid, such as water. To assist in reducing the complexities of these equations, certain frames of reference are utilised depending on the properties that each frame of reference possesses. In order to make use of these different reference frames for different purposes, the process of transforming information from one frame to another must be conducted.

2.3.1 Frames of reference

Within the context of control systems, the two main reference frames used are the n-frame and the b-frame. Both contain three translational components and three rotational components, yet the origin of each frame differs. This difference in origin can lead to useful properties which contain certain advantages when designing a control system.

2.3.1.1 N-Frame

The n-frame is a co-ordinate space usually defined as a plane oriented at a tangent to the Earth's surface. The most common of these frames for underwater vehicle control design is the North-East-Down (NED) frame. As its name suggests, the three axes of the translational components of this frame have the x-axis pointing towards true North, the y-axis pointing towards East, and the z-axis in the downward direction perpendicular to the Earth's surface. In general, waypoints are defined with reference to a fixed point on the earth, and therefore it is convenient to conduct guidance and navigation in this frame.

2.3.1.2 B-Frame

The b-frame, also known as the body frame, is a moving reference frame that has its origin fixed to the body of a vehicle. Due to various properties that exist at different points within the body of the vehicle, it is convenient to place the origin of this frame at one of these points to take advantage of, for example, body symmetries, centre of gravity, or centre of buoyancy. As a general rule, the x-axis of this frame points from aft to fore along the longitudinal axis of the body, the y-axis points from port to starboard, and the z-axis points from top to bottom. Due to the orientation of this frame, it is appropriate to express the velocities of the vehicle in this frame.

2.3.2 Kinematic equation

As mentioned in 2.3.1, both the NED and body frames have properties that are useful for underwater vehicle control design. Because both are used for different purposes, a means of converting information from one frame to the other is required. The kinematic equation, (1), achieves this task (Fossen, 1994, 2002).

$$\dot{\eta} = J(\eta)\upsilon \qquad (1)$$

Here, the 6 degree-of-freedom (DoF) position and orientation vector in (1), decomposed in the NED frame, is denoted by (2).

$$\eta = \begin{bmatrix} p^n & \Theta \end{bmatrix}^T \qquad (2)$$

Within (2), the three position components are given in (3),

$$p^n = \begin{bmatrix} x_n & y_n & z_n \end{bmatrix}^T \qquad (3)$$

and the three orientation components, also known as Euler angles, are given in (4).

$$\Theta = \begin{bmatrix} \phi & \theta & \psi \end{bmatrix}^T \qquad (4)$$

The 6 DoF translational and rotational velocity vector in (1), decomposed in the body frame, is denoted by (5).

$$\nu = \begin{bmatrix} v_o^b & \omega_{nb}^b \end{bmatrix}^T \qquad (5)$$

Here, the three translational velocity components are given in (6),

$$v_o^b = \begin{bmatrix} u & v & w \end{bmatrix}^T \tag{6}$$

and the three rotational velocity components are given in (7).

$$\omega_{nb}^b = \begin{bmatrix} p & q & r \end{bmatrix}^T \tag{7}$$

In order to rotate from one frame to the other, a transformation matrix is used in (1). This transformation matrix is given in (8).

$$J(\eta) = \begin{bmatrix} R_b^n(\Theta) & 0_{3\times3} \\ 0_{3\times3} & T_\Theta(\Theta) \end{bmatrix} \tag{8}$$

Here, the transformation of the translational velocities from the body frame to the NED frame are achieved by rotating the translational velocities in the body frame, (6), using the Euler angles (4). Three principal rotation matrices are used in this operation, as shown in (9).

$$R_{x,\phi} = \begin{bmatrix} 1 & 0 & 0 \\ 0 & \cos\phi & -\sin\phi \\ 0 & \sin\phi & \cos\phi \end{bmatrix}, R_{y,\theta} = \begin{bmatrix} \cos\theta & 0 & \sin\theta \\ 0 & 1 & 0 \\ -\sin\theta & 0 & \cos\theta \end{bmatrix}, R_{z,\psi} = \begin{bmatrix} \cos\psi & -\sin\psi & 0 \\ \sin\psi & \cos\psi & 0 \\ 0 & 0 & 1 \end{bmatrix} \tag{9}$$

The order of rotation is not arbitrary, due to the compounding effect of the rotation order. Within guidance and control, it is common to use the *zyx*-convention where rotation is achieved using (10).

$$R_b^n(\Theta) = R_{z,\psi} R_{y,\theta} R_{x,\phi} \tag{10}$$

Overall, this yields the translational rotation matrix (11).

$$R_b^n(\Theta) = \begin{bmatrix} \cos\psi\cos\theta & -\sin\psi\cos\phi + \cos\psi\sin\theta\sin\phi & \sin\psi\sin\phi + \cos\psi\sin\theta\cos\phi \\ \sin\psi\cos\theta & \cos\psi\cos\phi + \sin\psi\sin\theta\sin\phi & -\cos\psi\sin\phi + \sin\psi\sin\theta\cos\phi \\ -\sin\theta & \cos\theta\sin\phi & \cos\theta\cos\phi \end{bmatrix} \tag{11}$$

The transformation of rotational velocities from the body frame to the NED frame is achieved by again applying the principal rotation matrices of (9). For ease of understanding, firstly consider the rotation from the NED frame to the body frame in which $\dot{\psi}$ is rotated by $R_{y\,\theta}$, added to $\dot{\theta}$, and this sum then rotated by $R_{x,\,\varphi}$ and finally added to $\dot{\phi}$. This process is given in (12).

$$\omega_{nb}^{b} = \begin{bmatrix} \dot{\phi} \\ 0 \\ 0 \end{bmatrix} + R_{x,\phi}^{T} \left(\begin{bmatrix} 0 \\ \dot{\theta} \\ 0 \end{bmatrix} + R_{y,\theta}^{T} \begin{bmatrix} 0 \\ 0 \\ \dot{\psi} \end{bmatrix} \right) \tag{12}$$

By expanding (12), the matrix for transforming the rotational velocities from the NED frame to the body frame is defined as (13),

$$T_{\Theta}^{-1}(\Theta) := \begin{bmatrix} 1 & 0 & -\sin\theta \\ 0 & \cos\phi & \cos\theta\sin\phi \\ 0 & -\sin\phi & \cos\theta\cos\phi \end{bmatrix} \tag{13}$$

and therefore the matrix for transforming the rotational velocities from the body frame to the NED frame is given in (14).

$$T_{\Theta}(\Theta) = \begin{bmatrix} 1 & \sin\phi\tan\theta & \cos\phi\tan\theta \\ 0 & \cos\phi & -\sin\phi \\ 0 & \sin\phi/\cos\theta & \cos\phi/\cos\theta \end{bmatrix} \tag{14}$$

Overall, (1) achieves rotation from the body frame to the NED frame, and by taking the inverse of (8), rotation from the NED frame to the body frame can be achieved, as shown in (15).

$$\nu = J^{-1}(\eta)\dot{\eta} \tag{15}$$

2.3.3 Kinetic equation

The 6 DoF nonlinear dynamic equations of motion of an underwater vehicle can be conveniently expressed as (16) (Fossen, 1994, 2002).

$$M\dot{\nu} + C(\nu)\nu + D(\nu)\nu + g(\eta) = \tau + \omega \tag{16}$$

Here, M denotes the 6×6 system inertia matrix containing both rigid body and added mass, as given by (17).

$$M = M_{RB} + M_{A} \tag{17}$$

Similar to (17), the 6×6 Coriolis and centripetal forces matrix, including added mass, is given by (18).

$$C(\nu) = C_{RB}(\nu) + C_{A}(\nu) \tag{18}$$

Linear and nonlinear hydrodynamic damping are contained within the 6×6 matrix $D(v)$, and given by (19).

$$D(\nu) = D + D_n(\nu) \tag{19}$$

Here, *D* contains the linear damping terms, and Dn(ν) contains the nonlinear damping terms.

The 6 × 1 vector of gravitational and buoyancy forces and moments are represented in (16) by g(η), and determined using (20).

$$g(\eta) = \begin{bmatrix} (W-B)\sin\theta \\ -(W-B)\cos\theta\sin\phi \\ -(W-B)\cos\theta\cos\phi \\ -(y_g W - y_b B)\cos\theta\cos\phi + (z_g W - z_b B)\cos\theta\sin\phi \\ (z_g W - z_b B)\sin\theta + (x_g W - x_b B)\cos\theta\cos\phi \\ -(x_g W - x_b B)\cos\theta\sin\phi - (y_g W - y_b B)\sin\theta \end{bmatrix} \tag{20}$$

Here, *W* is the weight of the vehicle, determined using W=mg where m is the dry mass of the vehicle and g is the acceleration due to gravity. B is the buoyancy of the vehicle which is due to how much fluid the vehicle displaces while underwater. This will be determined by the size and shape of the vehicle. Vectors determining the locations of the centre of gravity and the centre of buoyancy, relative to the origin of the body frame, are given by (21) and (22) respectively.

$$r_g^b = \begin{bmatrix} x_g & y_g & z_g \end{bmatrix}^T \tag{21}$$

$$r_b^b = \begin{bmatrix} x_b & y_b & z_b \end{bmatrix}^T \tag{22}$$

The 6 × 1 vector of control input forces is denoted by τ, and is given by (23).

$$\tau = \begin{bmatrix} X & Y & Z & K & M & N \end{bmatrix}^T \tag{23}$$

Here, the translational forces affecting surge, sway and heave are *X*, *Y*, and *Z* respectively, and the rotational moments affecting roll, pitch and yaw are *K*, *M* and *N* respectively.

The 6 × 1 vector of external disturbances is denoted by ω.

Overall, (16) provides a compact representation for the nonlinear dynamic equations of motion of an underwater vehicle, formulated in the body frame. By applying the rotations contained within (8), (16) can be formulated in the NED frame as given in (24).

$$M_\eta(\eta)\ddot{\eta} + C_\eta(\nu,\eta)\dot{\eta} + D_\eta(\nu,\eta)\dot{\eta} + g_\eta(\eta) = \tau_\eta(\eta) + \omega \tag{24}$$

Within (24), the equations in (25) contain the rotations of the various matrices from the body frame to the NED frame.

$$\begin{aligned} M_\eta(\eta) &= J^{-T}(\eta) M J^{-1}(\eta) \\ C_\eta(\nu,\eta) &= J^{-T}(\eta)\left[C(\nu) - M J^{-1}(\eta)\dot{J}(\eta)\right]J^{-1}(\eta) \\ D_\eta(\nu,\eta) &= J^{-T}(\eta) D(\nu) J^{-1}(\eta) \\ g_\eta(\eta) &= J^{-T}(\eta) g(\eta) \\ \tau_\eta(\eta) &= J^{-T}(\eta)\tau \end{aligned} \qquad (25)$$

The presence of nonlinearities contained within $D_n(\nu)$, combined with the coupling effect of any non-zero off-diagonal elements within all matrices, can lead to a highly complex model containing a large number of coefficients.

2.4 Control laws

Diverse control strategies and corresponding control rules have been employed for AUV systems. The standard for control systems is the classical proportional-integral-derivative (PID) control, which has been effectively employed to manage many plants, including autonomous vehicles. PID schemes, however, are not particularly effective in managing the nonlinear dynamics of AUVs with uncertain models operating in unknown settings characterized by significant wave and current disturbances. PID methods are predominantly employed for rudimentary AUVs operating in areas devoid of external disturbances.An alternative control scheme known as sliding mode control (SMC), which is a form of variable structure control, has proven far more effective and robust at handling nonlinear dynamics with modelling uncertainties and nonlinear disturbances. SMC is a nonlinear control strategy which uses a nonlinear switching term to obtain a fast transient response while still maintaining a good steady-state response. Consequently, SMC has been successfully applied by many researchers in the AUV community. One of the earliest applications of using SMC to control underwater vehicles was conducted by Yoerger and Slotine wherein the authors demonstrated through simulations studies on an ROV model, the SMC controller's robustness properties to parametric uncertainties (Yoerger & Slotine, 1985). A multivariable sliding mode controller based on state feedback with decoupled design for independently controlling velocity, steering and diving of an AUV is presented in Healey and Lienard (1993). The controller design was successfully implemented on NPS ARIES AUV as reported in Marco and Healey (2001).

2.4.1 PID control

The fundamentals of PID control is that an error signal is generated that relates the desired state of the plant to the actual state (26),

$$e(t) = x_d(t) - x(t) \quad (26)$$

Where $e(t)$ is the error signal, $x_d(t)$ is the desired state of the plant, and $x(t)$ is the current state of the plant, and this error signal is manipulated to introduce a corrective action, denoted $\tau(t)$, to the plant.

PID control is named due to the fact that the three elements that make up the corrective control signal are: *proportional* to the error signal by a factor of KP, a scaled factor, K_I, of the *integral* of the error signal, and a scaled factor, K_D, of the *derivative* of the error signal, respectively (27).

$$\tau(t) = K_P e(t) + K_I \int_0^t e(\lambda) d\lambda + K_D \frac{d}{dt} e(t) \quad (27)$$

PID control is best suited to linear plants, yet has also been adopted for use on nonlinear plants even though it lacks the same level of performance that other control systems possess.

However, due to the wide use and acceptance of PID control for use in controlling a wide variety of both linear and nonlinear plants, it is very much employed as the "gold standard" that control systems are measured against. An example of a PID-based control strategy applied to underwater vehicles is given in Jalving (1994).

2.4.2 Sliding mode control

As mentioned previously, sliding mode control is a scheme that makes use of a discontinuous switching term to counteract the effect of dynamics that were not taken into account at the design phase of the controller.

To examine how to apply sliding mode control to AUVs, firstly (24) is compacted to the form of (28),

$$M_\eta(\eta)\ddot{\eta} + f(\dot{\eta}, \eta, t) = \tau_\eta(\eta) \quad (28)$$

where $f(\eta, \eta, t)$ contains the nonlinear dynamics, including Coriolis and centripetal forces, linear and nonlinear damping forces, gravitational and buoyancy forces and moments, and external disturbances.

If a sliding surface is defined as (29),

$$s = \dot{\eta} + c\eta \quad (29)$$

where c is positive, it can be seen that by setting s to zero and solving for η results in η converging to zero according to (30)

$$\eta(t) = \eta(t_0) e^{ct_0} e^{-ct} \quad (30)$$

regardless of initial conditions. Therefore, the control problem simplifies to finding a control law such that (31) holds.

$$\lim_{t\to\infty} s(t) = 0 \quad (31)$$

This can be achieved by applying a control law in the form of (32),

$$\tau = -T(\dot{\eta},\eta)\operatorname{sign}(s), T(\dot{\eta},\eta) > 0 \quad (32)$$

with $T(\dot{\eta},\eta)$ being sufficiently large. Thus, it can be seen that the application of (32) will result in η converging to zero.

If η is now replaced by the difference between the current and desired states of the vehicle, it can be observed that application of a control law of this form will now allow for a reference trajectory to be tracked.

Two such variants of SMC are the *uncoupled* SMC and the *coupled* SMC.

2.4.2.1 Uncoupled SMC

Within the kinetic equation of an AUV, (16), simplifications can be applied that will reduce the number of coefficients contained within the various matrices. These simplifications can be applied due to, for example, symmetries present in the body of the vehicle, placement of centres of gravity and buoyancy, and assumptions based on the level of effect a particular coefficient will have on the overall dynamics of the vehicle. Thus, the assumption of body symmetries allows reduced level of coupling between the various DoFs. An uncoupled SMC therefore assumes that no coupling exists between the various DoFs, and that simple manoeuvring is employed such that it does not excite these coupling dynamics (Fossen, 1994). The effect this has on (16) is to remove all off-diagonal elements within the various matrices which significantly simplifies the structure of the mathematical model of the vehicle (Fossen, 1994), and therefore makes implementation of a controller substantially easier.

2.4.2.2 Coupled SMC

Although the removal of the off-diagonal elements reduces the computational complexity of the uncoupled SMC, it also causes some limitation to the control performance of AUVs, particularly those operating in highly dynamic environments and required to execute complex manoeuvres. Taking these two factors into account, these off-diagonal coupling terms will have an influence on the overall dynamics of the vehicle, and therefore cannot be ignored at the design phase of the control law.

Coupled SMC is a new, novel control law that retains more of the coupling coefficients present in (16) compared to the uncoupled SMC (Kokegei et al., 2008, 2009). Furthermore, even though it is unconventional to design a controller in this way, the body frame is selected as the reference frame for this controller. This selection avoids the transformations employed in (24) and (25) used to rotate the vehicle model from the body frame to the NED frame although it does require that guidance and navigation data be transformed from the NED frame to the body frame. By defining the position and orientation error in the NED frame according to (33),

$$\tilde{\eta} = \hat{\eta} - \eta_d \tag{33}$$

where $\hat{\eta}$ represents an estimate of the current position and orientation provided by the navigation system, and η_d represents the desired position and orientation provided by the guidance system, a single rotation is required to transform this error from the NED frame to the body frame.

In general, desired and current velocity and acceleration data are already represented in the body frame, and as such, no further rotations are required here for the purposes of implementing a controller in the body frame.

By comparing the number of rotations required to transform the vehicle model into the NED frame, as seen in (25), for the uncoupled control scheme with the single rotation required by the coupled control scheme to transform the guidance and navigation data into the body frame, it can be seen that the latter has less rotations involved, and is therefore less computationally demanding.

3. CONTROL ALLOCATION

The role of the control law is to generate a generalised force to apply to the vehicle such that a desired state is approached. This force, τ, for underwater vehicles consists of six components, one for each DoF, as seen in (23). The control allocation system is responsible for distributing this desired force amongst all available actuators onboard the vehicle such that this generalised 6 DoF force is realised. This means that the control allocation module must have apriori knowledge of the types, specifications, and locations, of all actuators on the vehicle.

3.1 Role

The role of the control allocation module is to generate the appropriate signals to the actuators in order for the generalised force from the control law to be applied to the vehicle. Since the vehicle under consideration is over-actuated, which means multiple actuators can apply forces to a particular DoF, the control allocation is responsible for utilising all available actuators in the most efficient way to apply the desired force to the vehicle. Power consumption is of particular importance for all autonomous vehicles, as it is a key factor in determining the total mission duration. The control allocation is therefore responsible for applying the desired forces to the vehicle, while minimising the power consumed.

3.2 Actuators

The force applied to a vehicle due to the various actuators of a vehicle can be formulated as (34),

$$\tau = TKu \tag{34}$$

where, for an AUV operating with 6 DoF with n actuators, T is the actuator configuration matrix of size $6 \times n$, K is the diagonal force coefficient matrix of size $n \times n$, and u is the control input of size $n \times 1$. Actuators are the physical components that apply the desired force to the vehicle, and the particular configuration of these actuators will determine the size and structure of T, K and u, with each column of T, denoted t_i, in conjunction with the corresponding element on the main diagonal of K, representing a different actuator.

A vast array of actuators are available to underwater vehicle designers, the more typical of which include propellers, control fins and tunnel thrusters, and each has their own properties that make them desirable for implementation within AUVs. For all the following actuator descriptions lx defines the offset from the origin of the actuator along the x-axis, l_y defines the offset along the y-axis, and l_z defines the offset along the z-axis.

3.2.1 Propellers

Propellers are the most common actuators implemented to provide the main translational force that drives underwater vehicles. These are typically located at the stern of the vehicle and apply a force along the longitudinal axis of the vehicle. The structure of t_i for a propeller is given in (35).

$$t_i = \begin{bmatrix} 1 & 0 & 0 & 0 & l_z & -l_y \end{bmatrix}^T \tag{35}$$

As can be seen from (35), if the propeller is positioned such that there is no y-axis or z-axis offset, the force produced will be directed entirely along the x-axis of the vehicle, with no rotational moments produced.

3.2.2 Control surfaces

Control surfaces, or control fins, are actuators that utilise Newton's Third Law of motion to apply rotational moments to the vehicle. These surfaces apply a force to the water which causes a deflection in the water's motion. Hence, the water must also apply a force to the control surface. Due to this force being applied at a distance from the centre of gravity of the vehicle, a rotational moment is produced that acts on the vehicle. The typical configuration for control surfaces on an AUV is to have four independently controlled fins arranged in two pairs orientated horizontally and vertically at the stern of the vehicle. The structure of t_i for the horizontal fins is given in (36),

$$t_i = \begin{bmatrix} 0 & 0 & 1 & l_y & -l_x & 0 \end{bmatrix}^T \quad (36)$$

and for vertical fins given in (37).

$$t_i = \begin{bmatrix} 0 & 1 & 0 & -l_z & 0 & l_x \end{bmatrix}^T \quad (37)$$

These structures of t_i show that horizontal surfaces produce a heave force as well as roll and pitch moments, while vertical surfaces produce a sway force as well as roll and yaw moments. What must be considered here is that the force being produced by control surfaces relies on the vehicle moving relative to the water around it. If the vehicle is stationary compared to the surrounding water, control surfaces are ineffective. However, if the vehicle is moving relative to the surrounding water, these actuators are capable of applying forces and moments to the vehicle while consuming very little power.

3.2.3 Tunnel thrusters

The previously mentioned limitation of control surfaces can be overcome by the use of tunnel thrusters. These thrusters are usually implemented by being placed in tunnels transverse to the longitudinal axis of the vehicle. Similar to control surfaces, the typical arrangement is to position two horizontal tunnel thrusters equidistant fore and aft of the centre of gravity, and two vertical tunnel thrusters also equidistant fore and aft of the centre of gravity.

The structure of ti for horizontal thrusters is given in (38),

$$t_i = \begin{bmatrix} 0 & 1 & 0 & -l_z & 0 & l_x \end{bmatrix}^T \quad (38)$$

while the structure for vertical thrusters is given in (39).

$$t_i = \begin{bmatrix} 0 & 0 & 1 & l_y & -l_x & 0 \end{bmatrix}^T \quad (39)$$

What can be observed here is that horizontal thrusters provide a sway force as well as a roll and yaw moment, while vertical thrusters provide a heave force as well as roll and pitch moment. In general, horizontal tunnel thrusters are located such that l_z is zero, and vertical thrusters are located such that l_y is zero. The result of this choice is that no roll moment is produced by these actuators.

The advantage of tunnel thrusters is that forces and moments can be produced even if the vehicle is stationary with respect to the surrounding water. This greatly increases the manoeuvrability of the vehicle, as control of the vehicle when moving at low speeds is possible. However, there are limitations associated with the use of these actuators. Firstly, these actuators consume more power when activated compared to control surfaces. This is due to force being produced by the thrusters only when the thruster itself is activated. In contrast, control surfaces consume power when the deflection angle is altered, but require very little power to hold the surface in place once the desired angle has been achieved.

Secondly, thruster efficiency is reduced when the vehicle is moving. Under certain conditions, an area of low pressure is produced at the exit of the tunnel, which has the effect of applying a force to the vehicle in the opposite direction to which the water jet from the tunnel thruster is attempting to provide. The result is less total force being applied to the vehicle, and therefore reduced performance when moving at non-zero forward speeds (Palmer et al., 2009).

3.3 Allocation methods

As previously stated in the evaluation of several actuators for underwater vehicles, control surfaces are ineffectual while the vehicle is stationary, whereas tunnel thrusters prove to be highly advantageous. In contrast, when the vehicle is in motion, control surfaces exhibit superior efficiency in generating force relative to power consumption compared to tunnel thrusters. The function of control allocation is to achieve a balance among all actuators that applies the requisite generalized forces and moments to the vehicle while minimizing power consumption. This equilibrium enables the vehicle to sustain the maneuverability offered by all actuators, while concurrently maximizing mission length.

3.3.1 Non-optimal scheme

One of the most straightforward methods for control allocation is application of the inverse of (34), i.e., (40) (Fossen, 2002).

$$u = (TK)^{-1}\tau \tag{40}$$

This approach is straightforward to execute, comprising solely a single matrix multiplication. Consequently, it is straightforward and effective to execute within the computational processing limitations of an AUV. Nonetheless, owing to its simplicity, no efforts are undertaken to reduce power consumption. When the vehicle is equipped with both control surfaces and tunnel thrusters, both actuator types will be employed equally, even at maximum velocity relative to the surrounding fluid. Nevertheless, given the superior efficiency of employing control surfaces over tunnel thrusters while the vehicle is in motion relative to the water, a more sophisticated strategy is required for executing control allocation to reduce power consumption.

3.3.2 Quadratic programming

The limitations of the aforementioned non-optimal scheme can be overcome by formulating a quadratic programming optimisation problem to solve for actuator inputs (Fossen, 2002, Fossen et al., 2009). By introducing a weighting matrix into the problem statement, actuator usage can be biased towards utilising control surfaces over tunnel thrusters. Therefore, the generalised force desired from the control law can be realised by the actuators while minimising power consumption. There are however, limitations associated with this scheme. Firstly, although it is possible to calculate an explicit solution to this problem, in the event of actuator reconfiguration, such as an actuator failure, this explicit solution would need to be recalculated, which can be computationally intensive. Iterative approaches, such as sequential quadratic programming, can be implemented that allow for actuator failures, but this method has the potential to require several iterations of the programming problem be solved at each control sample interval. Again, this can be a computationally intensive task.

3.3.3 2-Stage scheme

A third scheme that is proposed here for the implementation of the control allocation is to break the control allocation problem into two smaller sub-problems, as seen in Figure 2, where the first sub-problem addresses control allocation to the main propeller and control surfaces, and the second addresses control allocation to the tunnel thrusters. Using this type of scheme, the control surfaces can firstly be used to their full extent in order to realise the generalised force as closely as possible. Only after full utilisation of the control surfaces occurs will the tunnel thrusters be introduced to provide forces and moments that the surfaces alone cannot produce. Using this methodology, the low power consumption control surfaces will be used as much as possible, while the higher power consumption tunnel thrusters will be called upon only when required to provide the extra manoeuvring capabilities that they possess. Furthermore, in a situation where accurate manoeuvring is not required, such as traversing from one waypoint to the next with no concern for what trajectory the vehicle follows, the second stage can be disabled such that no tunnel thruster is used, and control is performed entirely by the main propeller and control surfaces. However, if trajectory tracking is desired, the thruster allocation module can still be enabled. This will allow for the situation when the control surfaces provide inadequate force to the vehicle, and therefore the thrusters can assist in providing the extra force required to maintain the vehicle tracking the desired trajectory.

Implementation of this scheme would look somewhat like a 2-stage non-optimal scheme, as seen in Figure 2. The first stage would require the matrix operation $(TK)^{-1}\tau$ for the main propeller and control surfaces in order to obtain as much force required from these actuators. An estimate of the force produced for this particular set of control values would then be calculated such that this force estimate can be subtracted from the total force required. Any residual force requirement would then become the input to the second stage of the control allocation, which would perform the matrix operation $(TK)^{-1}\tilde{\tau}$ for the tunnel thrusters in order for these actuators to provide any extra force that the control surfaces cannot deliver alone. Therefore, the computational requirement for this scheme is quite minimal compared to the quadratic programming scheme, yet still heavily biases the use of control surfaces over tunnel thrusters.

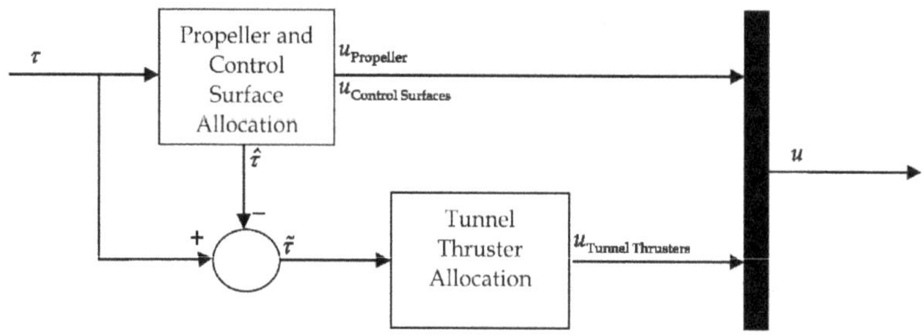

Fig. 2. 2-Stage Control Allocation Scheme Block Diagram.

4. OVERALL CONTROL SYSTEM OF AUTONOMOUS UNDERWATER VEHICLE

Overall, the control system of any autonomous underwater vehicle consists of a control law module and a control allocation module. The former is responsible for generating the generalised force in 6 DoF based on current and desired states, while the latter is responsible for distributing this generalised force amongst the actuators of the vehicle. Due to the modularity of such a control system, each of these subsystems can be designed and implemented independently, yet both must work cooperatively in order to accurately control the vehicle.

As discussed, the inputs to the control system are the desired state from the guidance system, and the current state from the navigation system. These are also the inputs to the control law. The outputs from the control system are the control signals that are sent to the individual actuators on the vehicle, which are the outputs generated by the control allocation. The intermediate signals connecting the control law to the control allocation are the 6 DoF generalised forces which are the outputs from the control law and therefore the inputs to the control allocation. Figure 3 shows the overall control system. As can be seen, the modularity of the system allows for control laws and control allocation schemes to be easily replaced, provided the inputs and outputs are of the same dimensions.

For comparison purposes, both the uncoupled SMC law and the coupled SMC law are simulated here. Both simulations consist of identical control allocation modules that contain the previously described 2-stage scheme.

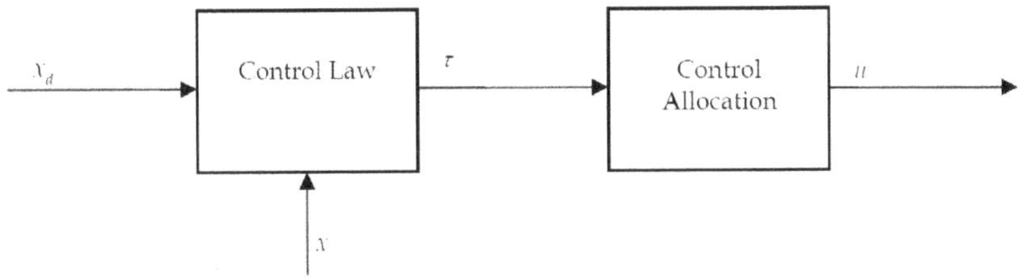

Fig. 3. Overall Control System Block Diagram.

The vehicle model used for these simulations is based on that which was proposed by Prestero (2001a, 2001b). This is a mathematical model for the REMUS (Remote Environmental Monitoring Unit) underwater vehicle developed by the Woods Hole Oceanographic Institute's Oceanographic Systems Laboratory. Alterations have been made to this model, based on current technology, which enable the vehicle to become fully actuated. These changes include the placement of four tunnel thrusters, two fore and two aft, such that sway, heave, pitch, and yaw motions are possible without any water flow over the control surfaces. Furthermore, the position of the vertical thrusters along the y-axis of the body, and the position of the horizontal thrusters along the z-axis of the body are set to 0m. Hence, there is no roll moment applied to the vehicle when these thrusters are activated. The thrusters introduced to the model are based on the 70mm IntegratedThrusterTM produced by TSL Technology Ltd. This particular device can provide a maximum thrust of 42N, and due to its compact size, is well suited to this particular application. Also, the propulsion unit has been altered within the model. The simulation model used contains a propulsion unit based on the Tecnadyne Model 540 thruster. This device is able to provide approximately 93.2N of thrust and, with a propeller diameter of 15.2cm, is therefore well suited for this application.

The trajectory that the vehicle is asked to follow here consists of a series of unit step inputs applied to each DoF. The translational DoFs experience a step input of 1 metre whereas the rotational DoFs experience a step input of 1 radian, or approximately 60°. All inputs are applied for a period of 20 seconds, such that both transient and steady state behaviour can be observed. These unit step inputs excite the vehicle in all combinations of DoFs, from a single DoF, through to all DoFs being excited at once. This simulation assumes no water current, and therefore no water flow over the vehicle when it is stationary. Hence, roll cannot be compensated for when stationary and for this reason this simulation does not excite the roll component of the model. All other DoFs are excited, however.

Due to the way the aforementioned trajectory is supplied to the vehicle model, no guidance system is present. The step inputs will be applied at set times regardless of the state of the vehicle, and therefore observation of the control performance can be observed without influence from unnecessary systems. Therefore total execution time is constant for all simulations. Performance metrics used here in evaluating each control system is the accumulated absolute error between the desired translation/rotation and actual translation/rotation for each DoF. By looking at the following plots, observations can be made regarding such time-domain properties as rise time, settling time and percentage overshoot.

Figures 4-13 show the desired and actual trajectories for each individual DoF when this complex set of manoeuvres is applied to both the uncoupled SMC and the coupled SMC. Figures 4, 6, 8, 10 and 12 show the complete trajectory for each DoF, while Figures 5, 7, 9, 11 and 13 show these same trajectories with the focus being on the last 300 seconds of the mission. This latter section of the mission is when multiple DoFs, particularly the rotational DoFs, are excited simultaneously.

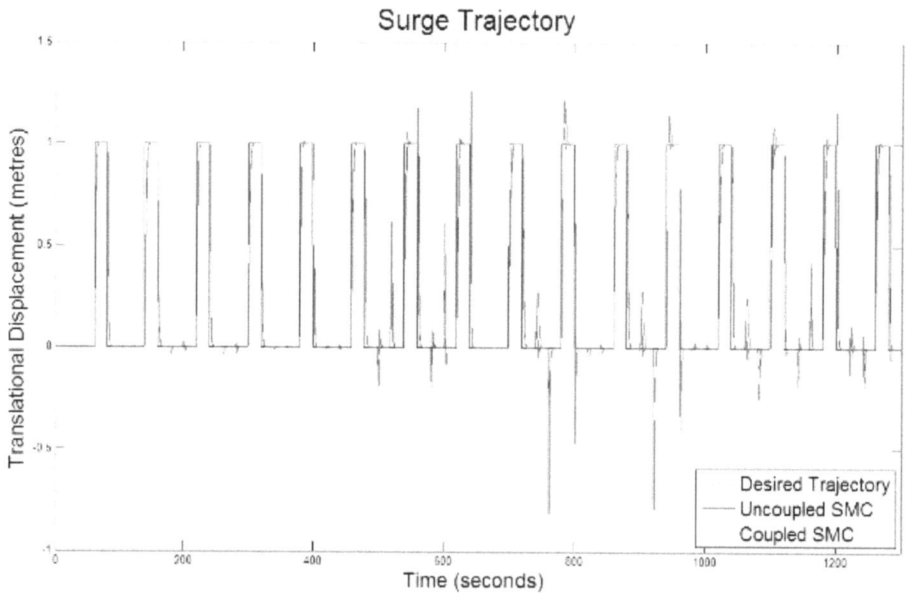

Fig. 4. Surge trajectory.

Figure 4 shows the surge motion of the vehicle for the two different control systems. As can be seen, both systems exhibit desired properties for the first 180 seconds. During this period, all manoeuvring is exciting only the translational DoFs which the main propeller and tunnel thrusters can handle independently. After 180 seconds, the other DoFs are also excited, and the effect of this combined motion produces significant overshoots, especially for the uncoupled system, within the surge motion of the vehicle. This is more easily seen in Figure 5 where the larger overshoots can be seen for the uncoupled system compared to the coupled system.

The sway motion of the AUV is shown in Figure 6. Minor overshooting is observed for both systems, especially when rotational DoFs are excited in combination with the translational DoFs. This can be observed in Figure 7. However, when the sway motion is excited, convergence to the desired set point is observed.

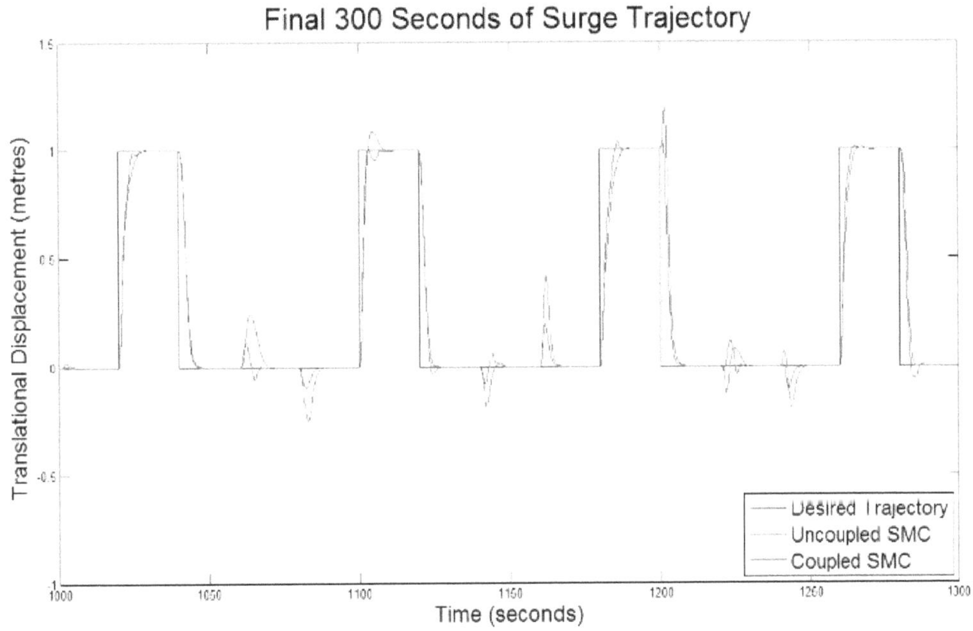

Fig. 5. Surge trajectory magnified.

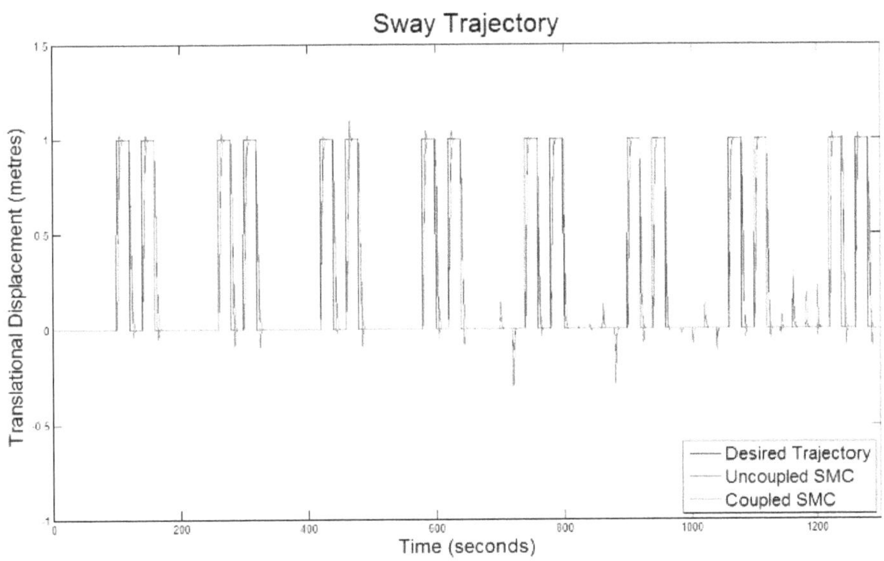

Fig. 6. Sway trajectory.

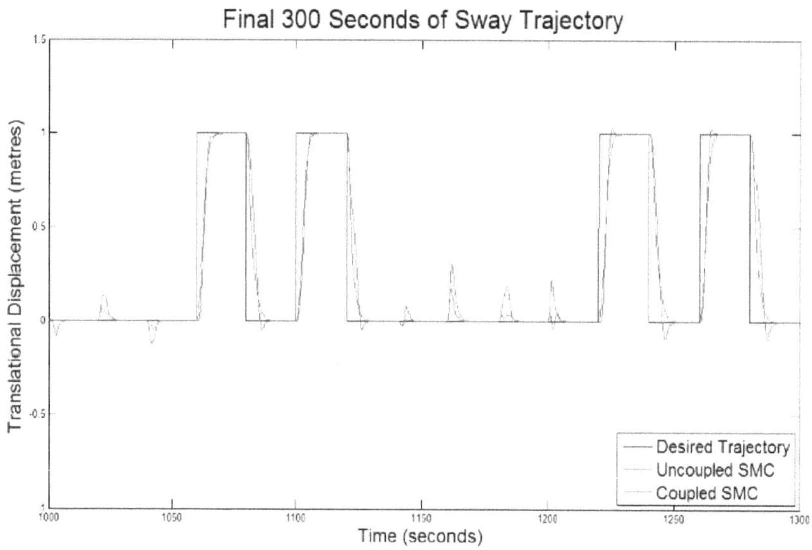

Fig. 7. Sway trajectory magnified.

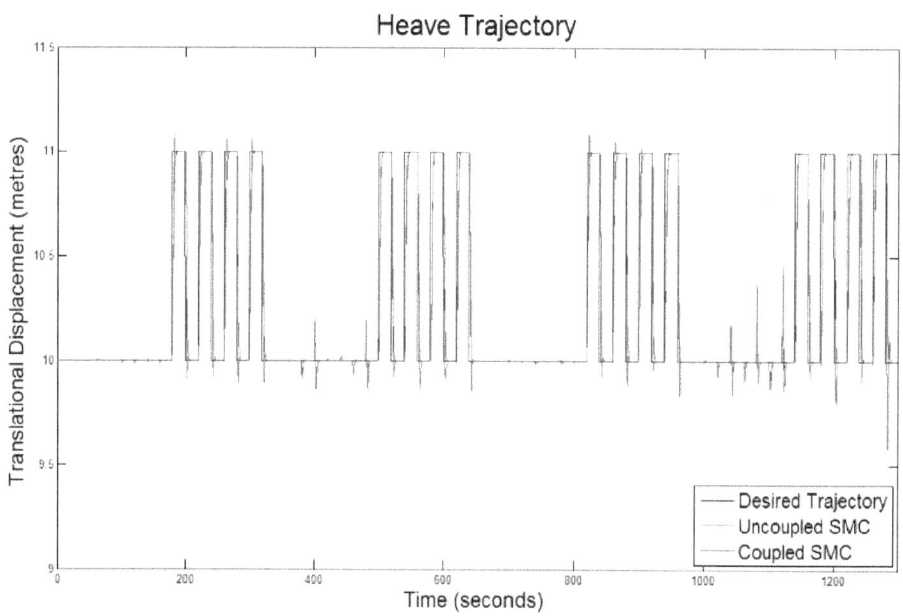

Fig. 8. Heave trajectory.

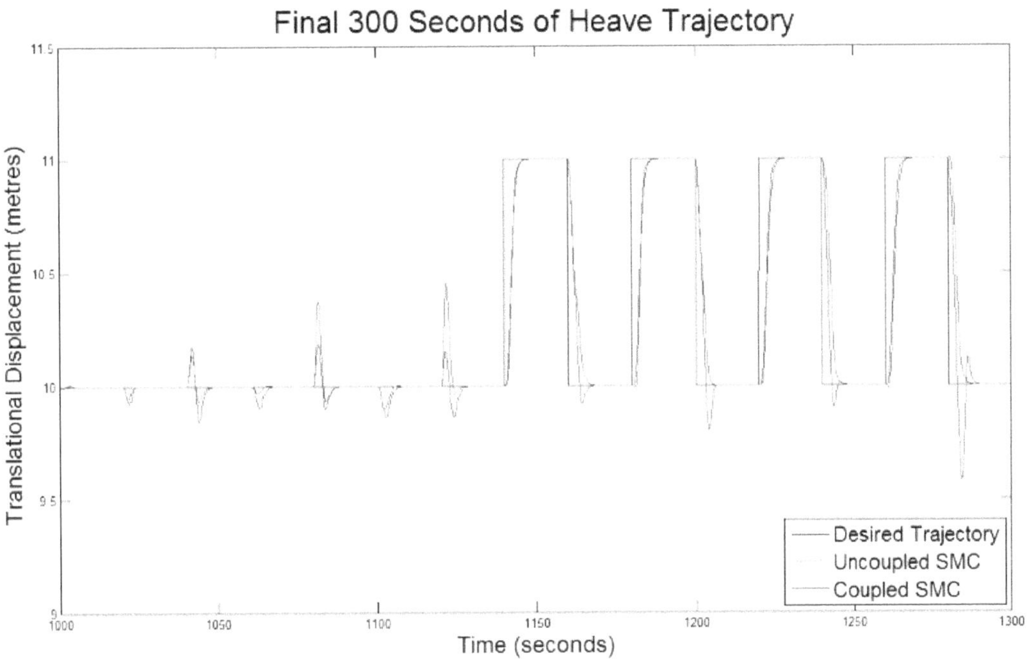

Fig. 9. Heave trajectory magnified.

Heave motion is shown in Figure 8. Here, the initial position of the vehicle is chosen to be 10 metres below the surface of the water, such that the vehicle is completely submerged for the entire trajectory. When observing the heave motion in Figure 9, it is clear that when heave is excited, convergence to the desired set point is achieved. However, it is also evident that significant coupling exists between this DoF and other DoFs as divergence from the desired set point occurs when the heave motion is not excited.

Pitch motion is observed in Figure 10. The first observation that can be made from this plot is the relatively large spikes in the motion for both the coupled and uncoupled systems between 800 seconds and 1000 seconds. The second observation is that a steady state error is observed when the uncoupled system attempts a pitch of 1 radian. This is more clearly observed in Figure 11. Due to the vertical offset of the vehicle's centre of buoyancy from its centre of gravity, a restorative moment will always be applied to the vehicle for non-zero pitch angles. This plot indicates that it is difficult for the uncoupled system to compensate for this effect. However, the coupled system is able to not only eliminate this steady state error, it also has significantly smaller overshoot in general, as can be seen in Figure 10.

Figure 12 shows the desired yaw motion of the vehicle. Here we can see significant overshooting from both systems, especially during the time period before the yaw motion is excited. By observing Figure 13, it can be seen that the coupled system achieves faster dynamics in terms of rise time, but the cost of this is overshoot. The uncoupled system has significantly less overshoot, and this is due to it taking slightly longer to react to trajectory changes.

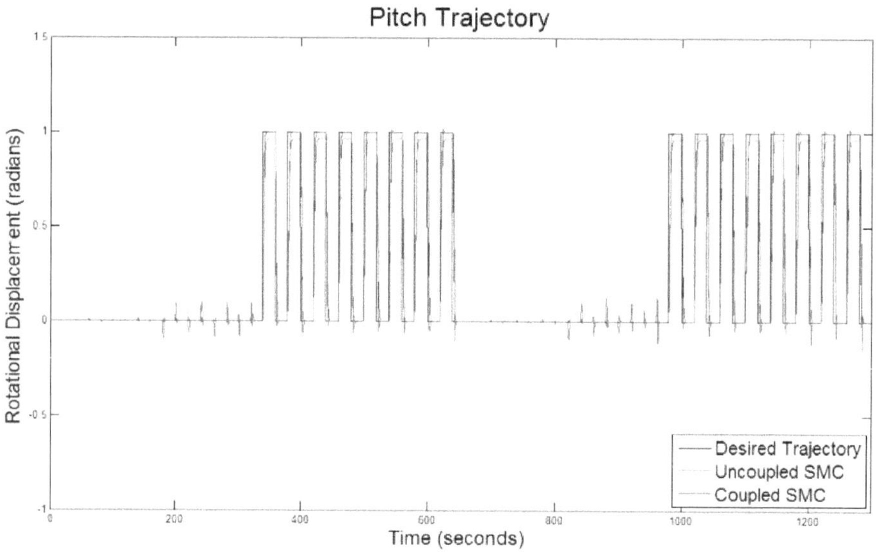

Fig. 10. Pitch trajectory.

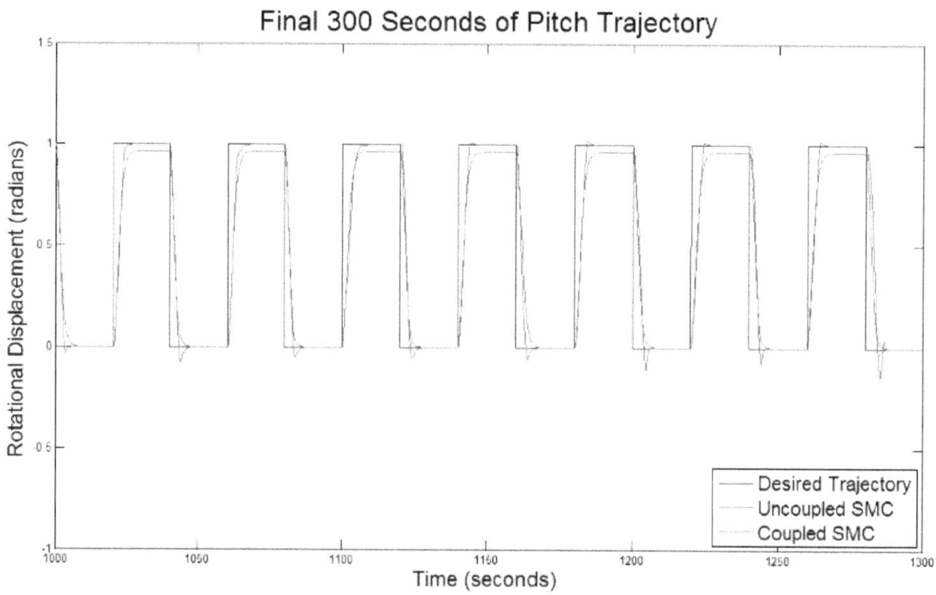

Fig. 11. Pitch trajectory magnified.

Overall, the simulation trajectory used here is extremely complex as it excites all possible combinations of DoFs without any water current present. This level of complexity allows for the coupling that exists between DoFs within an AUV model to be highlighted. For example, looking at the period of 0 to 660 seconds in Figure

12, the yaw motion is not being excited, yet motion in this DoF is observed. This coupling between multiple DoFs is the reason why controlling AUVs is a complex and challenging task.

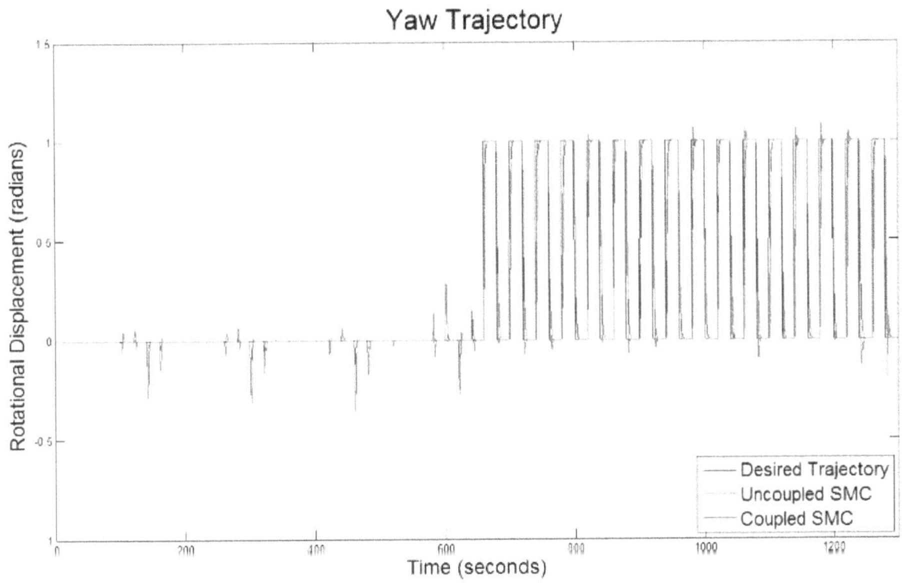

Fig. 12. Yaw trajectory.

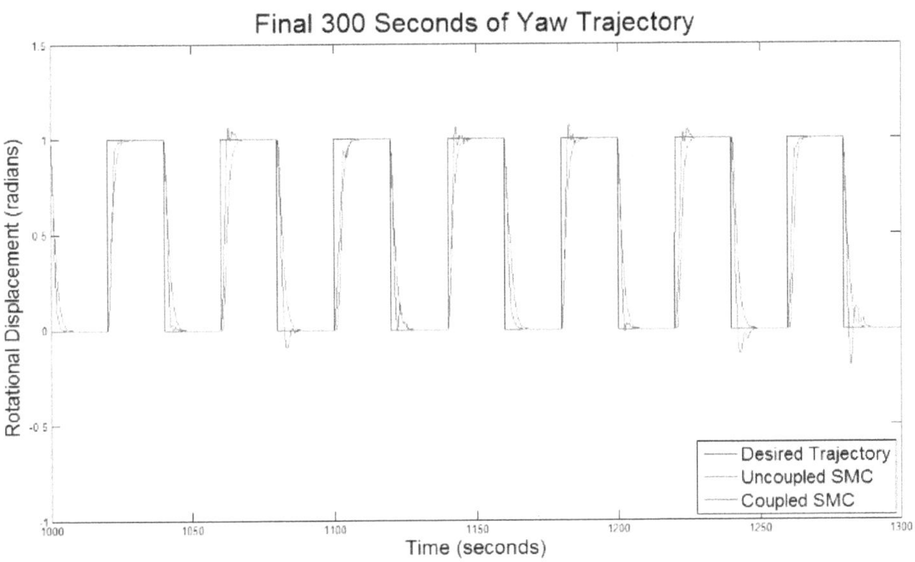

Fig. 13. Yaw trajectory magnified.

Furthermore, if the absolute error between actual and desired position is accumulated over the entire mission for each DoF, a measure of the accuracy of the control system can be obtained. A comparison can be made using Table 1. Here, we can see that by summing together the values obtained from the three translational

DoFs, the coupled system improves upon the uncoupled system by 9.75%, while by summing together the values obtained from the three rotational DoFs, the coupled improves upon the uncoupled system by 23.87%. This indicates that the coupled system is superior to the uncoupled system for both translational and rotational motion, when looking at the accumulated absolute error.

Table 1. Accumulated Absolute Translational and Rotational Errors.

Control System	Accumulated Absolute Translational Error	Accumulated Absolute Rotational Error
Uncoupled System	1.3173x104 (metres)	8.3713x103 (radians)
Coupled System	1.2003x104 (metres)	6.7580x103 (radians)

5. CONCISE SUMMARY

The growing utilization of AUVs for both civilian and defense purposes necessitates precision and reliability as essential variables for mission accomplishment. The control system is but one component of the autonomy architecture of an AUV that contributes to reaching this objective. The control law within the system must exhibit robustness against external disturbances and uncertainty in model parameters, while the control allocation should effectively employ the vehicle's actuators to exert the appropriate forces with minimal power consumption. PID control has been successfully applied across diverse systems to deliver effective compensation. Nonetheless, as PID control is more appropriate for linear models, its performance is inferior to that of other control schemes, notably nonlinear ones, when implemented in complicated nonlinear systems. Sliding mode control has demonstrated robustness against parameter uncertainty, making it an ideal candidate for application in this context, given the very complicated coupled nonlinear underwater vehicle model. The linked SMC aims to effectively utilize the model's interconnected structure, ensuring that motion induced in one degree of freedom is properly compensated by motion in another degree of freedom. Coupled SMC possesses a significant advantage over uncoupled SMC in trajectory tracking applications when many degrees of freedom are simultaneously activated. Diverse methodologies are available for control allocation, aiming to apply the requisite generalized forces while minimizing power consumption, encompassing both actuator utilization and computational requirements. Suboptimal schemes exist that utilize a generalized inverse of the force generated by all actuators as the allocation method, with the constraint that there is no capability to prioritize actuators under specific operating conditions, such as employing control surfaces over thrusters during high-speed maneuvers. Quadratic programming utilizes a weighting matrix that may influence the utilization of control surfaces relative to tunnel thrusters, and has been executed in both online and offline modes, each presenting distinct advantages and disadvantages. Online optimization facilitates modifications to the actuator configuration, including failures or altered saturation limits, however it is computationally intensive. Offline optimization is less computationally intensive during mission execution, but it cannot accommodate changes in actuator dynamics. The suggested two-stage strategy represents a balance between these approaches, utilizing control surfaces to their maximum capacity while employing tunnel thrusters solely as necessary. The primary objective of the control system is to ensure sufficient compensation for the vehicle despite unknown and unmodeled variables, while simultaneously minimizing power consumption to prolong mission length. Carefully selecting both the control rule and the control allocation scheme within the overall control system is essential for attaining these objectives.

REFERENCES

Fossen, T. I. (1994). *Guidance and Control of Ocean Vehicles,* John Wiley & Sons, Inc., ISBN 0- 471-94113-1, Chichester, England

Fossen, T. I. (2002). *Marine Control Systems: Guidance, Navigation and Control of Ships, Rigs and Underwater Vehicles* (1), Marine Cybernetics, ISBN 82-92356-00-2, Trondheim, Norway

Fossen, T. I., Johansen, T. A. & Perez, T. (2009). A Survey of Control Allocation Methods for Underwater Vehicles, In: *Underwater Vehicles,* Inzartsev, A. V., pp. (109-128), In- Tech, Retrieved from <http://www.intechopen.com/articles/show/title/a_survey_of_control_allocatio n_methods_for_underwater_vehicles>

Healey, A.J. & Lienard, D. (1993). Multivariable Sliding Mode Control for Autonomous Diving and Steering of Unmanned Underwater Vehicles. IEEE *Journal of Oceanic Engineering,* Vol. 18, No. 3, (July), pp. (327-339), ISSN 0364-9059

Jalving, B. (1994). The NDRE-AUV Flight Control System. IEEE *Journal of Oceanic Engineering,* Vol. 19, No. 4, (October), pp. (497-501), ISSN 0364-9059

Kokegei, M., He, F. & Sammut, K. (2008). Fully Coupled 6 Degrees-of-Freedom Control of Autonomous Underwater Vehicles, *MTS/IEEE Oceans '08,* Quebec City, Canada, September 15-18

Kokegei, M., He, F. & Sammut, K. (2009), Nonlinear Fully-Coupled Control of *AUVs, Society of Underwater Technology Annual Conference,* Perth, Australia, 17-19 February

Lammas, A., Sammut, K. & He, F. (2010). 6-DoF Navigation Systems for Autonomous Underwater Vehicles, In: *Mobile Robots Navigation*, Barrera, A., pp. (457-483), In-Teh, Retrieved from <http://www.intechopen.com/articles/show/title/6-dof-navigation-systems-for-autonomous-underwater-vehicles>

Lammas, A., Sammut, K. & He, F. (2008), Improving Navigational Accuracy for AUVs using the MAPR Particle Filter, *MTS/IEEE Oceans '08*, Quebec City, Canada, 15-18 September

Marco, D. B. & Healey, A. J. (2001). Command, Control, and Navigation Experimental Results with the NPS ARIES AUV. *IEEE Journal of Oceanic Engineering*, Vol. 26, No. 4, (October), pp. (466-476), ISSN 0364-9059

Palmer, A., Hearn, G. E. & Stevenson, P. (2009). Experimental Testing of an Autonomous Underwater Vehicle with Tunnel Thrusters, *First International Symposium on Marine Propulsors*, Trondheim, Norway, 22-24 June

Prestero, T. (2001a), Development of a Six-Degree of Freedom Simulation Model for the REMUS Autonomous Underwater Vehicle, 12th *International Symposium on Unmanned Untethered Submersible Technology*, University of New Hampshire, Durham, NH, 26-29 August

Prestero, T. (2001b). *Verification of a Six-Degree of Freedom Simulation Model for the REMUS Autonomous Underwater Vehicle.* Master of Science in Ocean Engineering and Master of Science in Mechanical Engineering, Massachusetts Institute of Technology and the Woods Hole Oceanographic Institution, Cambridge and Woods Hole

Yoerger, D. R. & Slotine, J.-J. E. (1985). Robust Trajectory Control of Underwater Vehicles. *IEEE Journal of Oceanic Engineering*, Vol. 10, No. 4, (October), pp. (462-470), ISSN 0364-9059

8 Chapter SHORT-RANGE NETWORKS FOR UNDERWATER SOUND COMMUNICATION

1. INTRODUCTION: AN OVERVIEW

This chapter talks about how a short-range acoustic communication channel model was made and how its properties can be used to build and test MAC (Medium Access Control) and routing protocols that will help network-enabled autonomous underwater vehicles (AUVs). As underwater activities have grown, they have needed to be able to share data between a wide range of underwater and surface-based communication nodes. Another type of node is an AUV, but in the future, they will likely be used in groups to work as an ad hoc sensor network. In this case, the swarm network will be built with uniform nodes, which means that each one is the same, as shown in Figure 1. The swarm network will then connect to other fixed underwater communication nodes. This chapter is mostly about how to make sure that AUVs can reliably send and receive data, which is needed to take advantage of the way a swarm network works as a whole. A simple 2-dimensional (2D) topology, as shown in Figure 1(b), will be used to investigated swarm based operations of AUVs. The vehicles within the swarm will move together, in a decentralised, self organising, ad-hoc network with all vehicles hovering at the same depth. Figure 1(b) shows the vehicles arranged in a 2D horizontal pattern above the ocean floor giving the swarm the maximum coverage area at a single depth, while forming a multi-hop communication network. The coverage area will depend on application. For example, the exploration of oil and gas deposits underwater using hydrocarbon sensing would initially require a broad structure scanning a large ocean footprint before narrowing the range between vehicles as the sensing begins to target an area. Thus vehicles may need to work as closely as 10 m with inter-node communication distance extending out to 500 m. The operational distances are far less than the conventional operations of submarines and underwater sensors to surface nodes, which typically function at range exceeding 1 km. Consequently, the modeling and equipment development for the communication requirements of these operations has concentrated on extended-range data transmission and channel modeling.

To fully capitalize on the advantages of short-range communication systems, it is essential to examine the characteristics of short-range communication channels.

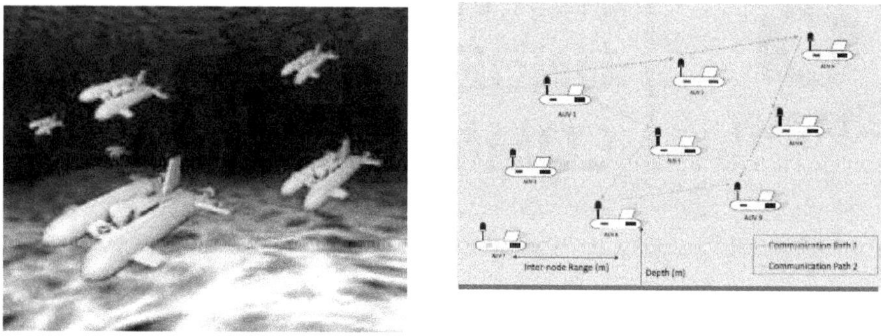

(a) AUV Swarm demonstrating stylised SeaVision©vehicles

(b) 2D AUV Swarm Topology

Fig. 1. Swarm Architecture

Most AUV development work has concentrated on the vehicles themselves and their operations as a single unit (Dunbabin et al., 2005; Holmes et al., 2005), without giving much attention to the development of the swarm architecture which requires wireless communication networking infrastructure. To develop swarm architectures it is necessary to research effective communication and networking techniques in an underwater environment. Swarm operation has many benefits over single vehicle use. The ability to scan or 'sense' a wider area and to work collaboratively has the potential to vastly improve the efficiency and effectiveness of mission operations. Collaboration within the swarm structure will facilitate improved operations by building on the ability to operate as a team which will result in emergent behaviours that are not exhibited by individual vehicles. A swarm working collaboratively can also help to mitigate the problem of high propagation delay and lack of bandwidth available in underwater communication environments. Swarm topology will facilitate improved communication performance by utilising the inherent spatial diversity that exists in a large structure. For example, information can be transmitted more reliably within a swarm architecture by using multi-hop networking techniques. In such cases, loss of an individual AUV, which can be expected at times in the unforgiving ocean environment, will have less detrimental effect compared to a structure where multiple vehicles operate on their own. (Stojanovic, 2008).

The underwater acoustic communication channel is recognised as one of the harshest environments for data communication, with long range calculations of optimal channel capacity of less than 50kbps for SNR (Signal-to-Nosie Ratio) of 20dB (Stojanovic, 2006) with current modem capacities of less than 10kbps (Walree,

2007). Predictability of the channel is very difficult with the conditions constantly changing due to seasons, weather, and the physical surroundings of sea floor, depth, salinity and temperature. Therefore, it must be recognised that any channel model needs to be adaptable so that the model can simulate the channel dynamics to be able to fully analyse the performance of underwater networks.

In general, the performance of an acoustic communication system underwater is characterised by various losses that are both range and frequency dependent, background noise that is frequency dependent and bandwidth and transmitter power that are both range dependent. The constraints imposed on the performance of a communication system when using an acoustic channel are the high latency due to the slow speed of the acoustic signal propagation, at 0.67 ms/m (compared with RF (Radio Frequency) in air at 3.3 ns/m), and the signal fading properties due to absorption and multipath. Specific constraints on the performance due to the mobility of AUV swarms is the Doppler effect resulting from any relative motion between a transmitter and a receiver, including any natural motion present in the oceans from waves, currents and tides.

Noise in the ocean is frequency dependent. There are three major contributors to noise underwater: ambient noise which represents the noise in the far field; self noise of the vehicle (considered out of band noise); and intermittent noise sources including noises from biological sources such as snapping shrimp, ice cracking and rain. Ambient noise is therefore the component of noise taken into account in acoustic communication performance calculations. It is characterised as a Gaussian Distribution but it is not white as it does not display a constant power spectral density. For the frequencies of interest for underwater acoustic data communication, from 10 to 100 kHz, the ambient noise value decreases with increasing frequency. Therefore, using higher signal frequencies, which show potential for use in shorter range communication, will be less vulnerable to the impact of ambient noise.

Short range underwater communication systems have two key advantages over longer range operations; a lower end-to-end delay and a lower signal attenuation. End-to-end propagation at 500 m for example is approximately 0.3 sec which is considerable lower than the 2 sec at 3 km but still critical as a design parameter for shorter range underwater MAC protocols. The lower signal attenuation means potentially lower transmitter power requirements which will result in reduced energy consumption which is critical for AUVs that rely on battery power. Battery recharge or replacement during a mission is difficult and costly. The dynamics associated with attenuation also changes at short range where the spreading component dominates over the absorption component, which means less dependency on temperature, salinity and depth (pressure). This also signifies less emphasis on frequency as the frequency dependent part of attenuation is in the absorption component and thus will allow the use of higher signal frequencies and higher bandwidths at short ranges. This potential needs to be exploited to significantly improve the performance of an underwater swarm network communication system.

A significant challenge for data transmission underwater is multipath fading. The effect of multipath fading depends on channel geometry and the presence of various objects in the propagation channel. Multipath's occur due to reflections (predominately in shallow water), refractions and acoustic ducting (deep water channels), which create a number of additional propagation paths, and depending on their relative strengths and delay values can impact on the error rates at the receiver. The bit error is generated as a result of inter symbol interference (ISI) caused by these multipath signals. For very short range single transmitter-receiver systems, there could be some minimisation of multipath signals (Hajenko & Benson, 2010; Waite, 2005). In swarm operations, it is essential to consider a distinct combination of multipath signals, particularly those produced by the other vehicles inside the swarm. Thorough examination of the physical layer parameters and their optimal design will enhance the benefits of a short-range communication system that must leverage the constrained resources present in an underwater acoustic networking context. This section will delineate the parameters related to underwater audio data transfer. Section 3 will describe the features of the underwater data transmission channel, along with a discussion of the pros and disadvantages of the short-range channel. Section 4 will demonstrate the effects on AUV swarm communications and the formulation of a short-range channel model for the design and assessment of MAC and routing algorithms. Section 5 thereafter presents a discourse on the protocol strategies necessary for AUV swarm network construction.

2. ACOUSTIC UNDERWATER COMMUNICATION NETWORK: AN INTRODUCTION

The underwater data communication link and networking environment significantly differ from the RF data communication channel in the terrestrial atmosphere. Figure 2 depicts a conventional underwater setting for data transmission utilizing a solitary transmitter-receiver pair.

A simple schematic of the data transmission scheme involving a projector (transmitter) and a hydrophone (receiver) is presented in Figure 3. The projector takes the collected sensor and navigational data and formats it into packets at the Data Source and this is then modulated with the carrier frequency. The modulated signal is amplified to a level sufficient for signal reception at the receiver. There is an optimum amplification level as there is a trade-off between error free transmission and conservation of battery energy. The acoustic power radiated from the projector as a ratio to the electrical power supplied to it, is the efficiency η_{tx} of the projector and represented by the Electrical to Acoustic conversion block. On the receiver side, the sensitivity of the hydrophone converts the sound pressure that hits the hydrophone to electrical energy, calculated in dB/V. Signal detection, includes amplification and shaping of the input to determine a discernible signal. Here a detection threshold needs to be reached and is evaluated as the ratio of the mean signal power to mean noise power (SNR). The carrier frequency is then supplied for demodulation, before the transmitted data is available for use within the vehicle for either data storage or for input into the vehicles control and navigation requirements.

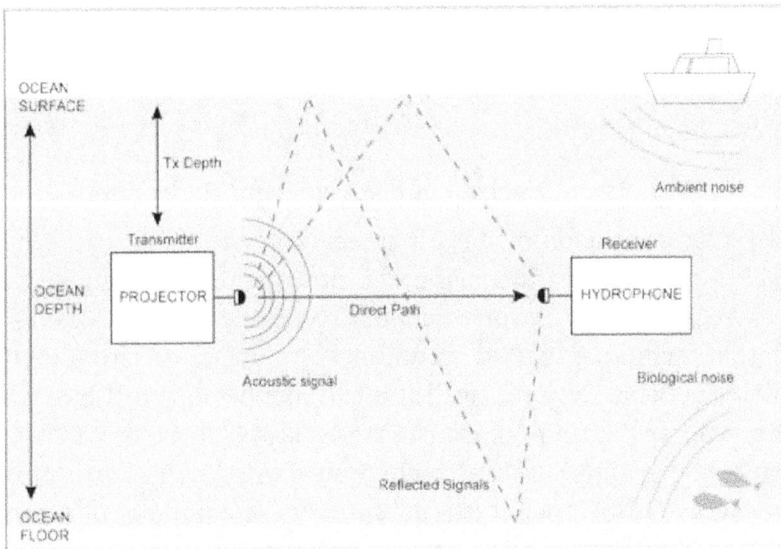

Fig. 2. Underwater Acoustic Environment

Underwater data communication links generally support low data rates mainly due to the constraints of the communication channel. The main constraints are the high propagation delay, lower effective SNR and lower bandwidth. The effects of these constraints could be reduced by using short distance links and the use of multi-hop communication techniques to cover longer transmission ranges. For an AUV swarm network, use of the above techniques could be crucial to design an effective underwater network. To develop a multi-node swarm network it is necessary to manage all point to point links using a medium access control (MAC) protocol. In a multi-access communication system like a swarm network a transmission channel is shared by many transceivers in an orderly fashion to transmit data in an interference free mode. Figure 2 shows a point to point communication link with two AUVs. When a network is scaled up to support N number of AUVs then it becomes necessary to control multiple point to point or point to multi-point links.

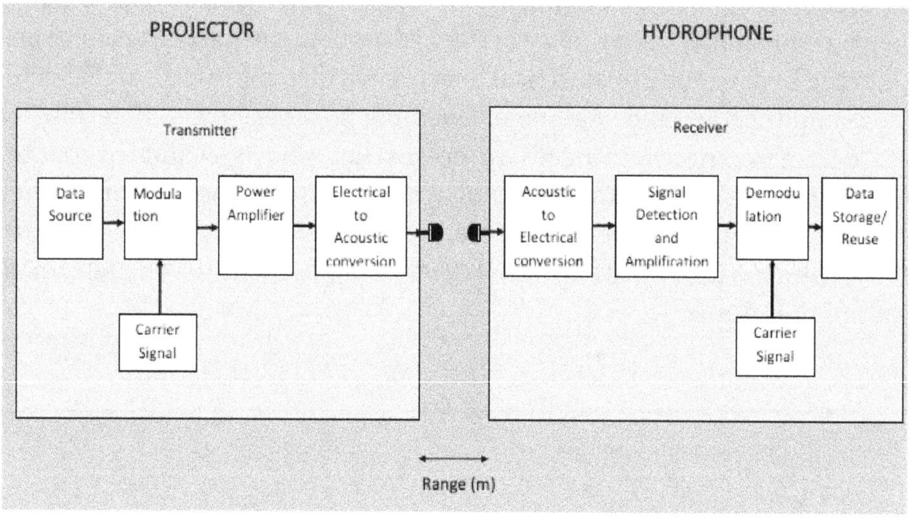

Fig. 3. Block Diagram of Projector and Hydrophone

To control the transmission of data it is necessary to design an effective MAC protocol which can control transmission of information from different AUVs. The design of a MAC protocol in a swarm network could be more complex if a multi-hop communication technique is used. The multi-hop communication technique will allow a scalable network design as well as it can support long distance transmission without the need of high power transmitter and receiver circuits. For example, using a multi-hop communication technique if AUV3 in Figure 1(b) wants to transmit packets to AUV7 then it can potentially use a number of communication paths to transmit packets. Some of the possible paths from AUV3 to AUV7 are: AUV3-AUV2-AUV1-AUV4-AUV7 or AUV3-AUV6-AUV9-AUV8-AUV7. The path selection in a network is controlled by the routing protocols. Optimum routing protocols generally select transmission paths based on a number of factors. However, the main factor used to select an optimum path in a wireless network is the SNR which indicates the quality of a link. Likewise, the MAC protocol will utilize the transmission channel status information to construct an optimal packet access strategy. To build these protocols successfully, it is essential to comprehend the features of short-range underwater channels. Prior to addressing the protocol design difficulties, we will first assess the features of the short-range underwater channel in the subsequent sections.

3. CHARACTERISTICS OF UNDERWATER DATA TRANSMISSION CHANNELS

This section will examine the parameters of the ocean channel that influence the transmission of acoustic signals from the projector to the hydrophone. Established underwater channel models will be utilized to generate and present the data transmission properties for a short-range link.

3.1 Acoustic signal level

The projector source level, $SL_{tprojector}$, is generally defined in terms of the sound pressure level at a reference distance of 1 m from its acoustic centre. The source intensity at this reference range is $I = P_{tx} / Area$ (W/m^2) and measured in dB're 1 μ Pa' but strictly meaning're the intensity due to a pressure of 1 μ Pa'. For an omni directional projector the surface area is a sphere ($4\pi r^2 = 12.6m^2$). Thus, $SL_{projector} = 10log((P_{tx}/12.6)/I_{ref})$ dB, where P_{tx} is the total acoustic power consumed by projector and the reference wave has an intensity:

$I_{ref} = (Pa_{ref})^2/\rho * c$ (Wm^{-2}) where reference pressure level; Pa_{ref} is 1 μ Pa, ρ is the density of the medium and; c is the speed of sound (averages for sea water: $\rho = 1025\ kg/m^3$ and c=1500 m/s) (Coates, 1989; Urick, 1967).

The equation for the transmitter acoustic signal level ($SL_{projector}$) at 1 m for an omni-directional projector can be written:

$$SL_{projector}(P) = 170.8 + 10logP_{tx} \qquad dB \qquad (1)$$

If the projector is directional, then the projector directivity index is $DI_{tx} = 10log(\frac{I_{dir}}{I_{omni}})$ where *Iomni* is the intensity if spread spherically and I_{dir} is the intensity along the axis of the beam pattern. Directivity can increase the source level by 20dB (Waite, 2005). The more general equation for the transmitter acoustic signal level ($SL_{projector}$) can be written:

$$SL_{projector}(P,\eta,DI) = 170.8 + 10logP_{tx} + 10log\eta_{tx} + DI_{tx} \qquad dB \qquad (2)$$

where the efficiency of the projector ηtx takes into account the losses associated with the electrical to acoustic conversion as shown in Figure 3, thus reducing the actual SL radiated by the projector. This efficiency is bandwidth dependent and can vary from 0.2 to 0.7 for a tuned projector (Waite, 2005).

3.2 Signal attenuation

Sound transmission in the ocean is affected by the physical and chemical characteristics of saltwater as well as the geometry of the channel. An underwater audio signal undergoes attenuation from dispersion and absorption. Furthermore, multipath fading may occur at the hydrophone, contingent upon channel geometry. Path loss quantifies the reduction in signal intensity from the projector to the hydrophone. Comprehending and formulating an accurate path loss model is essential for the computations of Signal-to-Noise Ratio (SNR).

3.2.1 Spreading loss

Spreading loss is due to the expanding area that the sound signal encompasses as it geometrically spreads outward from the source.

$$PL_{spreading}(r) = k * 10log(r) \qquad dB \tag{3}$$

where r is the range in meters and k is the spreading factor.

When the medium in which signal transmission occurs is unbounded, the spreading is spherical and the spreading factor k=2 whereas in bounded spreading, considered as cylindrical k=1. Urick (1967) suggested that spherical spreading was a rare occurrence in the ocean but recognised it may exist at short ranges. As AUV swarm operations will occur at short range it is likely that spherical spreading will need to be considered which means a higher attenuation value. Spreading loss is a logarithmic relationship with range and its impacts on the signal is most significant at very short range up to approximately 50m as seen in Figure 5(a). At these shorter ranges spreading loss plays a proportionally larger part compared with the absorption term (which has a linear relationship with range).

3.2.2 Absorption loss

The absorption loss is a representation of the energy loss in the form of heat due to the viscous friction and ionic relaxation that occurs as the wave generated by an acoustic signal propagates outwards and this loss varies linearly with range as follows:

$$PL_{absorption}(r, f) = 10log(\alpha(f)) * r \qquad dB \tag{4}$$

where r is range in kilometres and α is the absorption coefficient.

More specifically the absorption of sound in seawater is caused by three dominant effects; viscosity (shear and volume), ionic relaxation of boric acid and magnesium sulphate ($MgSO_4$) molecules and the relaxation time. The effect of viscosity is significant at high frequencies above 100 kHz, whereas the ionic relaxation effects of magnesium affect the mid frequency range from 10 kHz up to 100 kHz and boric acid at low frequencies up to a few kHz. In general, the absorption coefficient, α, increases with increasing frequency and decreases as depth increases (Domingo, 2008; Sehgal et al., 2009) and is significantly higher in the sea compared with fresh water due predominately to the ionic relaxation factor.

Extensive measurements of absorption losses over the last half century has lead to several empirical formulae which take into account frequency, salinity, temperature, pH, depth and speed of sound. A popular version is Thorp's expression (Thorp, 1965), Equation 5, which is based on his initial investigations in the 60's and has since been converted into metric units (shown here). It is valid for frequencies from 100Hz to 1MHz and is based on seawater with salinity of 35% ppt, pH of 8, temp of 4°C and depth of 0 m (atmospheric pressure) which is assumed but not stated by Thorp.

$$\alpha(f) = \frac{0.11f^2}{1+f^2} + \frac{44f^2}{4100+f^2} + 2.75\times10^{-4}f^2 + 0.0033 \qquad dB/km \qquad (5)$$

Fisher and Simmons (1977) and others (Francois & Garrison, 1982) have since proposed other variations of α. In particular, Fisher and Simmons in the late 70's found the effect associated with the relaxation of boric acid on absorption and provided a more detailed form of absorption coefficient α in dB/km which varies with frequency, pressure (depth) and temperature (also valid for 100 Hz to 1 MHz with salinity 35% ppt and acidity 8 pH)(Fisher & Simmons, 1977; Sehgal et al., 2009), given in Equation 6.

$$\alpha(f,d,t) = \frac{A_1 f_1 f^2}{f_1^2+f^2} + \frac{A_2 P_2 f_2 f^2}{f_2^2+f^2} + A_3 P_3 f^2 \qquad dB/km \qquad (6)$$

where d is depth in meters and t is temperature in °C. The 'A' coefficients represent the effects of temperature, while the 'P' coefficients represent ocean depth (pressure) and f_1, f_2 represent the relaxation frequencies of Boric acid and ($MgSO_4$) molecules. These terms were developed by Fisher and Simmons (1977) and presented more recently by (Domingo, 2008; Sehgal et al., 2009).

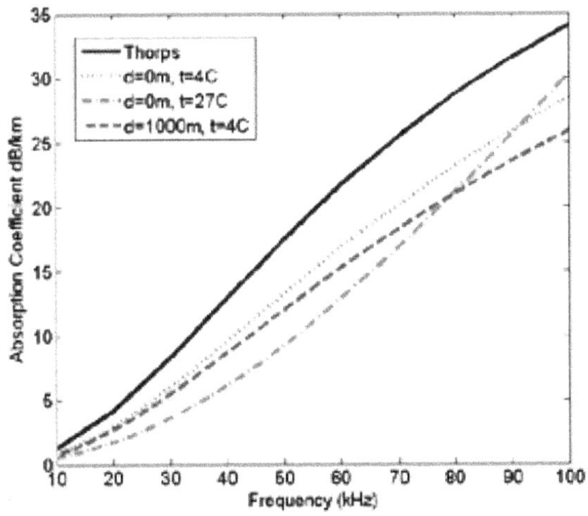

Fig. 4. Absorption Coefficient vs Frequency

Figure 4 shows the absorption coefficients in dB/km vs signal frequency for both Thorp and Fisher and Simmons coefficients and shows that in general α increases with increasing frequency at any fixed temperature and depth. Up until around 80kHz temperature change has a more significant affect on α than depth (Waite, 2005), but above these frequencies depth begins to dominate (Domingo, 2008; Sehgal et al., 2009). In any case, Thorps 'approximation' is quite close to Fisher and Simmons and is clearly more conservative at the frequencies shown. Sehgal (2009) shows that at higher frequencies above 300kHz, Thorps model predicts lower losses as it does not take into account the relaxation frequencies found by Fisher and Simmons.

If depth and frequency are fixed and temperature varied from 0 to 27°C, there is a decrease in α of approximately 4 dB/km for frequencies in the range of 30 to 60kHz which correlates to work presented by Urick (Urick, 1967). If we consider where AUV swarms are most likely to operate, in the 'mixed surface layer', where temperature varies considerable due to latitude (but has an average temp of 17°C (Johnson, 2011), temperature may be an important factor. It should be noted that if operating in lower temperatures α is higher and thus using 0°C will be a conservative alternative. At shorter ranges, the significance of α is expected to be less due to the linear relationship with range which will be discussed further in this chapter.

As mentioned, depth (pressure) has less of an effect on α than temperature at these lower frequencies. Domingo (Domingo, 2008) investigates the effect of depth (pressure) on absorption and confirms that for lower frequencies of less than 100kHz there is less change in α. More specifically Urick (1967) defined the variation by: $\alpha_d = \alpha * 10^{-3}(1 - 5.9 * 10^{-6}) * d$ dB/m (where d = depth in meters) but has also suggested an approximation of a 2% decrease for every 300 m depth. Thus, depth (pressure) variations are not expected to play a significant role in short range AUV swarm operations especially those that use a 2D horizontal topology as described in this chapter.

3.3 Path loss

Total path loss is the combined contribution of both the spreading and absorption losses. Urick (1967) established that this formula of spreading plus absorption yields a reasonable agreement with long range observations.

$$PathLoss(r,f,d,t) = k * 10log(r) + \alpha(f,d,t) * r * 10^{-3} \quad (7)$$

For very short range communication (below 50 m), see Figure 5(a), the contribution of the absorption term is less significant than the spreading term. It can be seen in Figure 5(b) that the Thorp model shows a conservative or worst case value for the ranges of interest up to 500 m. The Fisher and Simmons model for a particular frequency however provides some insight into the variations due to depth and temperature. However, the spreading factor k has the most significant affect on Path Loss, seen in Figure 5(a), at these shorter ranges according to these models.

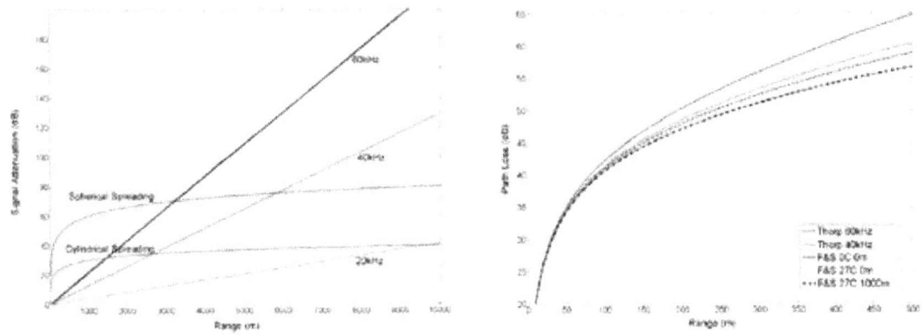

(a) Signal Attenuation showing spherical spreading and absorption factors

(b) Comparing Absorption Models using spherical spreading. Frequency change shown using Thorp Model and Temperature °'C' and Depth 'm' changes shown in Fisher and Simmons Model.

Fig. 5. Path Loss vs Range

As range increases and the absorption term begins to dominate, any variations in α also becomes more significant. For data communication, the changes in the attenuation due to signal frequency are particularly important as the use of higher frequencies will potentially provide higher data rates.

In summary using the two models, Thorp and Fisher and Simmons, the two important characteristics that can be drawn from Path Loss at the short ranges of interest for AUV swarm operation are:

- spreading loss dominates over absorption loss, and thus the 'k' term has a significant impact on the attenuation of the signal at shorter ranges as illustrated in Figure 5 (a). For AUV swarm operations and while the range between vehicles is much less than depth spherical spreading can be assumed, and
- at the ranges below 500 m the frequency component of absorption loss is most significant compared with the possible temperature and pressure (depth) changes as seen in Figure 5(b) and as range increases the difference also increases, effectively meaning that the communication channel is band-limited and available bandwidth is a decreasing function of range.

3.4 Characteristics of underwater multipath

Multipath signals, in general, represent acoustic energy loss, however, for communication systems it is the Inter Symbol Interference (ISI) that will also be detrimental at the receiver as it can significantly increase the error rate of the received signal. Multipath signals are created underwater through various mechanisms described in this section, so that, at the receiver many components of the original signal will arrive at different times due to the different length of propagation paths the multipath signals have taken. It is this delay spread of the signal component arrivals that can cause ISI to occur if they overlap with previous or future signal arrivals which will cause symbol corruption or loss and therefore bit errors. As the speed of sound propagation is very slow in an acoustic channel this delay spread can be significant.

There are two major mechanisms responsible for creating multi-path signals and these are: reverberation, which refers to the reflections and scattering of the sound signal; and ray bending, which is a result of the unique sound speed structure in the oceans which create temperature gradient channels that trap acoustic signals. Multi-path signal formation is therefore determined by the geometry of the channel in which transmission is taking place, the location of the transmitter and receiver, and importantly the depth at which it is occurring. In shallow water, multi-path is due predominately to reverberation whereas in deep water it is dominated by ray bending, although reverberation will occur in deep water if the transmitter and receiver are located near the surface or bottom (Coates, 1989; Domingo, 2008; Etter, 2003; Urick, 1967).

There are several physical effects which create reverberation underwater;

- Multi-path propagation caused by boundary reflections at the sea-floor or sea-surface, seen in Figure 2.
- Multi-path propagation caused by reflection from objects suspended in the water, marine animals or plants or bubbles in the path of the transmitted signal
- Surface scattering caused by sea-surface (waves) or sea-floor roughness or surface absorption, particularly on the sea bottom depending on material
- Volume scattering caused by refractive off objects suspended in the signal path

Ray bending, causes various propagation path loss mechanisms in deep water depending on the placement of transmitter and receivers. The propagating acoustic signal bends according to Snell's Law, to lower signal speed zones. Figure 6 shows a typical ocean Sound Speed Profile, although variations occur with location and seasons. The profile is depth dependent, where sound speed is influenced more by temperature in the surface layers and by pressure at greater depths.

The various path loss mechanisms include; (Domingo, 2008)

- Surface duct, Figure 7(a) occurs when the surface layer has a positive temperature gradient, the acoustic signals can bend back towards the surface, then reflect back into the layer off the surface

- Deep Sound Channel, sometimes referred to as the SOFAR (Sound Fixing and Ranging) channel, where acoustic propagation occurs above and below the level of minimum sound speed, when the sound rays continually are bent towards the depth of minimum speed, shown in Figure 7(a)
- Convergence zone, in deep water areas when the transmitter is located quite close to the surface and the sound rays bend downwards as a result of decreasing temperatures until the increase in pressure forces the rays back towards the surface, as shown in Figure 7(b)

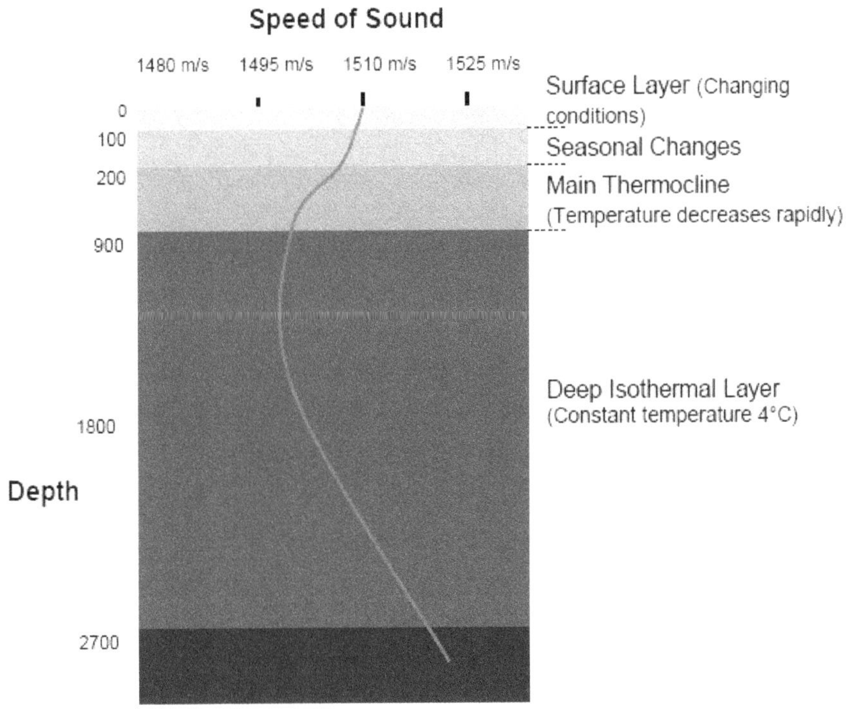

Fig. 6. Typical Sound Speed Profile in the Ocean Cox (1974)

Reliable acoustic path, which occurs when the transmitter is located in very deep water and receiver in shallow water. Referred to as reliable as it is not generally affected by bottom or surface reflections, as shown in Figure 7(b) and Shadow zones that are considered a special case, as these 'zones' are void from any signal propagation. This means that in Shadow zones a hydrophone may not receive any signal at all.

Thus the geometry of the channel being used is a major determinate of the number of significant propagation paths and their relative strengths and delays. Apart from the Shadow Zones where no signal or multipath components of the signal can reach the hydrophone, the hydrophone may receive the direct signal and a combination of various multipath signals that have been reflected, scattered or bent. It is these multiple components of the signal that are delayed in time due to the various path lengths that may create ISI and errors in symbol detection.

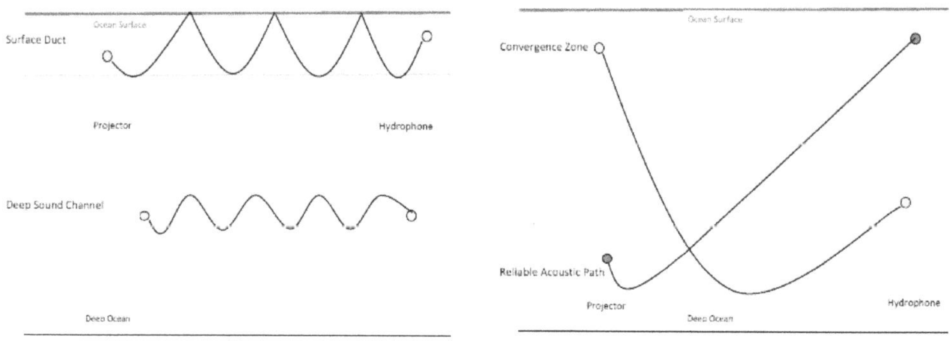

(a) Surface Duct and Deep Sound Channel (b) Convergence Zone and Reliable Acoustic Path

Fig. 7. Ray Bending Path Loss Mechanisms

For very short range channels that will be used in AUV swarm operations, multipath will be influenced also by the range-depth ratio, which is expected to produce fewer multipath signals at the hydrophone (Hajenko & Benson, 2010; Parrish et al., 2007). In addition some improvement can be gained through directing the beam of the transmitted signal and the directional properties of the receiver (Essebbar et al., 1994), however this will require an additional level of complexity for mobile AUV's due to the need for vehicle positioning before sending or when receiving a signal.

Most of the discussion so far has focused on time-invariant acoustic channel multipath where deterministic propagation path models have been developed for the various reflective and ray bending path options. These are significant in themselves with multipath spreads in the order of 10 to 100 ms. Take Figure 2, where projector and hydrophone are separated by 100m and are at a depth of 100m, the delay spread between the direct path and the first surface reflection is ≈ 28 ms. Multipath in an underwater channel, however, also has time-varying components caused by the surface or volume scattering or by internal waves in deep water that are responsible for random signal fluctuations. Unlike in radio channels, the statistical characterisation of these random processes in the underwater channel are in their early development stages. Experimental results have shown that depending on the day, the location and the depth of communication link, the results of multipath can follow one of the deterministic models discussed here to worst case coherence times in the order of seconds(Stojanovic, 2006). Another source of time variability in an underwater communication channel occurs when there is relative motion between the transmitter and receiver as will be briefly discussed in the following sub-section.

3.5 The doppler effect

The motion of AUV's relative to each other will cause two possible forms of Doppler distortion in the received signal, Doppler Shifting caused by an apparent shift in frequency as the vehicles move towards or away from each other and Doppler Spreading or its time domain dual coherence time, which is the measure of the time varying nature of the frequency dispersiveness in the doppler spectrum (Rappaport, 1996). The doppler shift (Δf) of a received signal is $f_c \frac{\Delta v}{c}$ where f_c is the original signal frequency and Δv is the relative velocity between the moving vehicles. As an example, if the vehicles were moving at a moderately slow speed of 1 m/s (2 knots) relative to each other and f_c = 40kHz the $\Delta f \approx$ 27Hz. Doppler spread or coherence time measurements as mentioned above can be as long as 1 s. Thus doppler shifting and spreading cause complications for the receiver to track the time varying changes in the channel which need to be designed into the channel estimation algorithms and explicit delay synchronisation approach within communication protocols. As swarm operations for exploration require rigid topology where there is minimal relative speed differences between vehicles, the impact of doppler effects diminished somewhat in this context and thus will not be considered further.

3.6 Noise

Three primary sources of underwater noise are: ambient oceanic noise, vehicle self-noise, and intermittent noises, which encompass biological sounds like snapping shrimp, ice cracking, and rainfall. An accurate noise model is essential for assessing the SNR at the hydrophone, enabling the determination of bit error rates (BER) to evaluate protocol performance.

3.6.1 Ambient noise

Ambient Noise in the ocean has been well defined (Urick, 1967). It can be represented as Gaussian and having a continuous power spectrum density (psd). It is made up of four components (outlined below), each having a dominating influence in different portions of the frequency spectrum.

For the frequency region of interest for AUV swarm communication systems (10 kHz to 100 kHz), the ambient noise psd decreases with increasing frequency, refer to Figure 8. At a frequency over 100kHz the ambient thermal noise component begins to dominate and the overall noise psd begins to increase, but this point moves further away from the frequencies of interest for AUV communication particularly as the wind speed increases.

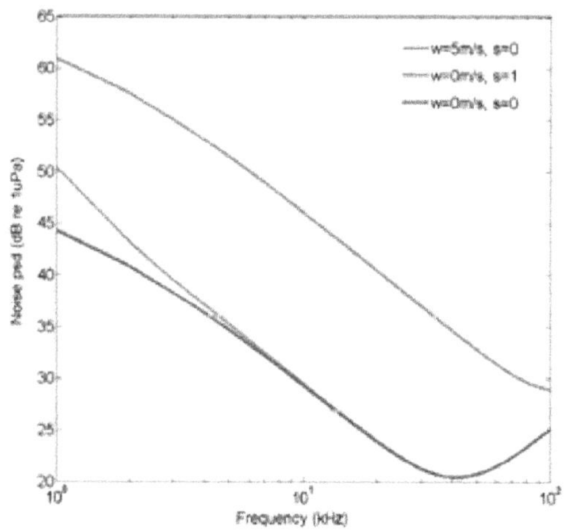

Fig. 8. Power Spectral density of the Ambient Noise, W - wind, S - shipping

Turbulence noise influences only the very low frequency regions $f < 10Hz$ $10logN_{turb}(f) = 17 - 30log(f)$

Shipping noise dominates the 10 - 100Hz region and has defined a shipping activity factor of s whose value ranges from 0 to 1 for low to high activity respectively: $10logN_{ship}(f) = 40 + 20(s - 0.5) + 26log(f) - 60log(f + 0.03)$

Wave and other surface motion caused by wind and rain is a major factor in the mid frequency region of 100Hz - 100kHz where wind speed is given by w in m/s: $10logN_{wind}(f) = 50 + 7.5w^{1/2} + 20log(f) - 40log(f + 0.4)$

Thermal noise becomes dominate over 100kHz: $10logN_{th}(f) = -15 + 20log(f)$

where wind speed is given by w in m/s (1m/s is approximately 2 knots) and f is in kHz. Ambient Noise power also decreases with increasing depth as the distance from the surface and therefore shipping and wind noise becomes more distant. Ambient noise has been shown to be 9dB higher in shallow water than deep water (Caruthers, 1977). Swarm operations and other underwater networking activities will necessitate that communication nodes, including AUVs, operate in close proximity to one another, thereby introducing an additional layer of ambient noise from the other vehicles in the swarm, regardless of the operating depth. The forthcoming section on Self Noise will address the idea that this supplementary 'ambient noise,' associated with the 'Shipping Noise' component, will have a little impact on auditory communication, which typically operates at frequencies beyond 10 kHz.

3.6.2 Self noise

Self noise is defined as the noise generated by the vehicle itself as the platform for receiving signals. This noise can reach the hydrophone mounted on the AUV either through the mechanical structure or through the water passing over the hydrophone. The degree to which turbulent flows cause transducer self noise depends on the location (mounting) of the transducer and its directivity characteristics (Sullivan & Taroudakis, 2008). Self Noise can also be seen as an equivalent isotropic noise spectrum as presented by Urick from work done during WWII on submarines. In general, as with ambient noise, there is decreasing levels of self noise with increases in frequency however self noise is also significantly affected by speed with decreasing noise spectra when the vessels are travelling at slower speeds or are stationary (Eckart, 1952; Kinsler et al., 1982; Urick, 1967).

Kinsler (1982) notes that at low frequencies (<1kHz) and slow speeds machinery noise dominates and at very slow speeds self noise is usually less important than ambient noise. Whereas at higher frequencies (10kHz) propeller and flow noise begins to dominate and as speed increases the hydrodynamic noise around the hydrophone increases strongly and becomes more significant than the machinery noise. This is due to the cavitation from the propeller due to the entrainment of air bubbles under or on the blade tip of the propeller. At higher speeds, self noise can be much more significant than ambient noise and can become the limiting factor.

The self noise of different size and types of vehicles are as varied as there are vehicle designs and there is little recent published values. Each vehicle itself produces large variations in self noise with speed and operating conditions (Eckart, 1952). Self noise can be controlled by selection of motor type, configuration, mounting and motor drivers. The trend for most AUV's will be the use of small brush-less DC electric motors which have been used on the development of the SeaVision vehicle (Mare, 2010). Preliminary testing of self noise on these vehicles shows an increase in noise due to increases in speed, as has been predicted, but there was no way to distinguish between machinery and hydrodynamic affects. Higher frequency components (up to 20kHz) were present as the speed increased due to the increased work from the thrusters. When the SeaVision vehicle hovered in a stationary position the frequency of the noise psd centred around 2kHz, which is out of band noise.

Holmes (Holmes et al., 2005) at WHOI recently investigated the self noise of REMUS, their torpedo shape AUV, used as a towed array. At the maximum RPM of the AUV, the 1/3rd octave noise level, when converted to source level by the calibrated transmission factor, was 130 dB re 1μPa at 1m directly aft of the vehicle for a centre frequency of 1000 Hz. This would represent the radiated noise source level for a vehicle moving at 3 knots (1.5m/s). Vehicles typically radiate less noise in free operating conditions than in tethered conditions, so the second trail on the REMUS was measuring the radiated noise of the vehicle to examine the power spectral density of the noise as recorded on the hydrophone array as it was towed behind the vehicle. The results showed the RPM dependent radiated noise in the aft direction at a distance of 14.6 m behind the vehicle looking at frequency range up to 2500Hz which is again out of band noise.

Given that the operating frequencies of the communication system are anticipated to exceed the majority of self-noise, and considering the relatively sluggish operation of the vehicles, the projected impact of self-noise on hydrophone reception will be minimal.

3.6.3 Intermittent sources of noise

The origins of intermittent noise can become highly consequential in areas or periods when they coincide with the operation of AUV swarms. Research has mostly focused on marine bioacoustics and the impact of rain and water bubbles generated by raindrops.

Major contributors to underwater bio-acoustic noise include;

- Shellfish - Crustacea - most important here are the snapping scrimp who produce a broad spectrum of noise between 500Hz and 20kHz
- Fish - toadfish 10 to 50Hz
- Marine mammals - Cetacea - porpoises 20 to 120Hz

Rain creates different noise spectrum to wind and needs to be dealt with separately as it is not a constant source of noise. Urick (1967) showed examples of increases of almost 30dB in the 5 to 10kHz portion of spectrum in heavy rain, with steady rain increasing noise by 10dB or sea state equivalent increase from 2 to 6. Eckart (1952) presented average value of rain at the surface from 100Hz to 10kHz of -17 to 9dB.

These main contributors to intermittent sources predominate in the lower frequency ranges up to 20kHz. Thus, interference in the operating frequencies of communication data signals is considered low.

4. SHORT RANGE CHANNEL MODELLING

Maximizing the capacity of the underwater acoustic channel is crucial due to its demanding and restricted resources, as previously described. Short-range data transmission operations offer several advantages compared to existing long-range underwater sound transmission methods. This will now be further examined regarding the design and development of data communication protocols. The evaluation will focus on determining the optimal signal frequency and bandwidth across various ranges and channel conditions, utilizing the highest possible Signal-to-Noise Ratio (SNR) at the hydrophone. An analysis of channel capacity and Bit Error Rate (BER) for various modulation schemes will be conducted to establish the context for the issues in Medium Access Control (MAC) and routing protocol design.

4.1 Frequency dependent component of SNR

The narrowband Signal-to-Noise-Ratio (SNR) observed at the receiver, assuming no multi-path or doppler losses is given by:

$$SNR(r, f, d, t, w, s, P_{tx}) = \frac{SL_{projector}(P_{tx}, \eta, DI)}{PathLoss(r, f, d, t) \sum Noise(f, w, s) \times B} \tag{8}$$

where B is the receiver bandwidth and the Signal Level (SL), PathLoss and Noise terms have been previously developed.

Taking the frequency dependent portion of the SNR from Equation 8, as developed by Stojanovic (2006), is the *PathLoss(r, f , d, t)* Σ *Noise(f , w, s)* product. Since SNR is inversely d,t proportional to *PathLoss(r, f , d, t)* Σ *Noise(f , w, s)* factor, the $\frac{1}{PathLoss(r,f,d,t)Noise(f,w,s)}$ term is illustrated in the Figures 9 (a) for longer ranges and Figure 9 (b) for shorter ranges. The first of these figures, shows various ranges up to 10km using Thorps absorption model (spherical spreading) and has been presented by several authors (Chen & Mitra, 2007; Nasri et al., 2008; Sehgal et al., 2009; Stojanovic, 2006). Figure 9(b) highlights shorter ranges of 500 m and 100 m and also illustrates the variation between the Thorp and Fisher and Simmons absorption loss models developed in Section 3.2.2.

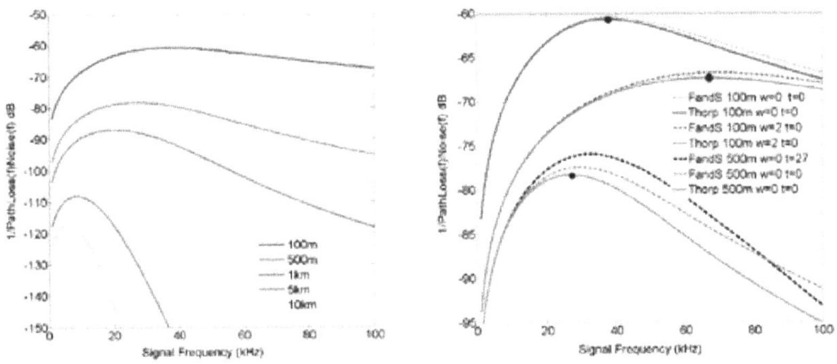

(a) Longer Ranges

(b) Short Ranges with channel parameter variations

Fig. 9. Frequency Dependent component of narrowband SNR

These figures show that there is a signal frequency where the frequency dependent component of SNR is optimised assuming that the projector parameters, including transmitter power and projector efficiency behave uniformly over the frequency band. The black dot at the apex of each of the three curves based on the Thorp absorption model indicate this optimum point. The two absorption models present similar responses and optimum frequencies. There is a minor variation in optimum frequency at 100 m where the absorption coefficient has a significantly lower contribution. The impact that the Ambient Noise component has on optimum signal frequency is seen most dramatically at the 100 m range when the wind state is changed from 0 to 2 m/s, the optimal signal frequency changes from 38 kHz to 68 kHz. From a communication perspective, if two vehicles were operating at 100 m and 38kHz and the wind state changed from 0 m/s to 2 m/s, there is a reduction of 9dB in the frequency dependent component of SNR. This is not an absolute reduction in SNR as the projector parameters and in particular the transmitter power level has not been considered here. It does however indicate the significant impact that wind and wave action can play with data communication underwater, and in addition this reduced SNR value does not include any increased losses associated with the increase scattering that wave action can generate. The impact of shipping, found in the Ambient Noise term, is not included here as its effect is minor on signal frequencies above 10kHz. In Section 3.2.2 Figure 4, temperature variations were seen to have the most impact on the signal frequency associated with the ranges of interest. Figure 9(b) illustrates this difference in terms of the frequency dependent component of SNR. This is further explored in Figure 10(a) in terms of the signal frequency variations over range.

The impact of changes in range can be seen if the vehicles moved from 100 m to 500 m (at wind state 0 m/s), the optimum signal frequency to maintain highest SNR decreases from 38 kHz to ≈ 28 kHz, Figure 9(b). Reduction in signal frequency implies a potential reduction in absolute bandwidth and with that a reduction in data rate which needs to be managed. This will be investigated further in the next sub sections. Figure 10 (a) and (b) show the optimum signal frequency verses range up to 500 m for the various parameters; temperature and depth, within the Thorp and Fisher and Simmons Absorption Loss models as well as the wind in the Ambient Noise model. The optimum frequency, decreases with increasing range due to the dominating characteristic of the absorption loss. It can be seen in Figure 10(a) that as the range increases there is an increasing deviation between the two models and between the parameters within the Fisher and Simmons model. There is approximately a 2.5 kHz difference between the models themselves at 500 m and up to 6 kHz when temperature increases are included. When wind is included, Figure 10(b), there is a dramatic change in optimum signal frequency at very short ranges and this difference reduces substantially over the range shown. This is due to the increasing significance of the Absorption Loss term relative to the constant Ambient Noise term (as it is not range dependent), which reduces the affect of the Noise term and therefore the wind parameter. In both Figure 10(a) and (b), the Fisher and Simmons model provides higher optimum frequencies due to the more accurate inclusion of the relaxation frequencies of boric acid and magnesium sulphate.

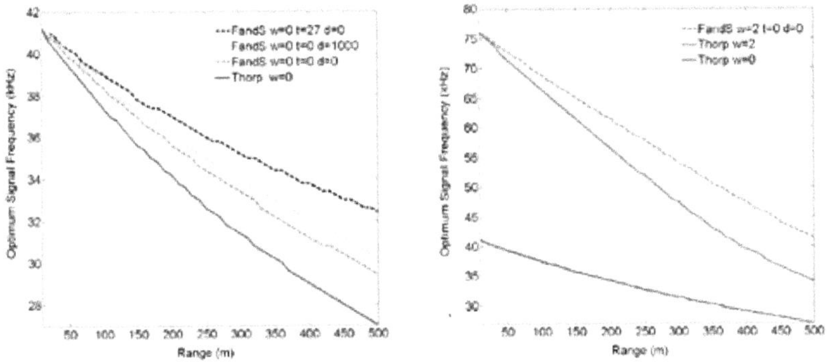

(a) Comparison of Absorption Loss Parameters (b) Comparison with changes in wind (from Ambient Noise Characteristics)

Fig. 10. Optimum Frequency determined from frequency dependent component of narrowband SNR

4.2 Channel bandwidth

Having established that at different ranges there is an optimum signal frequency that provides a maximum SNR, assuming constant transmitter power and projector efficiency, there is therefore an associated channel bandwidth with these conditions for different ranges. To determine this bandwidth a heuristic of ±3dB around the optimum frequency is used. Following a similar approach to Stojanovic (2006) the bandwidth is calculated according to the frequency range using 3dB around the optimum signal frequency $f_0(r)$ which has been chosen as the centre frequency. Therefore, the $f_{min}(r)$ is the frequency when $PathLoss(r, d, t, f_0(r))N(f_0(r)) - PathLoss(r, d, t, f))N(f) \geq 3dB$ holds true and similarly for $f_{max}(r)$ when $PathLoss(r, d, t, f))N(f) - PathLoss(r, d, t, f_0(r))N(f_0(r)) \geq 3dB$ is true. The system bandwidth B(r,d,t) is therefore determined by:

$$B(r,d,t) = f_{max}(r) - f_{min}(r) \quad (9)$$

Consequently, for a specified range, an ideal frequency exists from which a range-dependent 3dB bandwidth can be ascertained, as demonstrated in Figure 11. Section 4.1 addresses modifications in the optimal signal frequency corresponding to variations in range and channel conditions, including temperature, depth, and wind. These fluctuations are mirrored in the alterations observed in channel bandwidth, which will subsequently influence the prospective data transmission speeds. Figure 11 illustrates that both the ideal signal frequency and the 3dB channel bandwidth diminish as range rises. The influence of fluctuating wind conditions on channel bandwidth is substantial; but, as previously mentioned, wind and wave dynamics will also encompass time-variant complications and losses not addressed herein. Temperature elevations result in enhanced channel bandwidth within relevant ranges, attributed to less absorption loss with rising temperatures, hence providing advantages in operating within surface layers. This discussion emphasizes that the underwater acoustic channel is significantly band-limited, necessitating bandwidth-efficient modulation to optimize data throughput. Additionally, substantial advantages can be realized when conducting data transmission over shorter distances or in multi-hop configurations.

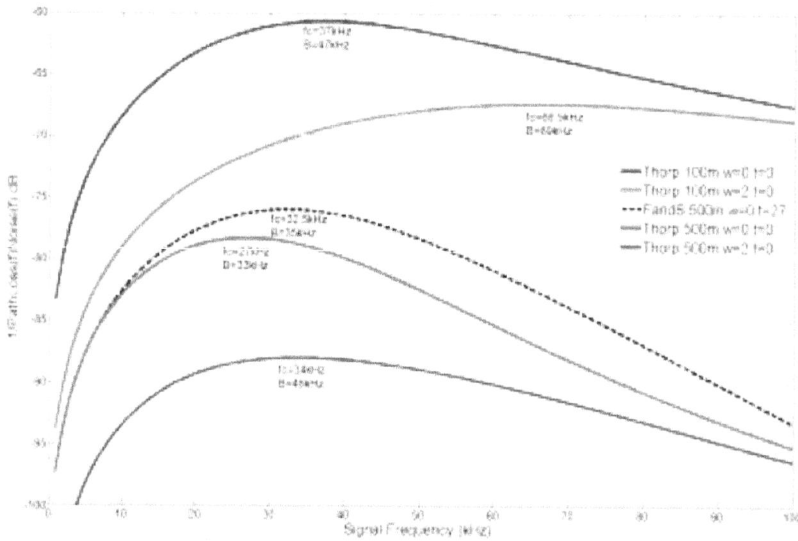

Fig. 11. Range dependent 3dB Channel Bandwidth shown as dashed lines. Where Y-axis is the frequency dependent component of the narrowband SNR

4.3 Channel capacity

Prior to evaluating the more realistic performance of the underwater data communication channel, the maximum achievable error-free bit rate C for various ranges of interest will be determined using the Shannon-Hartley expression, Equation 10. In these channel capacity calculations, all the transmitted power P_{tx} is assumed to be transferred to the hydrophone except for the losses associated with the deterministic Path Loss Models developed earlier.

The Shannon-Hartley expression using the Signal-to-Noise ratio, SNR(r), defined in Equation 8, is:

$$C = Blog_2(1 + SNR(r)) \quad (10)$$

where C is the channel capacity in bps and B is the channel bandwidth in Hz

Thus using the optimum signal frequency and bandwidths at 100 m and 500 m found in the Section 4.1 and 4.2, the maximum achievable error free channel capacities against range are shown in Figure 12. The signal frequency and channel bandwidth values for 100 m were f_o = 37kHz and B=47kHz and for 500 m were f_o =

27kHz and B=33kHz. These are significantly higher than values currently available in underwater operations(Walree, 2007), however they provide an insight into the theoretical limits. Two different transmitter power levels are used, 150dB re 1μPa which is approximately 10mW (Equation 1) and 140dB re 1 μ Pa is 1mW. Looking at the values associated with the same power level in Figure 12, the higher channel capacities are those associated with the determined optimum frequency and bandwidth for that range as would be expected. The change in transmitter power, however, by a factor of 10, does not produces a linear change in channel capacity across the range. These variations are important to consider as minimising energy consumption will be critical for AUV operations. In general, current modem specifications indicate possible data rate capacities of less than 10kbps (LinkQuest, 2008) for modem operations under 500 m, well short of these theoretical limits. This illustrates the incredibly severe data communication environment found underwater and that commercial modems are generally not yet designed to be able to adapt to specific channel conditions and varying ranges. The discussion here is to understand the variations associated with the various channel parameters at short range that may support adaptability and improved data transmission capacities.

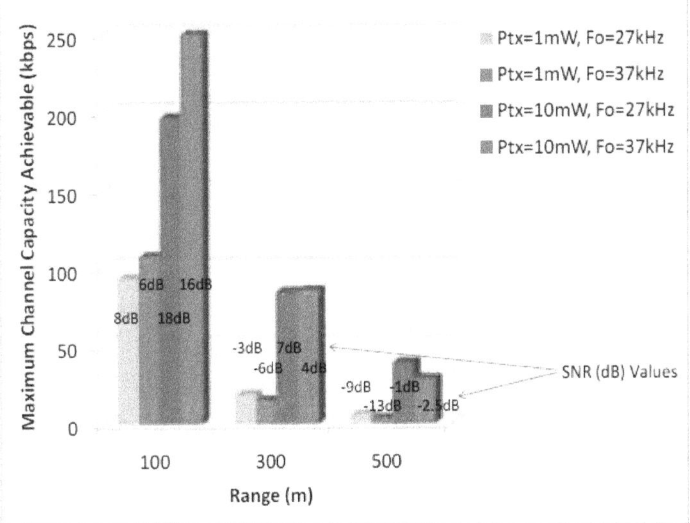

Fig. 12. Theoretical limit of Channel Capacity (kbps) verse Range

4.4 BER in short range underwater acoustic communication

Achieving close to the maximum channel capacities as calculated in the previous section is still a significant challenge in underwater acoustic communication. The underwater acoustic channel presents significant multipaths with rapid time-variations and severe fading that lead to complex dynamics at the hydrophone causing ISI and bit errors. The probability of bit error, BER, therefore provides a measure of the data transmission link performance. In underwater systems, the use of FSK (Frequency Shift Keying) and PSK (Phase Shift Keying) have occupied researchers approaches to symbol modulation for several decades. One approach is using the simpler low rate incoherent modulation frequency hopping FSK signalling with strong error correction coding that provides some resilience to the rapidly varying multipath. Alternatively, the use of a higher rate coherent method of QPSK signalling that incorporates a Doppler tolerant multi-channel adaptive equalizer has gained in appeal over that time (Johnson et al., 1999).

The BER formulae are well known for FSK and QPSK modulation techniques (Rappaport, 1996), which require the Energy per Bit to Noise psd, $\frac{E_b}{N_o}$, that can be found from the SNR (Equation 8) by:

$$\frac{E_b}{N_o} = SNR(r) \times \frac{B_c}{R_b} \qquad (11)$$

where R_b is the data rate in bps and B_c is the channel bandwidth. Equation 12 and 13 are the uncoded BER for BPSK/QPSK and FSK respectively:

$$QPSK: \quad \text{BER} = \frac{1}{2} erfc[\frac{E_b}{N_o}]^{1/2} \qquad (12)$$

$$FSK: \quad \text{BER} = \frac{1}{2} erfc[\frac{1}{2}\frac{E_b}{N_o}]^{1/2} \qquad (13)$$

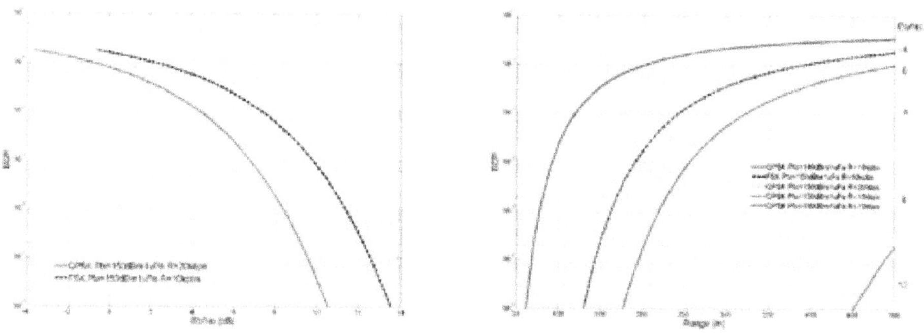

(a) BER vs $\frac{E_b}{N_o}$ (b) BER vs Range vs $\frac{E_b}{N_o}$ (for QPSK)

Fig. 13. Probability of Bit Error for Short Range Acoustic Data Transmission Underwater

The data rates R_b used are 10 and 20 kbps to reflect the current maximum commercial achievable levels. Figure 13 (a) and (b) show the BER for $\frac{E_b}{N_o}$ and Range respectively. Taking a BER of 10^{-4} or 1 bit error in every 10, 000 bits, the $\frac{E_b}{N_o}$ required for QPSK is 8dB for a transmitter power of 10mW and a data rate of 20kbps. This increases to 12dB if using FSK with half the data rate (10 kbps) and same Transmitter Power.

From Figure 13 (b), these settings will provide only a 150 m range. The range can be increased to 250 m using QPSK if the data rate was halved to 10 kbps or out to 500 m if the transmitter power was increased to 100mW in addition to the reduced data rate. Transmitter power plays a critical role, as illustrated here, by the comparison of ranges achieved from ≈ 75 m to 500 m with a change of transmitter power needed from 1mW to 100mW for this BER.

5. SWARM NETWORK PROTOCOL DESIGN TECHNIQUES

A short range underwater network, as shown in Figure 1(b) is essentially a multi-node sensor network. To develop a functional sensor network it is necessary to design a number of protocols which includes MAC, DLC (Data Link Control) and routing protocols. A typical protocol stack of a sensor network is presented in Figure 14. The lowest layer is the physical layer which is responsible for implementing all electrical/acoustic signal conditioning techniques such as amplifications, signal detection, modulation and demodulation, signal conversions, etc.. The second layer is the data link layer which accommodates the MAC and DLC protocols. The MAC is an important component of a sensor networks protocol stack, as it allows interference free transmission of information in a shared channel. The DLC protocol includes the ARQ (Automatic Repeat reQuest) and flow control functionalities necessary for error free data transmission in a non zero BER transmission environment. Design of the DLC functionalities are very closely linked to the transmission channel conditions. The network layers main operational control is the routing protocol; responsible for directing packets from the source to the destination over a multi-hop network. Routing protocols keep state information of all links to direct packets through high SNR links in order to minimise the end to end packet delay. The transport layer is responsible for end to end error control procedures which replicates the DLC functions but on an end to end basis rather than hop to hop basis as implemented by the DLL.

Application
Transport
Network
Data Link Layer (DLL)
Physical

Fig. 14. A typical protocol stack for a sensor network

The transport layer could use standard protocols such as TCP (Transmission Control Protocol) or UDP (User Datagram Protocol). The application layer hosts different operational applications which either transmit or receive data using the lower layers. To develop efficient network architectures, it is necessary to develop network and/or application specific DLL and network layers. The following subsections will present MAC and routing protocol design characteristics required for underwater swarm networking.

5.1 MAC protocol

Medium access protocols are used to coordinate the transmission of information from multiple transmitters using a shared communication channel. MAC protocols are designed to maximise channel usage by exploiting the key properties of transmission channels. MAC protocols can be designed to allocate transmission resources either in a fixed or in a dynamic manner. Fixed channel allocation techniques such as Frequency Division Multiplexing (FDM) or Time Division Multiplexing (TDM) are commonly used in many communication systems where ample channel capacity is available to transmit information (Karl & Willig, 2006). For low data rate and variable channel conditions, dynamic channel allocation techniques are generally used to maximise the transmission channel utilisation where the physical transmission channel condition could be highly variable. Based on the dynamic channel allocation technique it is possible to develop two classes of MAC protocols known as random access and scheduled access protocol. The most commonly used random access protocols is the CSMA (Carrier Sense Multiple Access) widely used in many networks including sensor network designs. Most commonly used scheduled access protocol is the polling protocol. Both the CSMA and polling protocols have flexible structures which can be adopted for different application environments. As discussed in this chapter, the underwater communication channel is a relatively difficult transmission medium due to the variability of link quality depending on location and applications. Also, the use of an acoustic signal as a carrier will generate a significant delay which is a major challenge when developing a MAC protocol. In the following subsection we discuss the basic design characteristics of the standard CSMA/CA protocol and its applicability for underwater applications.

5.1.1 CSMA/CA protocol

Carrier Sense Multiple Access with Collision Detection protocol is a distributed control protocol which does not require any central coordinator. The principle of this protocol is that a transmitter that wants to initiate a transmission, checks the transmission channel by checking the presence of a carrier signal. If no carrier signal is present which indicates the channel is free and the transmitter can initiate a transmission. For a high propagation delay network such a solution does not offer very high throughput due to the delay.

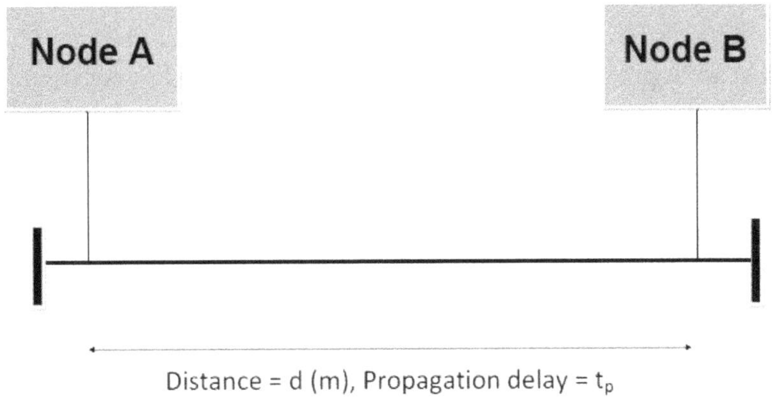

Fig. 15. CSMA/CA protocol based packet transmission example

Consider Figure 15, where two nodes are using CSMA/CA protocol, are spaced apart by 100 meters. In this case, if at t=0, Node A senses the channel then it will find the channel to be free and can go ahead with the transmission. If Node A starts transmission of a packet immediately then it can assume that the packet will be successfully transmitted. However, if Node B starts sensing the channel before the propagation delay time t_p then it will also find the channel is free and could start transmission. In this case both packet will collide and the transmission channel capacity will be wasted for a period of L+t_p where L is the packet transmission time. On the other hand, if Node B checks the channel after time t_p from the commencement of A's packet transmission, then it will find the channel is busy and will not transmit any packets. Now this simple example shows how the performance of random access protocol is dependent on the propagation delay. If propagation delay is small then there is much lower probability that a packet will be transmitted before the packet from A arrives at B. As the propagation delay increases the collision probability will also increase.

The CSMA/CA protocol is generally used in RF (Radio Frequency) networks where 100 m link delay will incur a propagation delay of 0.333 μ sec whereas an underwater acoustic link of same distance will generate a propagation delay of 0.29 sec which is about 875,000 times longer than the RF delay. One can easily see why an acoustic link will produce much lower throughput than is predicted by the Shannon-Hartley theorem as discussed in Section 4.3. If we assume that we are transmitting a 100 byte packet, then the packet will take about 0.08 sec to transmit on a 10 kbps RF link. The same packet will take 0.3713 sec on a 10 kbps acoustic link offering a net throughput of 2.154 kbps. This calculation is based on the assumption that the transmission channel is ideal i.e. BER=0. If the BER of the channel is non zero then the throughput will be further reduced.

Previous sections have shown that the BER of a transmission link is dependent on the link parameters, geometry of the application environment, modulation techniques, and presence of various noise sources. Non zero BER conditions introduce a finite packet error rate (PER) on a link which is described by Equation 14, where K represents the packet length. The PER will depend on the BER and the length of the transmitted packet. For a BER of 10^{-3} using a packet size of 100 bytes, the link will generate a PER value of 0.55 which means that almost every second packet will be corrupted and require some sort of error protection scheme to reduce the effective packet error rate.

There are generally two types of packet error correction techniques used in communication systems, one is forward error correction (FEC) scheme which uses a number of redundant bits added with information bits to offer some degree of protection against the channel error. The second technique involves the use of packet retransmission techniques using the DLC function known as the ARQ. The ARQ protocol will introduce retransmissions when a receiver is unable to correct a packet using the FEC bits. The retransmission procedure could effectively reduce the throughput of a link further because the same information is transmitted multiple times. From this brief discussion one can see that standard CSMA/CA protocols used in sensor networks are almost unworkable in the underwater networking environment unless the standard protocol is further enhanced. This is a major research issue which is currently followed up by many researchers and authors. Readers can find some of the current research work on the MAC protocol in the following references (Chirdchoo et al., 2008; Guo et al., 2009; Pompili & Akyildiz, 2009; Syed et al., 2008).

$$\mathrm{PER} = 1 - (1 - \mathrm{BER})^K \qquad (14)$$

5.2 Packet routing

Packet routing is another challenging task in the underwater networking environment. Packet routing protocols are very important for a multi-hop network because the receivers and the transmitters are distributed in a geographical area where nodes can also change their positions over time. Each node maintains a routing table to forward packets through multi-hop links. Routing tables are created by selecting the best cost paths from transmitters to receivers. The cost of a path can be expressed in terms of delay, packet loss, BER, real monetary cost $, etc.. For underwater networks, the link delay could be used as a cost metric, to transmit packets with a minimum delay. Routing protocols are generally classified into two classes: distance vector and link state routing protocols (LeonGarcia & Widjaja, 2004). The distance vector algorithms generally select a path from a transmitter to receiver based on shortest path through neighbouring networks. When the status of a link changes, for example, if the delay or SNR of a link is increased then the node next to the link will detect and inform its neighbour about the change and suggest a new link. This process will continue until all the nodes in the network have updated their routing table. The link state routing protocols work in a different manner. In this case all the link state information is periodically transmitted to all nodes in the network. In case of any change of state of a link, all nodes get notification and modify their routing table. In a swarm network link qualities will be variable which will require regular reconfiguration of routing tables. The performance of routing algorithms is generally determined by a number of factors including the convergence delay. In the case of a swarm network the convergence delay will be a critical factor because of high link delays. For underwater swarm applications, each update within a network will take considerably longer time than a RF network, causing additional packet transmission delays. Hence, it is necessary to develop the network structure in different ways than a conventional sensor network. For example, it may be necessary to develop smaller size clustered networks where cluster heads form a second tier network. Within this topology, local information will flow within the cluster and inter-cluster information will flow through the cluster head network. Cluster based communication architectures are also being used in Zigbee based and wireless personal communication networks (Karl & Willig, 2006). Further research is necessary to develop appropriate routing algorithms to minimise packet transmission delay in swarm networks. Readers can consult the following references to follow some of the recent progress in the area (Aldawibio, 2008; Guangzhong & Zhibin, 2010; LeonGarcia & Widjaja, 2004; Zorzi et al., 2008).

This section demonstrates that MAC and routing protocol designs necessitate transmission channel state information to enhance their performance. The considerable propagation delay of an underwater channel means that any alteration in link quality, such as signal-to-noise ratio (SNR), would substantially impact network performance. Therefore, it is essential to establish a novel class of protocols capable of adapting to fluctuating channel conditions while providing substantial throughput in swarm networks.

6. CONCISE SUMMARY

The expanding capabilities of Autonomous Underwater Vehicle (AUV) swarm operations and the prospect of employing multi-hop networking underwater have resulted in an escalating demand for utilizing a short-range acoustic communication channel. Comprehending the channel characteristics for data transmission is crucial for the formulation and assessment of novel MAC and Routing Level protocols that can more effectively utilize the constrained resources inside this challenging and unpredictable channel. The performance limitations of a communication system utilizing an acoustic channel include significant latency resulting from the slow propagation speed of acoustic signals (in contrast to RF) and signal degradation caused by absorption and multipath effects, especially reflections from surfaces, the seabed, and obstacles within the signal trajectory. The short-range acoustic channel is demonstrated to leverage decreased latency and transmitter power, alongside increased received SNR, signal frequencies, and bandwidths, albeit still within the kHz range. Each of these characteristics influences the methodology required for formulating suitable protocol designs and error control approaches while preserving the necessary network throughput and independent functioning of each node in the swarm. Substantial advantages will be realized when AUVs function as an intelligent swarm of cooperative nodes, achievable only through rapid and unambiguous communication among them inside an underwater short-range ad-hoc mobile sensor network.

REFERENCES

Aldawibio, O. (2008). A review of current routing protocols for ad hoc underwater acoustic networks, *First International Conference on the Applications of Digital Information and Web Technologies.* ICADIWT, pp. 431 –434.

Caruthers, J. (1977). *Fundamentals of Marine Acoustics,* Elsevier Scientific Publishing.

Chen, W. & Mitra, U. (2007). Packet scheduling for multihopped underwater acoustic communication networks, IEEE OCEANS'07, pp. 1–6.

Chirdchoo, N., Soh, W. & Chua, K. (2008). Ript: A receiver-initiated reservation-based protocol for underwater acoustic networks, IEEE *Journal on Selected Areas in Communications* 26(9): 1744 –1753.

Coates, R. (1989). *Underwater Acoustic Systems,* John Wiley and Sons.

Cox, A. W. (1974). *Sonar and Underwater Sound,* Lexington Books.

Domingo, M. (2008). Overview of channel models for underwater wireless communication networks, *Physical Communication* pp. 163 – 182.

Dunbabin, M., Roberts, J., Usher, K., Winstanley, G. & Corke, P. (2005). A hybrid auv design for shallow water reef navigation, Proc. *International Conference on Robotics and Automation (ICRA),* pp. 2117–2122.

Eckart, C. (1952). *Principles of Underwater Sound,* Research Analysis Group, National Research Council, California University.

Essebbar, A., Loubet, G. & Vial, F. (1994). Underwater acoustic channel simulations for communication, *IEEE OCEANS '94. 'Oceans Engineering for Today's Technology and Tomorrow's Preservation.'*, Vol. 3, pp. III/495 –III/500 vol.3.

Etter, P. (2003). *Underwater Acoustic Modeling and SImulation,* third edn, Spon Press.

Fisher, F. & Simmons, V. (1977). Sound absorption in sea water, *Journal of the Acoustical Society of America* 62(3).

Francois, R. & Garrison, G. (1982). Sound absorption based on ocean measurements: Part 1 and 2, *Journal of the Acoustical Society of America* 72(3,6): 896–907, 1879 - 1890.

Guangzhong, L. & Zhibin, L. (2010). Depth-based multi-hop routing protocol for underwater sensor network, 2nd *International Conference on Industrial Mechatronics and Automation (ICIMA),* Vol. 2, pp. 268 –270.

Guo, X., Frater, M. & Ryan, M. (2009). Design of a propagation-delay-tolerant mac protocol for underwater acoustic sensor networks, *IEEE Journal of Oceanic Engineering* 34(2): 170 –180.

Hajenko, T. & Benson, C. (2010). The high frequency underwater acoustic channel, *IEEE OCEANS 2010, Sydney,* pp. 1 –3.

Holmes, J., Carey, W., Lynch, J., Newhall, A. & Kukulya, A. (2005). An autonomous underwater vehicle towed array for ocean acoustic measurements and inversions, *IEEE Oceans 2005 - Europe,* Vol. 2, pp. 1058 – 1061 Vol. 2.

Johnson, M., Preisig, J., Freitag, L.& Stojanovic, M. (1999). FSK and PSK performance of the utility acoustic modem, IEEE OCEANS '99 MTS. *Riding the Crest into the 21st Century,* Vol. 3, pp. 1512 Vol. 3.

Johnson, R. (2011). *University Corporation of Atmospheric Research, Window to the Universe,* University Corporation of Atmospheric Research, http://www. windows. ucar. edu/ tour/link/earth/Water/overview.html.

Karl, H. & Willig, A. (2006). *Protocols and Architecures for Wireless Sensor Networks,* John Wiley and Sons, Ltd.

Kinsler, L., Frey, A., Coppens, A. & Sanders, J. (1982). *Fundementals of Acoustics,* John Wiley and Sons.

LeonGarcia, A. & Widjaja, I. (2004). *Communication Networks: Fundamental Concepts and Key Architecture,* second edn, McGraw Hill.

LinkQuest (2008). *SoundLink Underwater Acoustic Modems, High Speed, Power Efficient, Highly Robust,* LinkQuest Inc., http://www.link-quest.com/.

Mare, J. (2010). Design considerations for wireless underwater communication transceiver, *OCEANS10, Sydney.*

Nasri, N., Kachouri, A., Andrieux, L. & Samet, M. (2008). Design considerations for wireless underwater communication transceiver, *International Conference on Signals, Circuits and Systems.*

Parrish, N., Roy, S., Fox, W. & Arabshahi, P. (2007). Rate-range for an fh-fsk acoustic modem, *Proceedings of the second workshop on Underwater networks, WuWNet '07*, pp. 93–96.

Pompili, D. & Akyildiz, I. (2009). Overview of networking protocols for underwater wireless communications, IEEE *Communications Magazine* 47(1): 97 –102.

Rappaport, T. (1996). Wireless *Communications, Principles and Practice,* Prentice Hall.

Sehgal, A., Tumar, I. & Schonwalder, J. (2009). Variability of available capacity due to the effects of depth and temperature in the underwater acoustic communication channel, *IEEE OCEANS 2009, EUROPE,* pp. 1–6.

Stojanovic, M. (2006). On the relationship between capcity and distance in an underwater acoustic communication channel, *International Workshop on Underwater Networks, WUWNet'06.*

Stojanovic, M. (2008). Underwater acoustic communications: Design considerations on the physical layer, *5th Annual Conference on Wireless on Demand Network Systems and Services, WONS,* pp. 1–10.

Sullivan, E. & Taroudakis, M. (2008). *Handbook of Signal Processing in Acoustics Volume* 2, RSpringer.

Syed, A., Ye, W. & Heidemann, J. (2008). Comparison and evaluation of the t-lohi mac for underwater acoustic sensor networks, *IEEE Journal on Selected Areas in Communications* 26(9): 1731 –1743.

Thorp, W. H. (1965). Deep-ocean sound attenuation in the sub- and low-kilocycle-per-second region, *Journal of the Acoustical Society of America* 38(4): 648–654.

Urick, R. (1967). *Principles of Underwater Sound for Engineers,* McGraw-Hill.

Waite, A. (2005). *Sonar for Practicing Engineers,* third edn, Wiley.

Walree, P. (2007). Acoustic modems: Product survey, *Hydro International Magazine* 11(6): 36–39. Zorzi, M., Casari, P., Baldo, N. & Harris, A. (2008). Energy-efficient routing schemes for underwater acoustic networks, *IEEE Journal on Selected Areas in Communications* 26(9): 1754 –1766.

9 Chapter

INCORPORATED KNOWLEDGE AND SELF-SUFFICIENT PLANNING

1. INTRODUCTION: AN OVERVIEW

Oceanographic observatories, year-round energy sector subsea field inspections, and continuous homeland security coastal patrols now necessitate the regular and permanent deployment of undersea sensing instruments. These applications necessitate underwater networks of stationary sensors that cooperate with fleets of unmanned undersea vehicles (UUVs). Technological hurdles in the undersea realm, including power supply constraints, communication and perception interference, navigation ambiguities, and insufficient user delegation, are hindering their current advancement and implementation. To address these issues, more advanced embedded technologies are required to enhance the platform's autonomy while preserving the operator's trust. Embedded decision-making agents equipped with reasoning and planning algorithms can enhance the long-term management of diverse assets and deliver rapid dynamic responses to events by autonomously integrating global mission requirements with resource capabilities in real time. The issue, however, is that currently, applications are mono-domain. Mission targets are exclusively mono-platform, and missions often consist of a static procedural set of commands predetermined by the operator. This situation isolates the platforms and restricts the capacity for coordinated activities among adaptive collaborative agents. In a conventional mission workflow, operators delineate the mission for each specific platform, data is gathered throughout the mission, and subsequently processed offline. Thus, the primary function of underwater platforms is to collect sensor data during missions that are static and unable to adapt to long-term environmental problems or resource fluctuations. Embedded service agents must possess the ability to navigate and comprehend the intricate and dynamic settings in which these networks will function to make informed judgments and interoperate effectively. These decision-making tools are limited by the quality and extent of the accessible information. Consequently, a shared knowledge representation across embedded service-oriented agents is essential to equip them with the requisite common situational awareness. Two sources can furnish this type of information: domain knowledge derived from the expert (orientation) and inferred knowledge obtained from the analyzed sensor data (observation). In both scenarios, it is essential for the information to be stored, accessed, and exchanged efficiently by the deliberative agents throughout mission execution. These agents, offering diverse skills and collaborating, may be spread across several platforms or share restricted resources.

1.1 Contribution

This chapter begins with a study of various techniques to addressing the decision-making process for UUV missions. We propose a semantic framework that offers a solution for the hierarchical distributed representation of knowledge for transdisciplinary agent interaction. This system employs a collection of hierarchical ontologies to describe the knowledge derived from expert input and processed sensor data. It offers a universal machine comprehension among embedded agents that is both generic and extensible. It also comprises a reasoning interface for deducing new knowledge from the observed data and ensuring knowledge stability by verifying discrepancies. This paradigm enhances local (machine-level) and global (system-level) situational awareness across all service capabilities, ranging from adaptive mission planning and autonomous target detection to deliberative collision avoidance and escape strategies. It serves as a facilitator for on-board decision-making. Service-oriented agents can access various levels of information and enhance knowledge based on their capabilities. In the absence of the requisite information, the framework allows for the solicitation of other agents possessing the necessary capabilities to produce the required data; for instance, a target classification algorithm may request the corresponding agent to deliver the essential object detection analysis prior to advancing with its classification task. Secondly, we introduce an algorithm for autonomous mission adaptability. Our technique, informed by the semantic framework, alleviates the operator from decision-making responsibilities. We demonstrate that adaptation is crucial for ensuring long-term autonomy, since it enables platforms to respond to environmental events while simultaneously reducing the need for communication with the operator. The objective is to be both successful and efficient, as preparing a plan requires time. After the first time has been allocated to develop the initial plan, it may be more efficient to amend it when changes arise, rather than starting anew. Additionally, commitments may have been established for the present plan, including the trajectory communicated to other intelligent agents, the allocation of mission plan segments to executors, or the distribution of resources, among others. Modifying an existing plan guarantees that minimal obligations are rendered invalid. Utilizing plan proximity measurements, we demonstrate that similar plans are more likely to be accepted and trusted by the operator compared to those that are possibly entirely dissimilar. Ultimately, we demonstrate through a series of in-water trials how the integration of a decision-making algorithm and a shared knowledge representation facilitates the necessary interoperability among embedded service-oriented agents to accomplish high-level mission objectives, relieve the operator from routine decision-making, and ultimately enable the continuous presence of dynamic sensing networks underwater.

2. UNMANNED DECISION MAKING LOOP

This section delineates the decision-making process employed by UUV systems and introduces the unmanned decision loop, wherein observations, orientations, decisions, and actions (OODA) occur in a cyclical manner, facilitating adaptive mission planning.In order to describe the unmanned decision loop, we need to start by modelling the mission environment. A mission environment is defined by the tuple $\Pi = (\Sigma, \Omega)$, where:

Σ is the mission domain model containing information about domain, i.e. the platform and the environment of execution, and

Ω is the mission problem model containing information about the problem, i.e. mission status, requirements, and objectives.

The set of all possible mission environments for a given domain is defined as the domain space (e.g., the domain space of the underwater domain). It is denoted by Θ. A mission environment Π is an element of one and only one Θ.

From this model, a mission plan π that tries to accomplish the mission objectives can be produced. However, this mission environment evolves over time *t* as new observations of the domain model Σ_t and the problem model Ω_t continuously modify it:

$$\Pi_t \leftarrow \Pi_{t-1} \cup \Sigma_t \cup \Omega_t \quad (1)$$

The decision making process to calculate a mission plan π_t for a given mission environment Π_t occurs in a cycle of observe-orient-decide-act. This process was termed by Boyd (1992) as the OODA-loop, and it was modelled on human behaviour. Inside this loop, the *Orientation* phase contains the previously acquired knowledge and initial understanding of the situation of the mission environment (Π_{t-1}). The *Observation* phase corresponds to new perceptions of the mission domain model (Σ_t) and the mission problem model (Ω_t) that modify the mission environment. The *Decision* component represents the level of comprehension and projection, the central mechanism enabling adaptation before closing the loop with the *Action stage*.

Note that it is possible to make decisions by looking only at orientation inputs without making any use of observations. In this case, Eq. 1 becomes $\Pi_t \leftarrow \Pi_{t-1}$. In the same way, it is also possible to make decisions by looking only at the observation inputs without making use of available prior knowledge. In this case, Eq. 1 becomes $\Pi_t \leftarrow \Sigma_t \cup \Omega t$.

In current UUVs implementations, the human operator constitutes the decision phase.

See Figure 1 for a schematic representation of the control loop. When high bandwidth communication links exist, the operator remains in the OODA-loop during the mission execution taking the decisions. For each update of the mission environment Π_t received, the operator decides on the correspondent mission plan π_t to be performed. From the list of actions in this mission plan, the mission executive issues the correspondent commands to the platform. Examples of the implementation of this architecture are existing Remotely Operated Vehicles (ROVs).

However, when communication is unreliable or unavailable, the operator must attempt to include all possible if-then-else cases to cope with execution alternatives before the mission starts. This is the case of current UUVs implementations that follow an orientation-only model. Figure 2 shows this model, where the OODA-loop is broken because observations are not reported to the human operator.

We will now discuss a few recent UUV implementations which show where the state-of-the-art is currently positioned. Most implementations rely on pre-scripted mission plan managers that are procedural and static and might not even consider conditional executions (Hagen, 2001). At this level, the mission executive follows a sequence of basic command primitives and issues them to the functional control layer of the platform. Description about how these approaches maintain control of underwater vehicles can be found in Fossen (1994), Ridao et al. (1999) and Yuh (2000). In this situation, decisions taken by the operator are made using only orientation inputs related to some previous experience and *a-priori* knowledge. This has unpredictable consequences, in which unexpected situations can cause the mission to abort and might even cause the loss of the vehicle (Griffiths, 2005; von Alt, 2010).

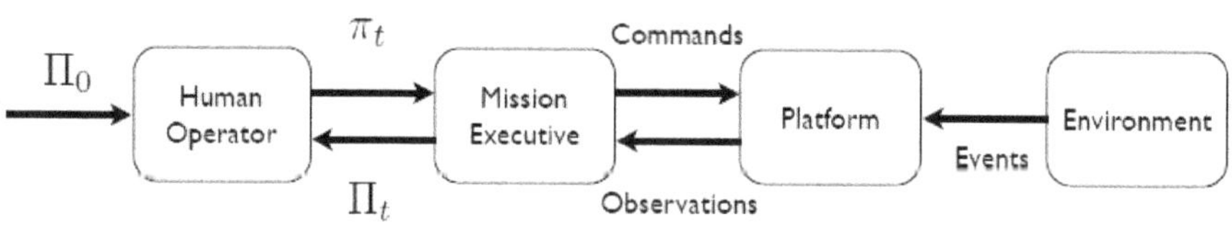

Fig. 1. Observation, Orientation, Decision and Action (OODA) loop for unmanned vehicle systems with decision making provided by the human operator.

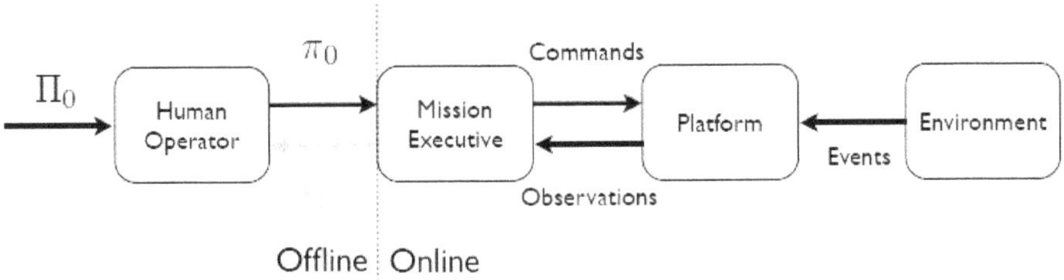

Fig. 2. Broken OODA-loop. Decision stage on the human operator based only on initial pre-mission orientation.

More modern approaches are able to mitigate this lack of adaptability by introducing sets of behaviours that are activated based on observations (Arkin, 1998). Behaviours divide the control system into a parallel set of competence-levels. They can be seen as manually scripted plans generated a-priori to encapsulate the decision loop for an individual task. Under this approach, the key factor is to find the right method for coordinating these competing behaviours.

The subsumption model, attributed to Brooks (1986), arbitrates behaviour priorities through the use of inhibition (one signal inhibits another) and suppression (one signal replaces other) networks. Most recent UUV control systems are a variant of the subsumption architecture.

This model was first applied to the control of UUVs by Turner (1995) during the development of the ORCA system. This system used a set of schemas in a case-based framework. However, its scalability remains unclear as trials for its validation were not conducted.

Later, Oliveira et al. (1998) developed and deployed the CORAL system based on Petri nets. The system was in charge of activating the vehicle primitives needed to carry out the mission. These primitives were chained by preconditions and effects.

The scaling problem was addressed by Bennet & Leonard (2000) using a layered control architecture. Layered control is a variant of the subsumption model that restricts of interaction between layers in order to keep it simple (Bellingham et al., 1990). The system was deployed for the application of adaptive feature mapping.

Another approach for coordinating behaviours is vector summation that averages the action between multiple behaviours. Following this principle, the DAMN system developed by Rosenblatt et al. (2002) used a voting-based coordination mechanism for arbitration implementing utility fusion with fuzzy logic.

The MOOS architecture developed by Newman (2002) was also able to guide UUVs by using a mission control system called Helm. Helm's mission plan was described by a set of prioritised primitive tasks. The most suitable action was selected using a set of prioritised mission goals. It used a state-machine for execution, a simplified version of a Petri net.

The O^2CA^2 system (Carreras et al., 2007) also used a Petri net representation of the mission plan (Palomeras et al., 2009). The system maintains the low level control (dynamics) from the guidance control (kinematics) uncoupled (Caccia & Veruggio, 2000). Although it contained a declarative mission representation, missions were programmed manually.

A detailed survey of other behaviour-based approaches applied to mission control systems for UUVs can be found in Carreras et al. (2006).

More recently, Benjamin et al. (2009) has applied multiple objective decision theory to provide a suitable framework for formulating behaviour-based controllers that generate Pareto-optimal and satisfying behaviours. This approach was motivated by the infeasibility of optimal behaviour selection for real-world applications. This approach has been implemented and deployed as part of the IvP Helm extension to MOOS. This method seems to be a more suitable for behaviour selection, although more computationally expensive. Also, the approach is limited to the control of only the direction and velocity parameters of the host platform.

After reviewing this related work, two problems affecting the effectiveness of the decision loop become evident. Firstly, orientation and observation should be linked together because it is desirable to place the new observations in context. Secondly, decision and action should be iterating continuously. These two problems have not been addressed together by previous approaches. These are the two of the goals that we address in this chapter. In order to achieve them autonomously, two additional components are required: a status monitor and a mission plan adapter. The status monitor reports any changes detected in the mission environment during the execution of a mission. When the mission executive is unable to handle the changes detected by the status monitor, the mission planner is called to generate a new modified mission plan that agrees with the updated mission environment. Figure 3 shows the OODA-loop for autonomous decision making. Comparing it to the previous Figure 2, the addition of status monitor and mission planner removes the need for human decisions in the loop. Note that the original mission plan π_0 could also be autonomously generated as long as the high-level goals are provided by the human operator in Π_0.

Adaptive mission planning enables a true unmanned OODA-loop. This autonomous decision making loop copes with condition changes in the mission environment during the mission execution. As a consequence, it releases the operator from decision making tasks in stressful environments containing high levels of uncertainty and dynamism.

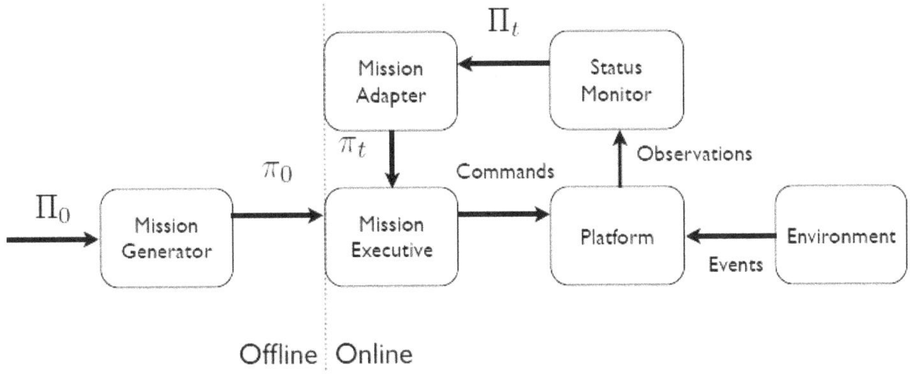

Fig. 3. Required OODA-loop for autonomous decision making in UUVs. Decision stage for adaptation takes place on-board based on initial orientation provided by the operator and observations provided by the status monitor.

The potential benefits of adaptive mission planning capabilities for autonomous decision making in UUVs were promoted by Turner (2005), Bellingham et al. (2006) and Patrón & Petillot (2008). Possibly the most advanced autonomous decision making framework for UUVs has been developed at the Monterey Bay Aquarium Research Institute. This architecture, known as T-REX, has been deployed successfully inside the Dorado AUV (Rajan et al., 2009). This is now providing adaptive planning capabilities to oceanographers for maximising the science return of their UUV missions (McGann et al., 2007; Rajan et al., 2007). Using deliberative reactors for the concurrent integration of execution and planning (McGann, Py, Rajan & Henthorn, 2008), live sensor data can be analysed during mission to adapt the control of the platform in order to measure dynamic and episodic phenomenon, such as chemical plumes (McGann, Py, Rajan, Henthorn & McEwen, 2008; McGann et al., 2009). Alternative approaches to adaptive plume tracing can also be found in the works of Farrell et al. (2005) and Jakuba (2007). Their research goals of all these approaches have been motivated by scientific applications and do not consider the needs of the human operators or the maritime industry.

However, autonomy cannot be achieved without humans, as it is necessary for this autonomy to be ultimately accepted by an operator. Our research is geared towards improving human access to UUVs in order to solve the maritime industry's primary requirement of improving platform operability (Patrón et al., 2007). We propose a goal-based approach to solving adaptive mission planning. The advantage of this approach is that it provides high levels of mission abstraction. This makes the human interface simple, powerful and platform independent, which greatly eases the operator's task of designing and deploying missions (Patrón, 2009). Ultimately, operators will not need any specialist training for an UUV specific platform, and instead missions will be described purely in terms of their goals. Apart from ease of use, we have also demonstrated using a novel metric (Patrón & Birch, 2009) that adaptive mission planners can produce solutions which are close to what a human planner would produce (Patrón et al., 2009a). This means that our solutions can be trusted by an operator.

Another advantage of our research over other state-of-the-art UUV implementations, is that we are industry focussed. Our service-oriented approach provides goal-based mission planning with discoverable capabilities, which meets industry's need for platform independence (Patrón et al., 2009b). Finally, our plan repair approach optimises the resources required for adaptability and maximises consistency with the original plan, which improves human acceptance of autonomy. Resource optimisation and consistency are very important properties for real world implementations, as we demonstrate in our sea trials (Patrón, Miguelanez, Petillot & Lane, 2008).

Section 3 describes how do we link together orientation and observation. Section 4 presents an approach to the continuous iteration of decision and action.

3. KNOWLEDGE-BASED SEMANTIC SITUATIONAL AWARENESS

Unmanned vehicle situation awareness S_{AV} consists in enabling the vehicle to autonomously understand the *'big picture'* (Adams, 2007). This picture is composed of the experience gained from previous missions (orientation) and the information obtained from the sensors while on mission (observation). Ontologies allow the representation of knowledge of these two components.

Ontologies are models of entities and interactions, either generically or in some particular practice of knowledge (Gruber, 1995). The main components of an ontology are concepts and axioms. A concept represents a set or class of entities within a domain (e.g., a *fault* is a concept within the domain of diagnostics). Axioms are used to constrain the range and domain of the concepts (e.g., a *driver* is a *software* that has a *hardware*). The finite set of concept and axiom definitions is called the *Terminology Box TBox* of the ontology.

Instances are the individual entities represented by a concept of the ontology (e.g. a remus is an instance of the concept *UUV*). Relations are used to describe the interactions between individuals (e.g. the relation *isComponentOf* might link the individual *SensorX* to the individual *PlatformY*). This finite set of instances and relations about individuals is called the *Assertion Box ABox*. The combination of *TBox* and *ABox* is what is known as a *Knowledge Base*. *TBox* aligns naturally to the orientation component of S_{AV} while *ABox* aligns to the observation component.

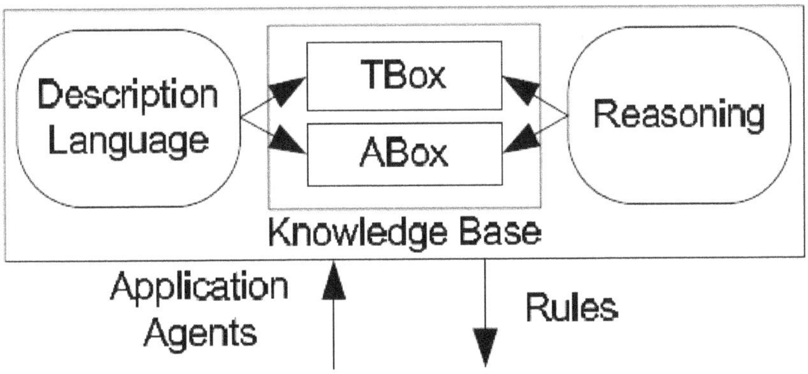

Fig. 4. Knowledge Base representation system including the TBox, ABox, the description language and the reasoning components. Its interface is made of orientation rules and agent queries.

In the past, authors such as Matheus et al. (2003) and Kokar et al. (2009) have used ontologies for situation awareness in order to assist humans during information fusion and situation analysis processes. Our work extends these previous works by using ontologies for providing unmanned situation awareness in order to assist autonomous decision making algorithms in underwater vehicles. One of the main advantages of using a knowledge base over a classical data base schema to represent S_{AV} is the extended querying that it provides, even across heterogeneous data systems. The meta-knowledge within an ontology can assist an intelligent agent (e.g., status monitor, mission planner, etc.) with processing a query. Part of this intelligent processing is due to the capability of reasoning. This enables the publication of machine understandable meta-data, opening opportunities for automated information processing and analysis.

For instance, a status monitor agent using meta-data about sensor location could automatically infer the location of an event based on observations from nearby sensors (Miguelanez et al., 2008). Inferences over the ontology are made by reasoners. A reasoner enables the domain's logic to be specified with respect to the context model and applied to the corresponding knowledge i.e., the instances of the model (see Fig. 4). A detailed description of how a reasoner works is outside of the scope of this article. For the implementation of our approach, we use the open source reasoner called Pellet (Sirin et al., 2007).

3.1 Semantic knowledge-based framework

A library of knowledge bases comprise the overall knowledge framework used in our approach for building SAV (Miguelanez et al., 2010; Patrón, Miguelanez, Cartwright & Petillot, 2008). Reasoning capabilities allow concept consistency providing reassurance that SAV remains stable through the evolution of the mission. Also, inference of concepts and relationships allows new knowledge to be extracted or derived from the observed data. In order to provide with a design that supported maximum reusability (Gruber, 1995; van Heijst et al., 1996), we adopt a three-level segmentation structure that includes the (1) Foundation, (2) Core and (3) Application ontology levels (see Fig. 5).

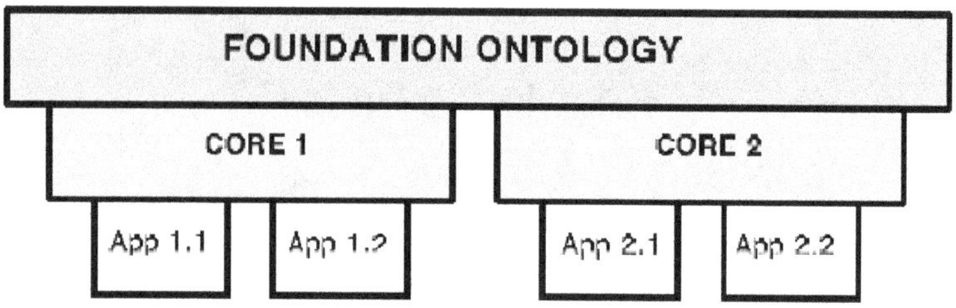

Fig. 5. Levels of generality of the library of knowledge bases for S_{AV}. They include the Foundation Ontology, the Core Ontology, and the Application Ontology levels.

Foundational Ontologies (FOs) represents the very basic principles and includes Upper and Utility Ontologies. Upper ontologies describe generic concepts (e.g., the Suggested Upper Merged Ontology or SUMO (Niles & Pease, 2001)) while Utility ontologies describe support concepts or properties (e.g. OGC_GML for describing geospatial information (Portele, 2007)). FOs meet the requirement that a model should have as much generality as possible, to ensure reusability across different domains.

The Core Ontology provides a global and extensible model into which data originating from distinct sources can be mapped and integrated. This layer provides a single knowledge base for cross-domain agents and services (e.g., vehicle resource / capabilities discovery, vehicle physical breakdown, and vehicle status). A single model avoids the inevitable combinatorial explosion and application complexities that results from pair-wise mappings between individual metadata formats and ontologies.

In the bottom layer, an Application Ontology provides an underlying formal model for agents that integrate source data and perform a variety of extended functions. As such, higher levels of complexity are tolerable and the design is motivated more by completeness and logical correctness than human comprehension. Target areas of these Application Ontologies are found in the status monitoring of the vehicle and its environment and the planning of the mission.

Figure 6 represents the relationship between the Foundation Ontologies (Upper and Utility), the Core Ontology and the Application Ontology for each service-oriented agent. Raw data gets parsed from sensors into assertions during the mission using a series of adapter modules for each of the sensing capabilities. It also shows that the knowledge handling by the agent during its decision making process is helped by the reasoner and the rule engine process.

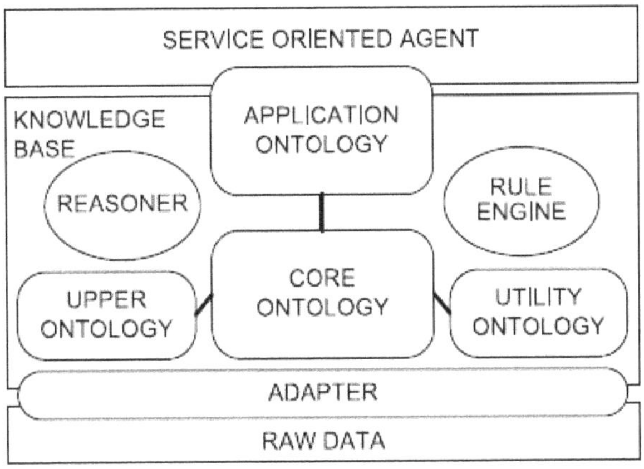

Fig. 6. SAV representation in the Knowledge Base using Core and Application ontologies supported by Upper and Utility ontologies. Generation of instances from raw data is performed by the Adapter. Handling of knowledge is done by the Reasoner, Rule Engine and the Service-Oriented Agent.

3.2 Foundation and core ontology

To provide the groundwork for the knowledge representation of unmanned vehicles, attention was directed towards the Joint Architecture for Unmanned Systems (JAUS) (SAE, 2008a). This standard was first formulated exclusively for the Unmanned Ground Vehicles (UGVs) domain but has lately been broadened to encompass all other settings, including aerial and aquatic, in an effort to establish a unified framework of architectural features and concepts. The JAUS paradigm categorizes four distinct types of Knowledge Stores: Status, World Map, Library, and Log. Our experience indicates that there is an overlap among these many knowledge repositories. The methodology presented in this work offers enhanced flexibility in information access and storage, while maintaining JAUS 'Message Interoperability' (SAE, 2008b) among agents. In the proposed paradigm, JAUS ideas are regarded as the foundational ontology for knowledge representation. The Core Ontology established in this study expands upon these notions while maintaining a concentration on unmanned systems. Several knowledge ideas identified within this domain are:

- Platform: Static or mobile (ground, air, underwater vehicles),
- Payload: Hardware with particular properties, sensors or modules,
- Agent: Software with specific capabilities,
- Sensor: A device that receives and responds to a signal or stimulus,
- Driver: Module for interaction with a specific sensor / actuator,

Additionally, the Standard Ontology for Ubiquitous and Pervasive Applications (SOUPA) (Chen et al., 2004) is used as an Utility Ontology. By providing generic context-aware concepts, it enables the spatio-temporal representation of concepts in the Core Ontology.

3.3 Application Ontology

Every service-oriented agent possesses a distinct Application Ontology. It signifies the agent's cognizance of the circumstances by including notions pertinent to the agent's experience. In the case study discussed in this chapter, the agents are the status monitor and the mission planner. Collectively, they furnish the status monitor and mission adapter components seen in Fig. 3, essential for completing the OODA loop and facilitating on-board decision-making adaption.

3.3.1 Status Monitoring Application Ontology

The Status Monitoring Application Ontology is used to express the SAV of the status monitor agent. To model the behaviour of all components and subsystems considering from sensor data to possible model outputs, the Status Monitoring Application Ontology is designed and built based on ontology design patterns (Blomqvist & Sandkuhl, 2005). Ontology patterns facilitate the construction of the ontology and promote re-use and consistency if it is applied to different environments. In this work, the representation of the monitoring concepts are based on a system observation design pattern. Some of the most important concepts identified for status monitoring are:

- Data: all internal and external variables (gain levels, water current speed),
- Observation: patterns of data (sequences, outliers, residuals,...),
- Symptom: individuals related to interesting patterns of observations (e.g., low gain levels, high average speed),
- Event: represents a series of correlated symptoms (low power consumption, position drift), Two subclasses of Events are defined: *CriticalEvent* for high priority events and IncipientEvent for the remaining ones.
- Status: links the latest and most updated event information to the systems being monitored (e.g. sidescan transducer),

Please note how some of these concepts are related to concepts of the Core Ontology (e.g. an observation comes from a *sensor*). These Core Ontology elements are the enablers for the knowledge exchange between service-oriented agents. This will be shown in the demonstration scenario of Section 5.3.

3.3.2 Planning Application Ontology

The Plan Application Ontology is used to express the SAV of the mission planner agent. It uses concepts originally defined by the *Planning Domain Definition Language* (PDDL). The PDDL language was originally created by Ghallab et al. (1998) to standardise plan representation. Concepts are extracted from the language vocabulary and the language grammar is used for describing the relationships and constraints between these concepts.

For the adaptation mission planning process, the Planning Application Ontology also requires concepts capable of representing the diagnosis of incidents or problems occurring in some parts of the mission plan (van der Krogt, 2005). Some of the most important concepts identified for mission plan adaptability are:

- Resource: state of an object (physical or abstract) in the environment (vehicle, position, sensor, etc.),
- Action: Modification of the state of resources (calibrate, classify, explore, etc.),
- Gap: A non-executable action,
- Execution: When an action is executed successfully,
- Failure: An unsuccessful execution of an action,

Please note how some of these concepts are also related to concepts of the Core Ontology (e.g. a list of capability concepts is required to perform a mission *action*).

4. ADAPTIVE MISSION PLANNING PROCESS

The adaptive mission planning process encompasses event detection, the impact of these events on the mission plan, and the subsequent response phase. Event detection is executed by the status monitoring agent. The adaptive mission planner agent conducts the diagnostic and repair of the mission plan.

4.1 Status Monitor Agent

The Status Monitor Agent evaluates all symptoms and observations from ambient and internal data to identify and classify events based on their priority and nature (critical or incipient). This agent can deduce fresh information about the vehicle's current condition concerning the operational status of its components based on internal events and contextual data. Similarly, environmental data is utilized for the detection and classification of external events to maintain the vehicle's situational awareness.

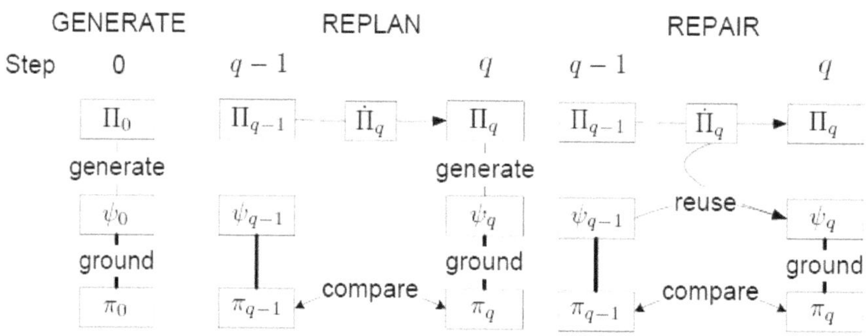

Fig. 7. Schematic representation of the autonomous mission generation, replan and repair processes using partial plan representation of the mission plans.

4.2 Mission plan adaptation agent

This section describes the mission environment model used by the mission plan repair techniques presented in this chapter. We continue using the mission environment model previously described. We generalise this model to any step q in the mission execution timeline. An instance of an UUV mission environment at a given step q can be simply defined as $\Pi_q = \{\Sigma_q, \Omega_q\}$. The mission domain model Σ_q contains the set of propositions defining the available resources in the system P_V and the set of actions or capabilities A_V . The mission problem model Ω_q contains the current platform state x_q and the mission requirements Q_O.

Based on this model, we analyse how to calculate mission plans on the plan space (Sacerdoti, 1975). A plan space is an implicit directed graph whose vertices are partially specified plans and whose edges correspond to refinement operations. In a real environment where optimality can be sacrificed by operability, partial plans are seen as a suitable representation because they are a flexible constrained-based structure capable of being adapted.

A partial plan ψ is a tuple containing a set of partially instantiated actions and a set of constraints over these partially grounded actions. Constraints can be of the form of ordering constraints, interval preservation constraints, point truth constraints and binding constraints. Ordering constraints indicate the ordering in which the actions should be executed. Interval preservation constraints link preconditions and effects over actions already ordered. Point truth constraints assure the existence of precondition facts at certain points of the plan. Binding constraints on the variables of actions are used to ground the actions to variables of the domain. Figure 8 shows a partial plan representation of an UUV mission. Partial plans are flexible to modification. They provide an open approach for handling extensions such as temporal and resource constraints. Due to nature of the constraints, it is easy to explain a partial plan to a user. Additionally, it is easily extensible to distributed multi-agent mission planning.

Figure 7 explains the processes of mission plan repair and mission plan replan for mission plan adaptation for UUVs using a partial plan representation of the mission plans. At the initial step, a partial ordered plan ψ0 is generated satisfying the original mission environment

Π_0. The ψ_0 is then grounded into the minimal mission plan π_0 including all constraints in ψ0. At step q, the semantic knowledge-based framework is updated by the diagnosis information Π' q providing a modified awareness of the mission environment Π_q. From here, two mission adaptation processes are possible: *Mission replan* generates a new partial plan ψ_q, as done at the first stage, based only on the knowledge of Π_q. On the other hand, *mission plan repair* re-validates the original plan by ensuring minimal perturbation of it. Given the partial plan at the previous step ψ_{q-1} and the diagnosis information $\Pi_{q'}$ the mission repair problem produces a solution partial plan ψ_q that satisfies the updated mission problem Π_q, by modifying ψ_{q-1}. The final step for both approaches is to ground ψ_q to its minimal mission plan π_q. It can be seen that mission repair better exploits the orientation capabilities for decision making: instead of taking the new mission environment as a given, it uses the diagnosis information about the changes occurred to guide the adaptation process.

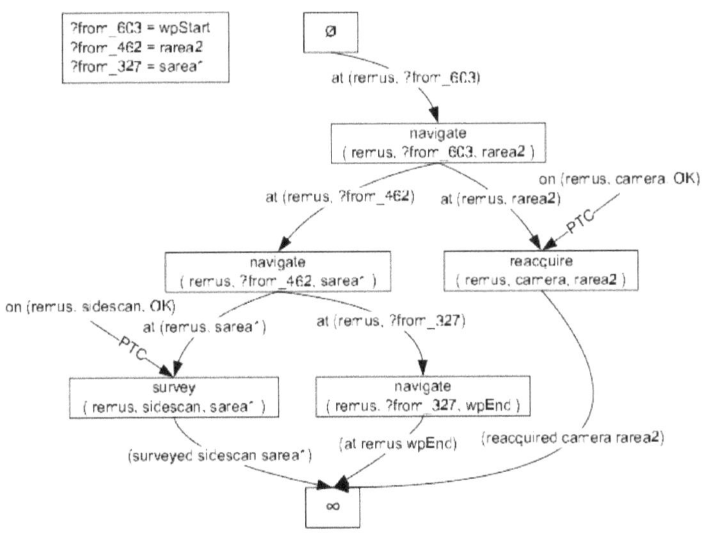

Fig. 8. Example of a partial ordered plan representation of an autonomously generated UUV mission. The ordering constraints are represented using the graph depth, interval preservation constraints are represented with black arrows, point truth constraints are represented with PTC-labelled arrows, and binding constraints are shown in the top left box.

We have now identified the benefits of mission plan repair over mission replan. Mission plan repair modifies the partial plan ψ_q, so that it uses a different composition, though it still maintains some of the actions and the constraints between actions from the previous partial plan. However, mission plan adaptation can also be achieved by mission execution repair by looking directly at the mission plan instantiation π_q. Execution repair modifies the instantiation of the mission plan πq such that a ground action g^{ah}_q that was previously instantiated by some execution e_q is newly bound by another action execution instance e'_q .

Executive repair is less expensive and it is expected to be handled directly by the mission executive agent. Plan repair, however, is computationally more expensive and requires action of the mission planner agent.

The objective is to maximise the number of execution repairs over plan repairs and, at the plan repair level, maximise the number of decisions reused from the previous mission instantiation. The information provided by the semantic-base knowledge base during the plan diagnosis phase is critical.

Executive repair fixes plan failures identified in the mission plan during the diagnosis stage. Our approach uses ontology reasoning in combination with an action execution template to adapt the mission plan at the executive level.

Once a mission plan πq is calculated by the mission planner, its list of ground actions is transferred to the executive layer. In this layer, each ground action g^{ah}_q of π_q gets instantiated into an action execution instance et using the action template for the action a_h available in the Core Ontology of the knowledge base. At the end of this phase, each e^t_q contains the script of commands required to perform its correspondent ground action. Flexibility in the execution of an action instance is critical in real environments. This is provided by a timer, an execution counter, a time-out register and a register of the maximum number of executions in the action execution instance. Additionally, three different outputs control the success, failure or time-out of its execution. These elements handle the uncertainty during the execution phase and enable the executive repair process. This minimise the number of calls to the adaptive mission planner agent and therefore the response time for adaptation.

Plan repair uses a strategy to repair with new partial plans the plan gaps identified during the plan diagnosis stage. Our approach uses an iteration of unrefinement and refinement strategies on a partial-ordered planning framework to adapt the mission plan.

Planning in the plan space is slower than in the state space because the nodes are more complex. Refinement operations are intended to achieve an open goal from the list of mission requirements or to remove a possible inconsistency in the current partial plan. These techniques are based on the least commitment principle, and they avoid adding to the partial plan any constraint that is not strictly needed. A refinement operation consists of one or more of the following steps: adding an action, an ordering constraint, a variable binding constraint or a causal link.

A partial plan is a solution to the planning problem if has no flaw and if the sets of constraints are consistent. Flaws are either subgoals or threats. Subgoals are open preconditions of actions that have not been linked to the effects of previous actions. Threats are actions that could introduce inconsistencies with other actions or constraints. We implemented a recursive non-deterministic approach based on the Partial ordered Planning (PoP) framework (Penberthy & Weld, 1992). This framework is sound, complete, and systematic. Unlike other Plan space planners that handle both types of flaws (goals and threats) similarly, each PoP recursive step first refines a subgoal and then the associated threats (Ghallab et al., 2004).

In our implementation, we introduce a previous step capable of performing an unrefinement of the partial plan when necessary. During the unrefinement strategy we remove refinements from the partial plan that are reported by the plan diagnosis phase to be affecting the consistency of the mission plan with the mission environment, i.e. to remove constraints and finally the actions if necessary.

In simple terms, when changes on the *ABox* Planning Application Ontology are sensed (Π_q) that affect the consistency of the current partial plan ψ_{q-1}, the plan repair process is initiated. The plan repair stage starts an unrefinement process that relaxes the constraints in the partial plan ψ_{q-1} that are causing the mission plan to fail.

The remaining temporal mission partial plan ψ_{q-1} is now relaxed to be able to cope with the new mission environment. However, this relaxation could open some subgoals and introduce threats in the partial plan that need to be addressed. The plan repair stage then executes a refinement process searching for a new mission plan ψ_q that is consistent with the new mission environment Π_q and removing these possible flaws. By doing this, it can be seen that the new mission plan ψ_q is not generated again from Π_q (re-planned) but recycled from ψ_q-1 (repaired). This allows re-use of the parts of the plan ψ_{q-1} that were still consistent with Π_q.

5. RESULTS

5.1 Architecture

The combination of the status monitor agent, the adaptive mission planner, the mission executive and the semantic knowledge-based framework is termed as the Semantic-based Adaptive Mission Planning system (SAMP). The SAMP system implements the four stages of the OODA-loop. Figure 9 represents the customised version of Figure 3 for SAMP.

The status monitor agent reports to the knowledge base the different changes occurring in the environment and the modifications of the internal status of the platform. The knowledge base stores the ontology-based knowledge containing the expert orientation provided *a priori* and the observations reported by the status monitor. A mission planner agent generates and adapts mission plans based on the situation awareness stored in the knowledge base. The mission executive agent executes mission commands in the functional layer based on the sequence of ground actions received from the mission planner. An Abstract Layer Interface (ALI) based on JAUS-like messages (SAE, 2008a) over UDP/IP packages implemented using the OceanSHELL protocol (Oce, 2005) provides independence from the platform's functional layer making the system generic and platform independent.

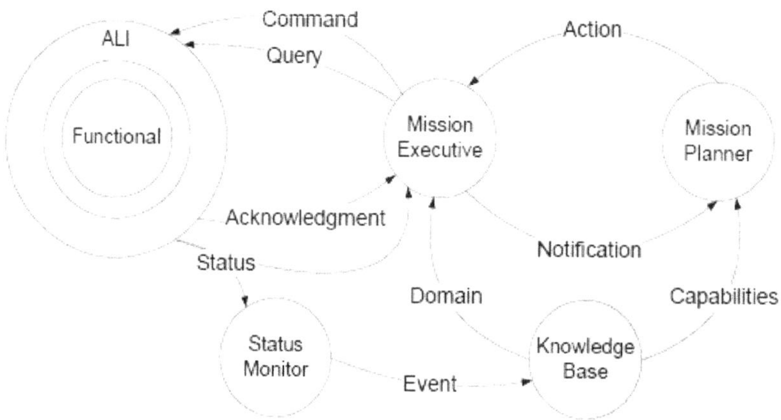

Fig. 9. Architecture of the SAMP system. The embedded agents are the planner, executive, monitor, and knowledge base. These agents interconnect via set of messages. The system integrates to the functional layer of a generic host platform by an abstract layer interface (ALI).

5.2 Simulation results

A set of synthetic simulated scenarios have been implemented to test the performance of the SAMP system. The tests are based on the mine counter measure (MCM) operation, where UUVs support and provide solutions for mine-hunting and neutralisation.

A set of 15 selected MCM scenarios were simulated covering the variability of missions described by the concepts of operations for unmanned underwater vehicles presented in the UUV (2004) and the JRP (2005). For each scenario, the detection of a failure in one of the components of the system was simulated. The mission plan was adapted to the new constraints using replanning methods and the mission plan repair approach based on partial plans introduced in Section 4.

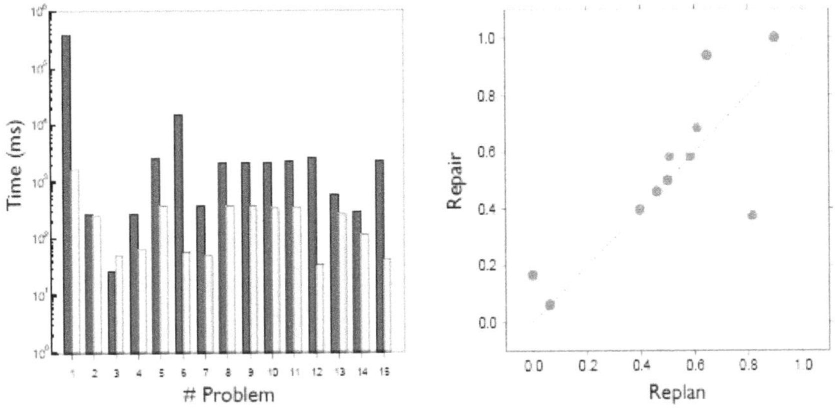

Fig. 10. Left: A semi-log plot displaying the computational time in miliseconds for replan (dark grey bars) and repair approaches (light grey bars). Right: Comparison of Plan Proximity ($PP_{0.5}$) of the replan and repair approaches to the original plan.

The performance of the two approaches was compared by looking at the computation time and the Plan Proximity (Patrón & Birch, 2009) of the adaptive mission plan provided to the original reference mission plan. Figure 10 left shows the computation time in milliseconds required for adapting the mission to the new constraints for replan (dark grey bars) and repair approaches (light grey bars). Note that a logarithmic scale is used for the time values. Figure 10 right displays the Plan Proximity to the original plan of the replan strategy result versus the repair strategy result. It can be seen that plans adapted using the mission repair strategy tend to be closer to the original plan than using the mission replan strategy. In these results, 14 out of 15 scenarios were computed faster by using mission plan repair. This computation was on average 9.1x times faster. Also, 14 out of 15 scenarios showed that mission plan repair had greater or equal Plan Proximity values as compared to mission replan. In general, our mission repair approach improves performance and time response while at the same time finds a solution that is closer to the original mission plan available before adaptation.

5.3 Experimental results

This section shows the performance of the SAMP system inside the real MCM application using a REMUS 100 UUV platform (see Fig. 11.left) in a set of integrated in-water field trial demonstration days at Loch Earn, Scotland (56° 23.1N,4° 12.0*W*). The REMUS UUV had a resident guest PC/104 1.4GHz payload computer where the SAMP system was installed. SAMP was capable of communicating with the vehicle's control and status modules and taking control of it by using an interface module that translated the ALI protocol into the manufacturer's *Remote Control* protocol (REM, 2008).

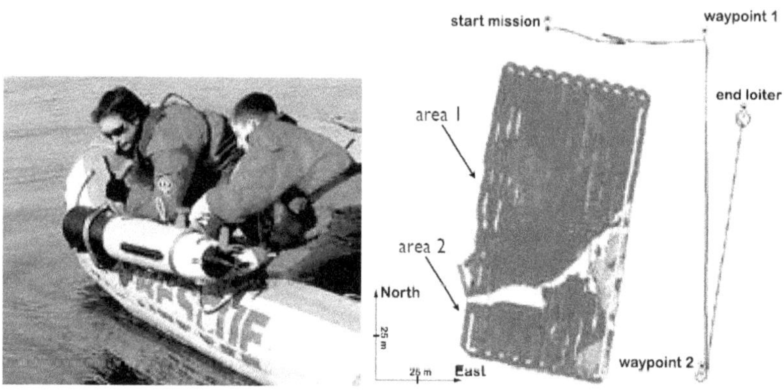

Fig. 11. Left: REMUS UUV deployment before starting one of its missions. Right: Procedural mission uploaded to the vehicle control module and a *priori* seabed classification information stored in the knowledge base. The two dark grey areas correspond to the classified seabed regions.

Figure 11.right shows the procedural waypoint-based mission as it was described to the vehicle's control module. This was known as the baseline mission. It was only used to start the vehicle's control module with a mission in the area of operation before taking control of it using the SAMP system. The baseline mission plan consisted on a start waypoint and two waypoints describing a North to South mission leg at an approximate constant Longitude (4o 16.2W). This leg was approximately 250 meters long and it was followed by a loiter pattern at the recovery location. The track obtained after executing this baseline mission using the vehicle control module is shown in Fig. 11 with a dark line. A small adjustment of the vehicle's location can be observed on the top trajectory after the aided navigation module corrects its solution to the fixes received from the Long Baseline (LBL) transponders previously deployed in the area of operations.

On the payload side, the SAMP system was oriented (in the OODA-loop sense) using *a priori* information about the environment and the platform and a declarative description of the goals of the mission. The *a priori* knowledge and the platform configuration capabilities was represented using Core Ontology concepts (see Fig. 12). Knowledge about the environment was provided based on automatic computer-aided seabed classification information generated from previous existent data (Reed et al., 2006). The two classified seabed areas are shown in Fig. 11. The declarative description of the mission requirements was represented using concepts from the Planning Application Ontology. They could be resumed as 'survey all known areas maximizing efficiency'.

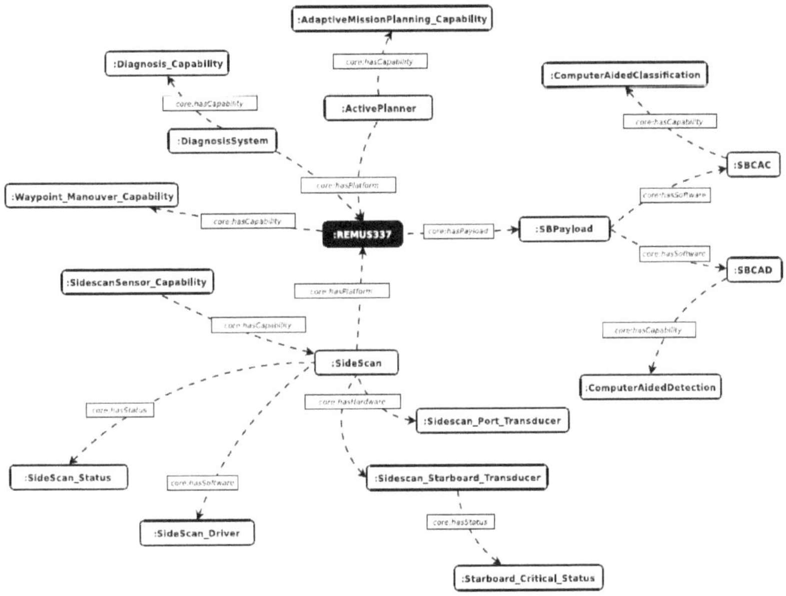

Fig. 12. Core Ontology instances for the demonstration scenario. The diagram represents the main platform, its components and their capabilities.

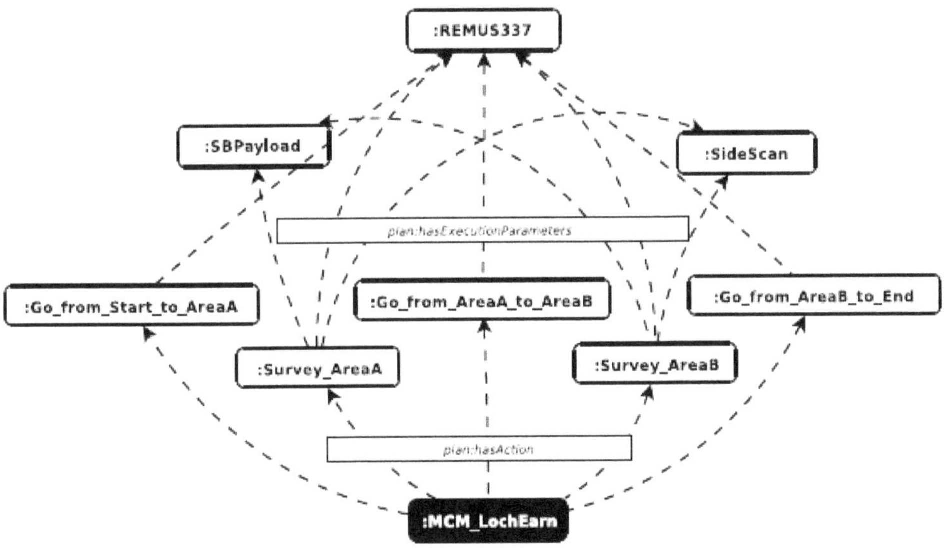

Fig. 13. Plan Application Ontology concepts representing the mission planning actions and their execution parameters and relationships.

5.3.1 Pre-mission reasoning

Please note that the previously described separation between Core knowledge and Planning knowledge gracefully aligns with the separation between platform engineers and mission scientists on current UUV operations. If the platform capabilities were described in Core Ontology terms by the engineers that manufactured the platform, it can be seen how, by using the SAMP approach, a scientific operator that only cares about the data should be able to describe the mission to the platform without knowing anything about the custom properties of the platform. It is, therefore, important to assist the operator in knowing if the platform capabilities can match the mission requirements before starting the mission:

Is this platform configuration suitable to successfully perform this mission?

In order to answer this question, new knowledge could be inferred from the initial Core Ontology orientation. The Core Ontology rule engine was executed providing with additional knowledge. A set of predefined rules helped orienting the knowledge base into inferring new relationships between instances. An example of a rule dealing with the transfer of payload capabilities to the platform is represented in Eq. 2.

$$\begin{aligned} & core : isCapabilityOf(?Capability, ?Payload) \wedge \\ & core : isPayloadOf(?Payload, ?Platform) \\ \rightarrow\ & core : isCapabilityOf(?Capability, ?Platform) \end{aligned} \quad (2)$$

Once all the possible knowledge was extracted, it was possible to query the knowledge base in order to extract the list of capabilities of the platform (see Eq. 3) and the list of requirements of the mission (see Eq. 4).

SELECT ?Platform ?Cap WHERE { rdf:type(?Platform, core:Platform) ∧ core:hasCapability(?Platform,?Cap) } (3)

SELECT ?Mission ?Req WHERE { plan:hasAction(?Mission, ?Action) ∧ plan:hasRequirement(?Action,?Req) } (4)

This way, it was possible to autonomously extract that the requirements of the mission of the experiment were[1]:

- *core:WaypointManeuver_Capability ∈ jaus:Maneuver_Capability*
- *core:ComputerAidedClassification_Capability ∈ jaus:Autonomous_RSTA-I_Capability*
- *core:ComputerAidedDetection_Capability ∈ jaus:Autonomous_RSTA-I_Capability*
- *core:SidescanSensor_Capability ∈ jaus:Environmental_Sensing_Capability*

which were a subset of the platform capabilities. Therefore, for this particular case, the platform configuration suited the mission requirements.

5.3.2 In mission adaptation

For these experiments, SAMP was given a static location in which to take control of the host vehicle. At this point, the mission planner agent generated a mission plan based on the knowledge available and the mission requirements. The instantiation of this mission plan is described in Fig. 13 using Planning Application Ontology concepts. The mission was then passed to the executive agent that took control of the vehicle for its execution.

While the mission was executed the status monitor agent maintained the knowledge base updated (in the OODA-loop sense) by reporting changes in the status of hardware components, such as batteries and sensors, and external parameters, such as water currents. When observations indicated that some of these changes were affecting the mission under execution, the mission planner was activated in order to adapt the mission to the changes. This indication was detected by the planner agent by querying the knowledge base with the following question:

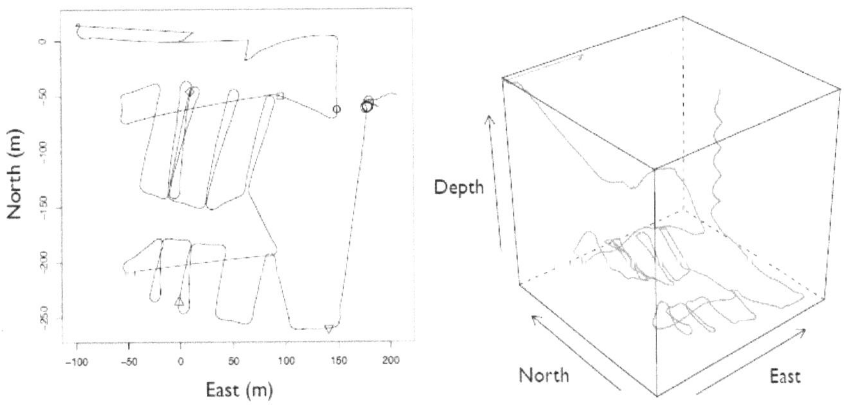

Fig. 14. Left: Vehicle's track during mission in a North-East coordinate frame projection with the origin at the starting point of the mission. Right: Three-dimensional display of the vehicle's track during the mission (Note that depth coordinates are not to scale).

Are the observations coming from the environment affecting the mission currently under execution?

In order to explain the reasoning process involved during the event detection, diagnosis and response phases of the mission adaptation process, a component fault as an internal event was temporarily simulated in the host vehicle. The fault simulated the gains of the starboard transducer of the sidescan sonar dropping to their minimum levels half way through the lawn mower survey of the first area.

For the detection phase, the low gain signals from the transducer triggered a symptom instance, which had an associated event level. This event level, represented in the Status Monitoring Application Ontology using a value partition pattern, plays a key role in the classification of the instances in the Event concept between being critical or incipient. This classification is represented axiomatically in the Eqs. 5 and 6.

$$\text{status:CriticalEvent} \sqsubseteq \text{status:Event} \sqcap \ni\text{status:causedBySymptom} \ldots \\ (\text{status:Symptom} \sqcap \ni\text{status:hasEventLevel} \ldots \\ (\text{status:Level} \sqcap \ni\text{status:High})) \tag{5}$$

$$\text{status:IncipientEvent} \sqsubseteq \ldots \\ \text{status:Event} \sqcap \ni\text{status:causedBySymptom} \ldots \\ (\text{status:Symptom} \sqcap \ni\text{status:hasEventLevel} \ldots \\ (\text{status:Level} \sqcap \ni\text{status:Med})) \tag{6}$$

After the *Event* individuals were re-classified, the *Status* property of the related component in the Core Ontology was updated.

During the diagnosis event phase, a critical status of a component is only considered to be caused by a critical event. Therefore, due to the fact that the sidescan sonar component is composed of two transducers, port and starboard, one malfunctioned transducer was only diagnosed as an incipient Status of the overall sidescan sonar component.

During the response phase, the Status property of the Core Ontology components were used by the mission planner to perform the plan diagnosis of the mission under execution. The query to the knowledge base shown in Eq. 7 reported that the two survey actions in the mission plan were affected by the incipient status of the sidescan sonar.

An incipient Status of the action parameters indicates that the action can still be performed by adapting the way it is being executed, an execution repair. If both transducers were down, a critical status of the sidescan sensor is diagnosed and a plan repair adaptation of the mission plan would have been required instead. In that case, the adaptive mission planner would have looked for redundant components or similar capabilities to perform the action or to drop the action from the plan.

The same procedure was used after the transducer recovery was reported to adapt the survey action to the normal pattern during the second lawn mower survey. In a similar process, SAMP adapted the lawnmower pattern survey of the areas to the detected water current Status at the moment of initialising the survey of the areas.

The timeline of the mission executed using the SAMP approach is described in the following figures: Figure 14 represents the final trajectory of the vehicle in 2D and 3D using a North-East coordinate frame projection with the origin at the starting point of the mission. Figure 15 displays the vehicle's telemetry recorded during the mission. It includes vehicle's velocity, compass heading, altitude, depth measurements and processed bathymetry over time. Figure 16 shows subset of the variables being monitored by the status monitor agent that were relevant to this experiment. These variables include direction of water current, remaining battery power, the availability of the transducers in the sidescan sensor and the mission execution status. Figure 17 represents the system activity of the payload computer recorded during the mission. The system activity logs show percentage of processor usage, memory usage, network activity and disk usage.

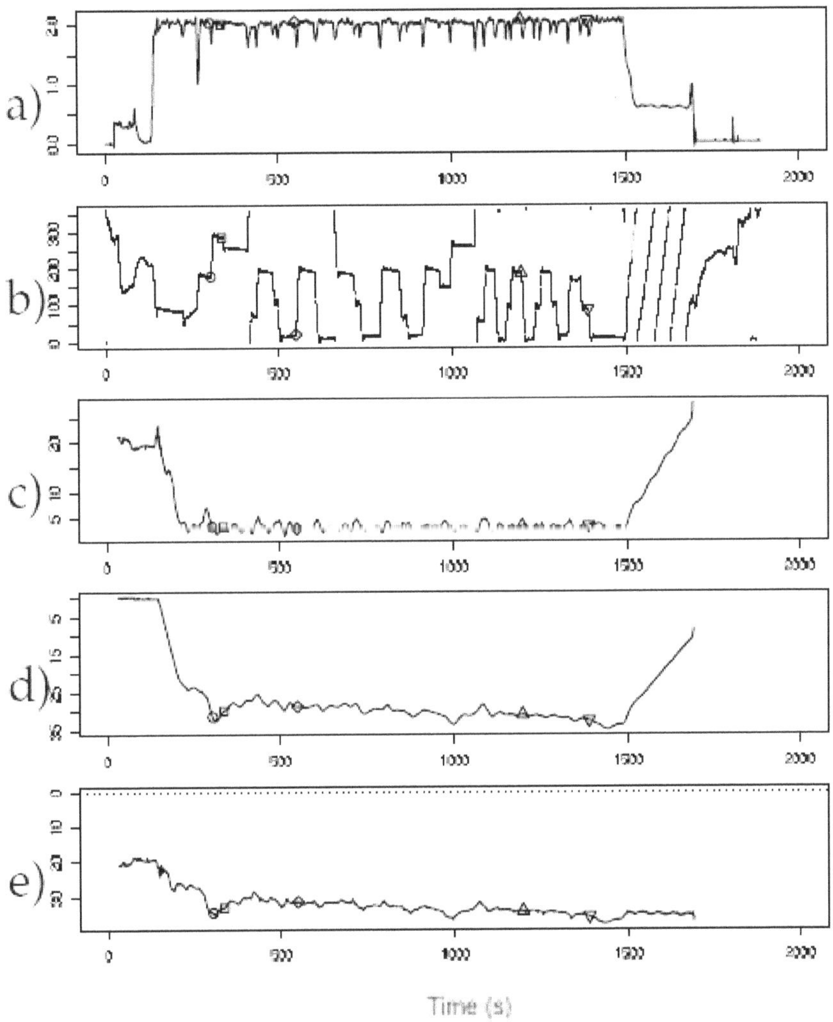

Fig. 15. Vehicle telemetry (top to bottom): a) vehicle velocity (m/s), b) compass heading (degrees), c) altitude (m), d) depth (m) and e) reconstructed profile of the seabed bathymetry (m) during the mission, all plotted against mission time (s).

$$\begin{aligned}&\mathit{SELECT}\quad ?Mission\ ?Action\ ?Param\ ?Status\\&\mathit{WHERE}\ \{\ plan{:}hasAction(\ ?Mission,\ ?Action)\wedge\\&\qquad plan{:}hasExecParam(\ ?Action,?Param)\wedge\\&\qquad plan{:}hasStatus(\ ?Param,\ ?Status)\ \}\end{aligned}\tag{7}$$

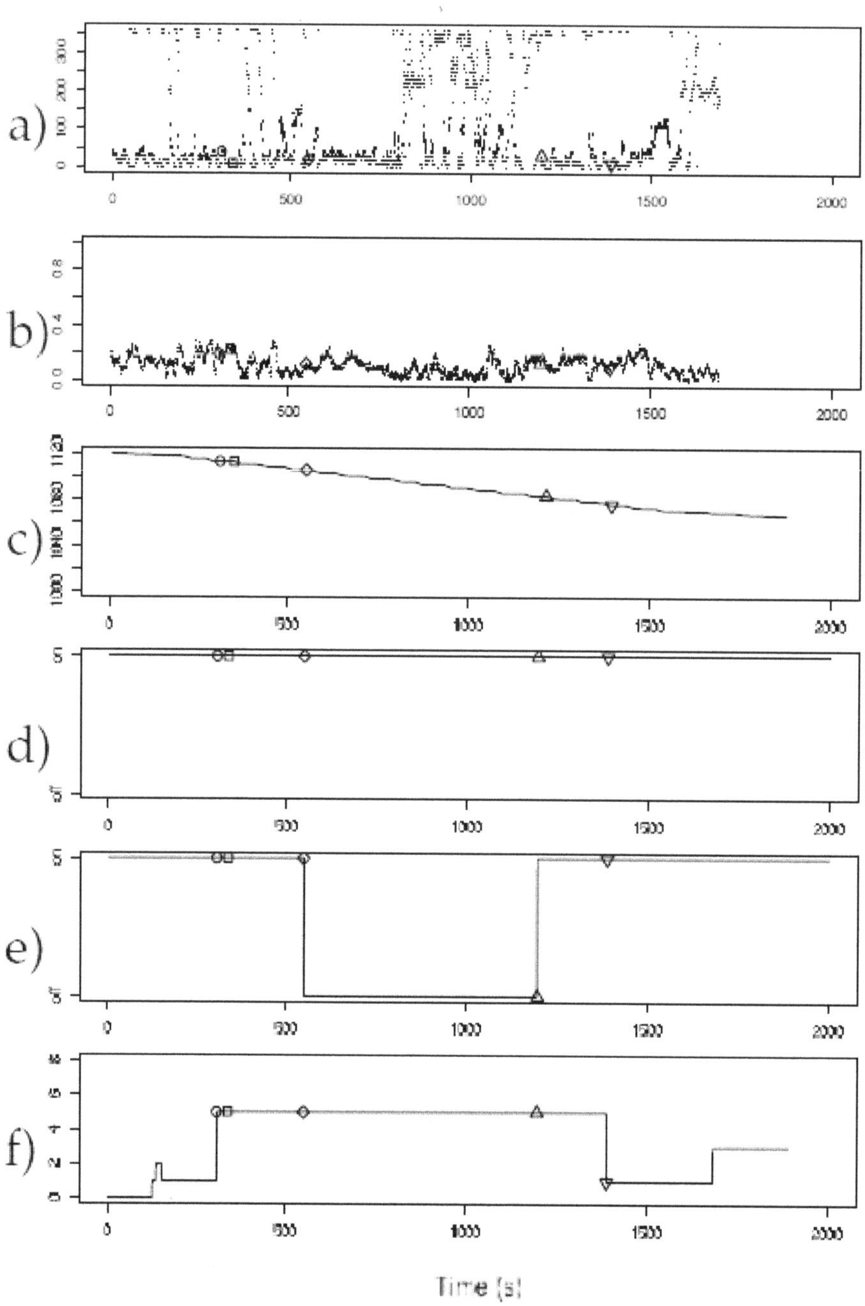

Fig. 16. Status monitoring (top to bottom): a) direction of water current (degrees), b) speed of water current (m/s), c) battery power (Wh), d) sidescan sensor port and e) starboard transducers availability (on/off) and f) mission status binary flag, all plotted against mission time (s).

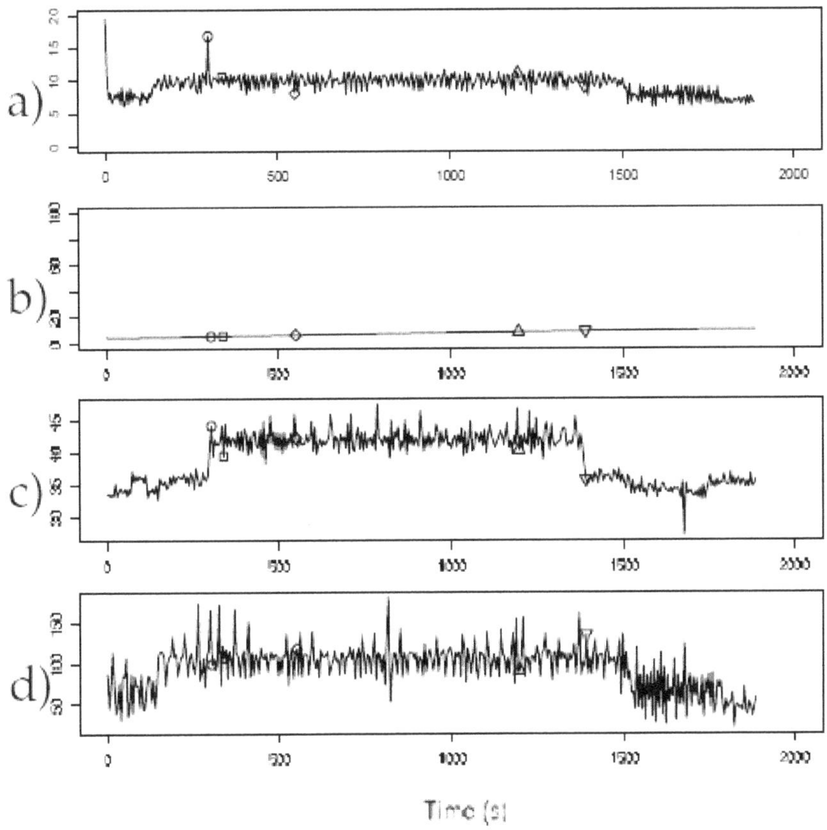

Fig. 17. System activity (top to bottom): a) % processor usage, b) % memory usage, c) network activity (packets/s) and d) disk usage (I/O sectors/s), all plotted against mission time (s).

Each of the symbols ○, □, ◇, △ and ▽on the aforementioned figures represents a point during the mission where an event occurred. Symbol ○represents the point where SAMP takes control of the vehicle. Note a change on the host platform mission status binary flag that becomes 0x05, i.e. the mission is active (0x01) and the payload is in control (0x04) (Figure 16.e).

Also, a peak on the CPU usage can be noted as this is the point where the mission partial plan gets generated (Figure 17.a).

Symbol □ represents the point where the vehicle arrives to perform the survey of the area. At this point, the action survey gets instantiated based on the properties of the internal elements and external factors. Although the Loch waters where the trials were performed were very still (see Figure 16.b), note how the vehicle heading during the lawnmower pattern performed to survey the areas follows the water current direction sensed at the arrival (approx. 12° , Symbol □ - Figure 16.a) in order to minimize drag and maximise battery efficiency. The heading of the vehicle during the survey can be observed in Figure 14 and Figure 15.b. The link between the vehicle heading in relation to the water current direction and its effect on the battery consumption was expert orientation knowledge captured by a relationship property between the two concepts in the Core Ontology.

Symbol ◇ represents the point when the status monitor agent detects and reports a critical status in the starboard transducer of the sidescan sonar (Figure 16.d). It can be seen how the lawnmower pattern was adapted to cope with the change and to use the port transducer to cover the odd and even spacing of the survey. This pattern avoids gaps in the sidescan data under the degraded component configuration and maximises sensor coverage for the survey while the transducer is down.

Symbol △ indicates the point where the starboard transducer recovery is diagnosed. It can be observed how the commands executing the action are modified in order to optimise the survey pattern and minimise distance travelled. Although also being monitored, the power status does not report any critical status during the mission that requires modification of the actions (Figure 16.c).

Symbol ▽ shows the location where all the mission goals are considered achieved and the control is given back to the mission control of the host vehicle (see Symbol ▽- Figure 16.e shows the mission is still active but the payload is not longer in control (0x01)). From this point the host vehicle's control module takes the control back and drives the vehicle to the loiter at the recovery location.

6. CONCISE SUMMARY

The undersea environment presents significant challenges for the operational maintenance of an unmanned underwater vehicle (UUV). The operability can be enhanced through the integrated modification of the mission plan. We establish a system that autonomously modifies mission plans in response to events occurring during a mission. We do this using a synthesis of ontological hierarchical knowledge representation and adaptive mission plan modification techniques. This strategy enhances robustness, system performance, and response time. The system's performance has been validated through simulation. The mission adaptation capability is demonstrated during an in-water field trial.In our fully integrated experiments we achieved the following:

Knowledge based framework: We have presented a semantic-based framework that provides the core architecture for knowledge representation for service oriented agents in UUVs. The framework combines the initial expert orientation and the observations acquired during mission in order to improve the situation awareness in the vehicle. This is currently unavailable in UUVs.

Goal-oriented plan vs. waypoint-based plan: The system uses a goal-oriented approach in which the mission is described in terms of 'what to do' instead of a 'how to do' it. The mission is parametrised and executed based on the available knowledge and vehicle capabilities. This is the first time that an approach to goal-based planning is applied to the adaptation of an underwater mission in order to maintain platform's operability.

Adaptation to environmental parameters and internal issues: The approach shows adaptability to environmental elements, such as water current flows in order to improve mission performance. The approach is also capable of dealing with the critical status of certain components in the platform and can react accordingly.

Platform agnostic: The approach is platform independent making it readily applicable to other domains, such as ground or air vehicles.

SAMP is receptive to event detections originating from other embedded service-oriented agents. We intend to implement the concept in more intricate scenarios requiring additional embedded agents, such as those for autonomous target recognition. We also intend to expand it to a fleet of cars executing a shared objective. In this situation, agents are allocated across many platforms. We are currently striving to achieve a collective situational awareness for a team of vehicles, ensuring that each member possesses the necessary awareness for their respective roles. The primary benefit of our semantic-based methodology is the minimal bandwidth necessary for transmitting id-coded ontological notions, hence addressing the constraints of underwater audio communication.

REFERENCES

Adams, J. A. (2007). Unmanned vehicle situation awareness: A path forward, *Proceedings of the 2007 Human Systems Integration Symposium,* Annapolis, Maryland, USA.

Arkin, R. C. (1998). *Behavior-based Robotics,* Massachusetts Institute of Technology Press, ISBN: 978-0262011655.

Bellingham, J., Consi, T., Beaton, R. & Hall, W. (1990). Keeping layered control simple, *Proceedings of the (1990) Symposium on Autonomous Underwater Vehicle Technology (AUV'90),* pp. 3–8.

Bellingham, J., Kirkwood, B. & Rajan, K. (2006). Tutorial on issues in underwater robotic applications, *16th International Conference on Automated Planning and Scheduling (ICAPS'06).*

Benjamin, M. R., Leonard, J. J., Schmidt, H. & Newman, P. M. (2009). An overview of MOOS-IvP and a brief users guide to the IvP Helm autonomy software,

Technical Report MIT-CSAIL-TR-2009-028, Computer Science and Artificial Intelligence Lab, MIT.

Bennet, A. A. & Leonard, J. J. (2000). A behavior-based approach to adaptive feature detection and following with autonomous underwater vehicles, *IEEE Journal of Oceanic Engineering* 25(2): 213–226.

Blomqvist, E. & Sandkuhl, K. (2005). Patterns in ontology engineering âĂŞ classification of ontology patterns, *7th International Conference on Enterprise Information Systems,* Miami, USA.

Boyd, J. (1992). A discourse on winning and losing - organic design for command and control, *Technical report,* http://www.d-n-i.net/dni/john-r-boyd/.

Brooks, R. (1986). A robust layered control system for a mobile robot, *Journal of Robotics and Automation* RA-2(1): 14–23.

Caccia, M. & Veruggio, G. (2000). Guidance and control of a reconfigurable unmanned underwater vehicle, *Control Engineering Practice* 8(1): 21 – 37.

Carreras, M., Palomeras, N., Ridao, P. & Ribas, D. (2007). Design of a mission control system for an AUV, *International Journal of Control* 80(7): 993–1007.

Carreras, M., Ridao, P., Garcia, R. & Batlle, J. (2006). *Behaviour Control of UUVs,* G. Roberts and R. Sutton, Eds Advances in Unmanned Maritime Vehicles (IEE Control Engineering), ISBN: 978-0863414503, chapter 4, pp. 67–86.

Chen, H., Perich, F., Finin, T. & Joshi, A. (2004). Soupa: Standard ontology for ubiquitous and pervasive applications, *International Conference on Mobile and Ubiquitous Systems: Networking and Services, Boston, MA.*

Farrell, J. A., Pang, S. & Li, W. (2005). Chemical plume tracing via an autonomous underwater vehicle, *IEEE Journal of Oceanic Engineering* 30, 2: 428–442.

Fossen, T. I. (1994). *Guidance and Control of Ocean Vehicles,* John Wiley and Sons, ISBN: 0-471-94113-1.

Ghallab, M., Howe, A., Knoblock, C., McDermott, D., Ram, A., Veloso, M., Weld, D. & Wilkins, D. (1998). PDDL: The planning domain definition language, *Technical report,* Yale Center for Computational Vision and Control.

Ghallab, M., Nau, D. & Traverso, P. (2004). *Automated Planning: Theory and Practice,* Morgan Kaufmann, ISBN: 1-55860-856-7.

Griffiths, G. (2005). AUTOSUB under ice, *Ingenia,* ISSN: 1472-9768 22: 30–32.

Gruber, T. R. (1995). Towards principles for the design of ontologies used for knowledge sharing, *International Journal Human-Computer Studies* 43: 907–928.

Hagen, P. E. (2001). AUV/UUV mission planning and real time control with the HUGIN operator systems, *Proceedings of the IEEE International Conference Oceans (Oceans'01)*, Vol. 1, Honolulu, HI, USA, pp. 468–473.

Jakuba, M. (2007). *Stochastic Mapping for Chemical Plume Source Localization with Application to Autonomous* Hydrothermal Vent Discovery, Doctor of philosophy in mechanical engineering, MIT/WHOI Joint Program in Applied Ocean Physics and Engineering.

JRP (2005). Joint robotics program master plan FY2005, *Technical report,* US Department of Defense.

Kokar, M. M., Matheus, C. J. & Baclawski, K. (2009). Ontology-based situation awareness, *Information Fusion* 10(1): 83–98.

Matheus, C., Kokar, M. & Baclawski, K. (2003). A core ontology for situation awareness, *Proceedings of the Sixth International Conference of Information Fusion* 1: 545 – 552.

McGann, C., Py, F., Rajan, K. & Henthorn, R. (2008). Adaptive control for autonomous underwater vehicles, *Association for the Advancement of Artificial Intelligence (AAAI'08)*, Chicago, IL, USA.

McGann, C., Py, F., Rajan, K., Henthorn, R. & McEwen, R. (2008). A deliberative architecture for AUV control, *International Conference on Robotic and Autonomation (ICRA'08)*, Pasadena, CA, USA.

McGann, C., Py, F., Rajan, K. & Olaya, A. G. (2009). Integrated planning and execution for robotic exploration, *International Workshop on Hybrid Control of Autonomous Systems*, Pasadena, CA, USA.

McGann, C., Py, F., Rajan, K., Thomas, H., Henthorn, R. & McEwen, R. (2007). T-REX: A model-based architecture for AUV control, *Workshop in Planning and Plan Execution for Real-World Systems: Principles and Practices for Planning in Execution, International Conference of Autonomous Planning and Scheduling (ICAPS'07).*

Miguelanez, E., Lewis, R., K.Brown, Roberts, C. & Lane, D. (2008). Fault diagnosis of a train door system based on the semantic knowledge representation, The *4th IET International Conference on Railway Condition Monitoring RCM 2008*, Derby Conference Centre, Derby, UK, pp. 1 –6.

Miguelanez, E., Patrón, P., Brown, K., Petillot, Y. R. & Lane, D. M. (2010). Semantic knowledge-based framework to improve the situation awareness of autonomous underwater vehicles, IEEE Transactions on Knowledge and Data Engineering (In Press) PP(99).

Newman, P. (2002). MOOS – a mission oriented operating suite, Technical report, Department of Ocean Engineering, Massachusetts Institute of Technology.

Niles, I. & Pease, A. (2001). Towards a standard upper ontology, Proceedings of the 2nd International Conference on Formal Ontology in Information Systems (FOIS-2001) .

Oce (2005). OceanSHELL: An embedded library for distributed applications and communications, Technical report, Ocean Systems Laboratory, Heriot-Watt University. Oliveira, P., Pascoal, A., Silva, V. & Silvestre, C. (1998). Mission control of the MARIUS AUV: System design, implementation, and sea trials, International Journal of Systems Science 29(10): 1065–1085.

Palomeras, N., Ridao, P., Carreras, M. & Silvestre, C. (2009). Using petri nets to specify and execute missions for autonomous underwater vehicles, Proceedings of the IEEE/RSJ International Conference on Intelligent Robots and Systems (IROS'09).

Patrón, P. (2009). Embedded knowledge and autonomous planning, Sea Technology Magazine, ISSN. 0093-3651 50(4): 101.

Patrón, P. & Birch, A. (2009). Plan proximity: an enhanced metric for plan stability, Workshop on Verification and Validation of Planning and Scheduling Systems, 19th International Conference on Automated Planning and Scheduling (ICAPS'09), Thessaloniki, Greece, pp. 74–75.

Patrón, P., Evans, J. & Lane, D. M. (2007). Mission plan recovery for increasing vehicle autonomy, Proceedings of the Conference of Systems Engineering for Autonomous Systems from the Defence Technology Centre (SEAS-DTC'07), Edinburgh, UK.

Patrón, P., Lane, D. M. & Petillot, Y. R. (2009a). Continuous mission plan adaptation for autonomous vehicles: balancing effort and reward, 4th Workshop on Planning and Plan Execution for Real-World Systems, 19th International Conference on Automated Planning and Scheduling (ICAPS'09), Thessaloniki, Greece, pp. 50–57.

Patrón, P., Lane, D. M. & Petillot, Y. R. (2009b). Interoperability of agent capabilities for autonomous knowledge acquisition and decision making in unmanned platforms, Proceedings of the IEEE International Conference Oceans Europe (Oceans Europe'09), Bremen, Germany.

Patrón, P., Miguelanez, E., Cartwright, J. & Petillot, Y. R. (2008). Semantic knowledge-based representation for improving situation awareness in service oriented agents of autonomous underwater vehicles, Proceedings of the IEEE International Conference Oceans (Oceans'08), Quebec, Canada.

Patrón, P., Miguelanez, E., Petillot, Y. R. & Lane, D. M. (2008). Fault tolerant adaptive mission planning with semantic knowledge representation for autonomous underwater Embedded Knowledge and Autonomous Planning: the Path Towards Permanent Presence of Underwater Networks 25 vehicles, Proceedings of the IEEE/RSJ International Conference on Intelligent Robots and Systems (IROS'08), Nice, France, pp. 2593–2598.

Patrón, P. & Petillot, Y. R. (2008). The underwater environment: A challenge for planning, Proceedings of the Conference of the UK Planning Special Interest Group (PlanSIG'08), Edinburgh, UK.

Penberthy, J. & Weld, D. S. (1992). UCPOP: A sound, complete, partial order planner for ADL, Proceedings of the National Conference on Knowledge Representation and Reasoning (KR), pp. 103–114.

Portele, C. (2007). OpenGISÂo″ geography markup language (GML) encoding standard v.3.2.1,

Technical report, Open Geospatial Consortium Inc.

Rajan, K., McGann, C., Py, F. & Thomas, H. (2007). Robust mission planning using deliberative autonomy for autonomous underwater vehicles, Workshop in Robotics in challenging and hazardous environments, International Conference on Robotics and Automation (ICRA'07).

Rajan, K., Py, F., McGann, C., Ryan, J., O'Reilly, T., Maughan, T. & Roman, B. (2009). Onboard adaptive control of AUVs using automated planning and execution, International Symposium on Unmanned Untethered Submersible Technology (UUST'09), Durham, NH, USA.

Reed, S., Ruiz, I., Capus, C. & Petillot, Y. (2006). The fusion of large scale classified side-scan sonar image mosaics, IEEE Transactions on Image Processing 15(7): 2049–2060.

REM (2008). REMUS remote control protocol v1.12, Technical report, Hydroid Inc.

Ridao, P., Batlle, J., Amat, J. & Roberts, G. (1999). Recent trends in control architectures for autonomous underwater vehicles, International Journal of Systems Science 30(9): 1033–1056.

Rosenblatt, J. K., Williams, S. B. & Durrant-Whyte, H. (2002). Behavior-based control for autonomous underwater exploration, International Journal of Information Sciences, 145(1–2): 69–87.

Sacerdoti, E. (1975). The nonlinear nature of plans, International Joint Conference on Artificial Intelligence (IJCAI'75), pp. 206–214.

SAE (2006). Society of automotive engineers AS-4 AIR5664 JAUS history and domain model, Technical report, SAE International Group.

SAE (2008a). Society of automotive engineers AS-4 AIR5665 JAUS architecture framework for unmanned systems, Technical report, SAE International Group.

SAE (2008b). Society of automotive engineers AS-4 ARP6012 JAUS compliance and interoperability policy, Technical report, SAE International Group.

Sirin, E., Parsia, B., Grau, B. C., Kalyanpur, A. & Katz, Y. (2007). Pellet: A practical owl-dl reasoner, Web Semantics: Science, Services and Agents on the World Wide Web 5(2): 51–53.

Turner, R. (2005). Intelligent mission planning and control of autonomous underwater vehicles, Workshop on Planning under uncertainty for autonomous systems, 15th International Conference on Automated Planning and Scheduling (ICAPS'05).

Turner, R. M. (1995). Context-sensitive, adaptive reasoning for intelligent auv control: Orca project update, In Proceedings of the Ninth International Symposium on Unmanned, Untethered Submersible Technology (UUST'95), Durham, NH, USA, pp. 426–435.

UUV (2004). The navy unmanned undersea vehicle (UUV) master plan, Technical report, US Department of the Navy.

van der Krogt, R. (2005). Plan repair in single-agent and multi-agent systems, PhD thesis, Netherlands TRAIL Research School.

van Heijst, G., Schreiber, A. & Wielinga, B. (1996). Using explicit ontologies in kbs development, International Journal of Human-Computer Studies 46(2-3).

von Alt, C. (2010). Remus technology 2020 and beyond, NATO Naval Armaments Group – Maritime Capability Group 3 on Mines, Mine Countermeasures and Harbour Protection – Industry Day, Porton Down, UK.

Yuh, J. (2000). Design and control of autonomous underwater robots: A survey, Journal of Autonomous Robots 8(1): 7–24.

10 Chapter

BEHAVIORAL RESPONSES OF DEEP-SEA FISH TO UNDERWATER VEHICLES: VARIATIONS AMONG VEHICLES, HABITATS, AND SPECIES

1. INTRODUCTION: AN OVERVIEW

Fish possess extensive perceptual capacities that enable them to respond behaviorally to diverse environmental inputs, including visual, auditory, mechanical, chemical, and electromagnetic signals. In today's cacophonous world, numerous artificially generated signals permeate aquatic environments, where fish detect and respond to them in frequently surprising ways. Accurate differentiation between natural and intentionally induced ("disturbed") behavior is crucial in ecological research aimed at identifying the dominant factors and mechanisms affecting fish abundance, distribution, and diversity. The necessity to account for anthropogenic behavioral disturbances as a significant element in ecological research, as noted by Beale (2007), extends to the residents of isolated aquatic environments, including the deep sea. *In situ* investigations employing diverse underwater vehicles (UVs) have profoundly altered the perception that the denizens of the deep, dark, and predominantly cold ocean are less behaviorally engaged and, hence, less vulnerable to anthropogenic disturbances. Although direct observation of deep-sea fauna dates back to William Beebe in the 1930s, in situ investigations of deep ocean species and their environments have been progressively more prevalent over the past 50 years. Following its original application for exploration and the finding of previously unknown habitats and creatures, UVs were utilized to rigorously examine the ecology of deep-sea organisms, particularly the larger and more easily observable fauna in the open ocean and near the seabed. Similar to census studies performed by divers in shallow waters, vertical or horizontal transects utilizing underwater vehicles were employed to acquire density or distribution data of fish (e.g., Yoklavich et al. 2007, Uiblein et al. 2010). Various fish species or closely related taxonomic groups were seen at quite high concentrations during these transects, facilitating quantitative behavioral studies. Initial in situ exploration revealed the first indications of pelagic and demersal fishes exhibiting behavioral activity at depths over 200 m, akin to that of shallow-water species (Beebe 1930, Heezen & Hollister 1971). Subsequent to these initial behavioral observations, comprehensive investigations into locomotion behavior and habitat utilization were conducted, primarily utilizing video equipment during bottom transects with manned submersibles (e.g., Lorance et al. 2002, Uiblein et al. 2002, 2003) and subsequently with remotely operated vehicles (ROVs) (e.g., Trenkel et al. 2004a, Lorance et al. 2006). Quantitative behavioral analyses performed with the submersible Nautile unequivocally demonstrated that fish species exhibit distinct variations in their swimming patterns and vertical location relative to the substrate (Uiblein et al. 2003). Furthermore, unique reactions to the approaching vehicle were seen, necessitating detailed analysis to differentiate natural behavior from responses to artificial disturbance. The disturbance effect of underwater vehicles on fish behavior significantly impacts fish density calculations derived from in situ transects, as the data may not accurately represent natural conditions when disturbance responses are pronounced and/or occur frequently (Trenkel et al. 2004b, Stone et al. 2008). Responses of deep-sea fishes to disturbances may result from various reasons, including noise generated by engines and thrusters, illumination light, movement, electromagnetic fields, or odor plumes emitted by the vehicle. Comprehensive inquiries into the true source(s) of disturbance are typically insufficient. This document presents a description and classification of disturbance responses, highlighting the distinctions among vehicles, ecosystems, and species. The data indicate that disturbance reactions are diverse and can independently provide valuable insights into the life styles of deep-sea fishes. Furthermore, if disturbance responses are recognized, natural behaviors (such as movement and vertical placement above the substrate) can be isolated and examined independently of artificial stimulation. This document presents nine case studies based on video transects from manned submersibles and remotely operated vehicles (ROVs) in the deep North Atlantic, focusing on variations in disturbance reactions across underwater vehicles, dive transects (habitats), and co-occurring species or species groupings. A distinct section is dedicated to the integrated studies of natural behavior and disturbance responses, providing a comprehensive overview. The results are examined in relation to (1) new insights regarding the artificially and naturally induced behavior of deep-sea fish, (2) the necessity of considering and integrating all influential factors in behavioral analysis and interpretation, and (3) prospective technological advancements and challenges in optimizing in-situ investigations of deep-sea fish behavior and ecology.

2. MATERIAL AND METHODS

Video recordings from the areas of the Bay of Biscay and the northern Mid-Atlantic Ridge made during six dives with four different underwater vehicles were studied.

The underwater vehicles were as follows (Table 1): the manned submersible Nautile and the ROV Victor 6000 (both at IFREMER, www.ifremer.fr), the ROV Aglantha (IMR, www.imr.no), and the ROV Bathysaurus (ARGUS, www.argus-rs.no). Each dive consisted of one to three horizontal transects close to the bottom which lasted between 10 and 174 minutes and covered various depth ranges between 812 and 1465 m (Table 1). During transects the respective vehicle moved slowly (ca. 0.5 to 1.0 knots on average) above the bottom, mostly in straight lines, sometimes interrupted by short stops.

Each of the 10 total transects crossed a distinct habitat within canyons or deep-sea terraces of the Bay of Biscay (Nautile, Victor 6000) and slopes or valleys of the northern Mid-Atlantic Ridge (Aglantha, Bathysaurus) (Table 1). The Mid-Atlantic Ridge study area was divided in a southern investigation box, close to the Azores, and a northern box situated in the area south and north of the Charlie Gibbs Fracture Zone (Table 1).

The four species/species groups selected for detailed analysis were the roundnose grenadier *Coryphaenoides rupestris* (family Macrouridae; Fig.1), the orange roughy *Hoplostethus atlanticus* (family Trachichthyidae; Fig.1) the false boarfish *Neocyttus helgae* (family Oreosomatidae; Fig.1) and codling (family Moridae). The term "codling" includes the most common *Lepdion eques* (North Atlantic codling; Fig. 1), its congeners L. guentheri and L. *schmidti,* and the slender codling Halagyreus johnssonii. Identification of species/species groups was based on the size and form of the body, head and fins, and color patterns and distributional data from the respective area deriving from collected material.

The recording of all behaviors started immediately after a fish appeared on the video screen. Four main behaviors, overall activity level, disturbance response, locomotion, and vertical positioning above the bottom, each consisting of two or more categories, were recorded for subsequent statistical analysis (Table 2). Fishes visualized on video with high or increasing swimming speed indicating burst swimming in response to prior disturbance by the submersible ("arriving disturbed") were excluded from further-going behavioral analyses. During the subsequent behavioral recordings, the UV frequently got closer to the fishes, with increasing illumination intensity caused by the front lights. If a disturbance response was observed during this process (i.e. a marked change in activity level and/or locomotion behavior), the recordings of locomotion or vertical body positioning were stopped immediately before the occurrence of this behavioral change. The disturbance response during UV approach was split into two separate categories, depending if it happened still at far distance or at close distance to the UV and mostly within the highest illumination radius.

Table 1. Overview of dives, vehicles and video transects, with numbers of encountered fish per transect. Samples analyzed are highlighted. For further explanations see text.

Cruise	**OBSERVHAL**	**VITAL**				**MAR-ECO**				
Region		Bay of Bascay				Northern Mid-Atantic Ridge				
Vehicle	Submersible Nauble	ROV Vector 6000				ROV Bathyaunus		ROV Agiantha		
Dive	OB22	VT1		VT3		ME4-2 (85 44)	ME10 (88 60)			ME16 (88 70)
Transect	OB22-1	VT1-1	VT1-2	VT3-1	VT3-2	ME4-2-1	ME10-1	ME10-2	ME10-3	ME16-1
Area	Maridak Terrace			Belle Isle Canyon		Southern Box	Northern Box, south			Northern Box, north
Date	17.05.02	23.8.02	24.8.02	31.8.02	1.9.02	12.7.04	19.7.04			26.7.04
Duration (min)	126	174	111	91	85	50	21	10	19	105
Rounchose grender	29	1	1	19	20	3	7	33	20	20
Orange roughy	5	2	0	3	0	0	0	73	43	0
Coding	32	29	37	32	33	3	29	11	31	0
False boadifish	0	33	0	0	0	41	4	3	1	0

Table 2. Overview of the behavioral categories studied

Behaviour	Category			
Disturbance response	No response	Close distance	Far distance	Arriving disturbed
Vertical position in water column	Close to bottom		Well above bottom	Far above bottom
Locomotion behaviour	Inactive	Drifting	Station holding	Forward moving

Fig. 1. Photographs of studied fish species (North Atlantic codling was the most common species of the codling group)

For the analysis of undisturbed natural behavior, four locomotion activity categories were identified: "inactive" (Table 2) (= without any movement), "station holding" (= body stationary with active swimming against current), "drifting" (= movement in lateral or backward direction with or without swimming activity), and "forward movement" (= clear active forward swimming movements). Three categories for vertical body positioning in relation to the bottom surface were determined: "close to bottom" (= positioned at the bottom or at distances of less than one body length above the bottom), "well above bottom" (= distance from bottom exceeds one body length), and "far above bottom" (= distance from bottom exceeds three body lengths).

In order to reduce the number of influential factors comparisons between underwater vehicles and species/species groups were mostly restricted to the same transect or area and comparisons among habitats were restricted to single species. Only samples with 19 or more individuals per species/species group encountered per transect were analyzed to allow statistical comparisons in all instances. For statistical comparisons of categorical data among species/species groups and habitats, G-tests of independency were carried out (Sokal & Rohlf 1981).

3. RESULTS

The behavioral data of 501 fishes from the four selected species/species groups were analyzed. Apart from a single exception (codling in dive transect OB22-1) disturbance responses occurred during all transects and in all species/species groups. On average 44 % of all fishes showed disturbance and in 7 of the 15 total observational sets (= species-transect combinations) that were analyzed, more than 50 % of the fish displayed disturbance responses. While pre-arrival disturbance was relatively rare (14 % of all disturbed behavior registered), disturbance responses at far distance occurred most frequently (59 %). The disturbance responses were only rarely directed towards any of the four UV's used. No clear signs of attraction or aggressive responses triggered by the UV's could be observed in any of the four species/species groups.

Differences between underwater vehicles (Fig.2)

The codling showed a significant difference ($p<0.005$) in disturbance responses between two dive transects performed in the same area at the Mériadzek terrace, Bay of Biscay, one with the manned submersible Nautile (transect OB22-1, Table 1) and the other with the ROV Victor 6000 (transect VT-1, Table 1). While no disturbance response was registered during the dive with Nautile, 35 % of the individuals encountered during the ROV transect showed clear signs of disturbance. Among the disturbed fish 23 % showed pre-arrival disturbance, while 54 % responded at far distance and 23 % responded at short distance to the approaching vehicle. Regarding undisturbed natural behavior, no significant differences in both vertical positioning and locomotion behavior were found between the two transects.

The codling showed a significant decrease in disturbance responses ($p<0,005$; Fig. 3b) between the first and second transect of ROV dive VT1 (Table 1) on the Mériadzek Terrace, Bay of Biscay. These two transects covered different depth zones (1392-1454 vs. 1208-1228 m), the first (VT1-1) being clearly deeper. Neither vertical positioning nor locomotion behavior differed significantly between the two transects.

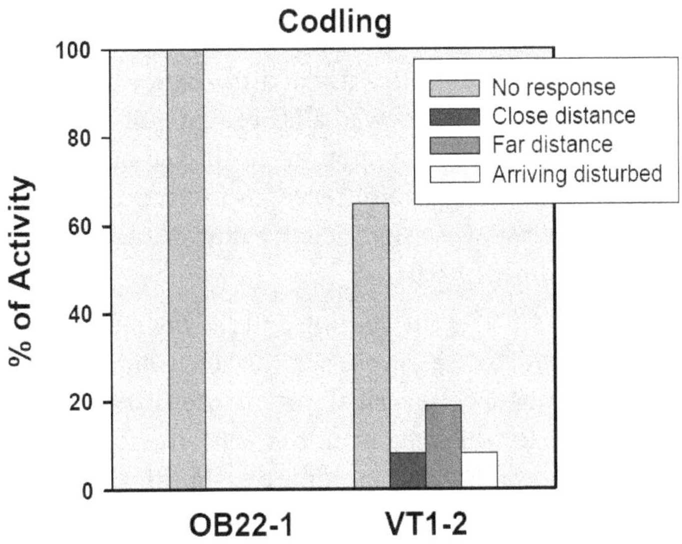

Fig. 2. Disturbance responses of codling during a manned submersible transect (left) and a ROV transect (right) in the area of Mériadzek Terrace, Bay of Biscay

Differences between dive transects and habitats (Fig. 3)

Orange roughy showed significant differences in disturbance responses ($p<0.01$; Fig. 3a) between two transects that crossed adjacent habitats at similar depths (812-879 m) during dive ME10 (Table 1) with the ROV Aglantha on the northern Mid-Atlantic Ridge, just south of the Charlie Gibbs Fracture zone. Each of the three categories of disturbance responses decreased in frequency between the first and the second transect thus indicating less responsiveness. Both vertical positioning and locomotion behavior did not differ significantly between transects.

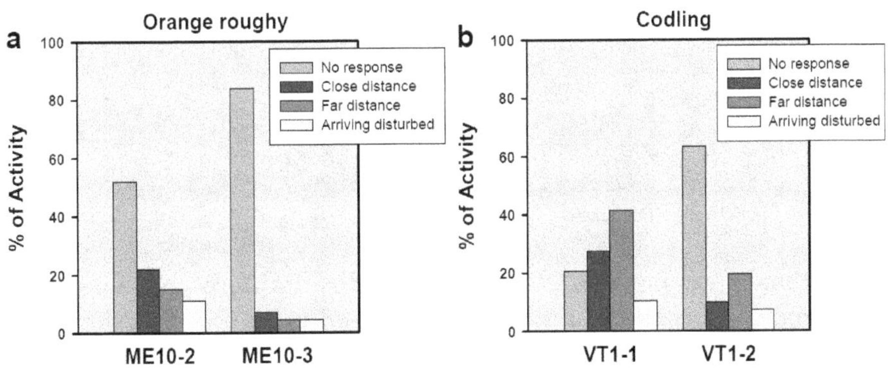

Fig. 3. Disturbance responses of (a) orange roughy during two subsequent ROV transects in the area of the northern Mid-Atlantic Ridge and (b) codling during two subsequent ROV transects in the area of Mériadzek Terrace, Bay of Biscay

Differences between co-occurring species/species groups (Fig. 4)

During the manned submersible transect OB22-1 on the Mériadzek terrace, roundnose grenadier differed significantly in disturbance responses ($p<0.0001$; Fig. 4a) from the codling. The former showed all three categories of disturbance, while the latter showed no disturbance responses at all (see also first case study; Fig. 2). Regarding natural behaviour, no differences in vertical positioning occurred, but roundnose grenadier showed significantly more forward movement and less station holding than codling ($p<0.01$).

During ROV dive transect VT1-1 the codling and the boarfish differed significantly from each other in disturbance responses ($p<0.005$; Fig. 4b) with the boarfish showing clearly less disturbed arrival and close-distance responses to the approaching vehicle. At far distance from the ROV, the frequency of disturbance responses was similar in both taxa. In addition, significant differences occurred both in vertical positioning and locomotion behavior which are dealt with at the end of the next section.

Variation in natural behavior and disturbance responses

Four different comparative data sets were selected (1) to exemplify situations with disturbance responses occurring at constant or variable rates between transects/habitats or between species/species groups and (2) to analyze in detail the undisturbed, natural vertical positioning and locomotion behavior of these species/species groups during the same dive transects.

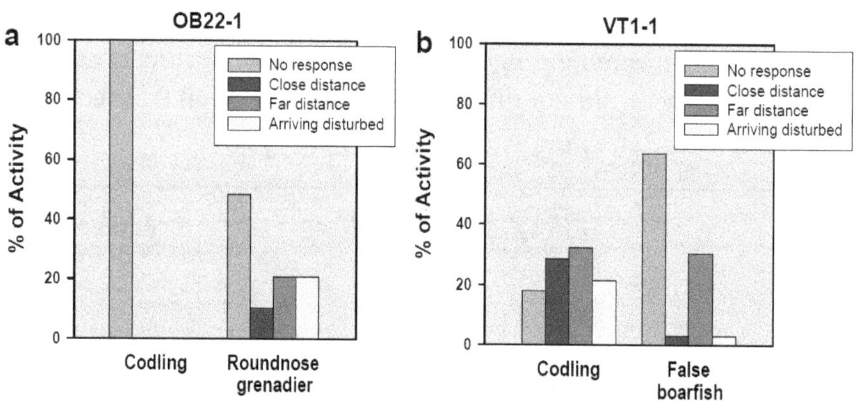

Fig. 4. Disturbance responses of (a) codling and roundnose grenadier during a manned submersible transect and (b) codling and false boarfish during a ROV transect in the area of Mériadzek Terrace, Bay of Biscay

Differences only in locomotion behavior (Fig. 5)

The disturbance responses of codling did not differ significantly between two transects (ME10-1, ME10-3, Table 1) of a dive with the ROV Bathysaurus on the Mid

Atlantic Ridge (Fig. 5a). There was however a significant difference in locomotion behavior ($p<0.05$). During the first transect all individuals encountered were active and mostly station holding, while several were inactively sitting on the bottom during the third transect, with less fish station holding. Drifting and forward moving occurred in both transects at rather similar rates. No significant differences in vertical positioning occurred.

During the ROV transect VT3-2 in the Bay of Biscay roundnose grenadier and codling did not differ significantly from each other in disturbance response and vertical positioning (Fig. 5b). However, they clearly differed in locomotion behavior ($p<0.005$), with roundnose grenadier showing less frequently station holding and more often drifting and forward movement than the codling.

Differences only in locomotion and vertical positioning (Fig. 6a)

In roundnose grenadier disturbance responses did not vary significantly between the ROV transects VT3-1 in the Bay of Biscay and ME16-1 on the Mid Atlantic Ridge (Table 1). In both cases only few individuals were recorded as being entirely undisturbed (10-21 %) and 67-87 % of all the disturbed individuals encountered responded to the vehicles at far distance. Both the locomotion behavior and the vertical positioning registered prior to disturbance responses differed significantly between the two habitats (locomotion: $p=0.0005$; vertical positioning: $p<0.025$). Roundnose grenadier occurred much higher above the bottom and showed a much higher rate of drifting on the ridge site. Station holding was frequently registered in the Bay of Biscay habitat, whereas it did not occur on the ridge site.

Differences in disturbance response and natural behavior between habitats and species (Fig. 6b)

The false boarfish showed significantly more disturbance responses ($p<0.05$), reacting more frequently at far distances during ROV transect VT1-1 in the Bay of Biscay compared to ROV transect ME4-2-1 on the Mid-Atlantic Ridge. In the latter habitat this species was positioned slightly higher above the bottom ($p=0.08$ for well- and far-above bottom categories combined) and showed a significant difference in locomotion behavior ($p<0.005$) with much less station holding and a higher rate of forward movement. Compared to the co- occurring codling in the Bay of Biscay transect, false boarfish showed significantly less disturbance ($p<0.005$), a much more frequent positioning well or far above the bottom ($p<0.0001$) and significantly more drifting and less station holding ($p<0.0001$)

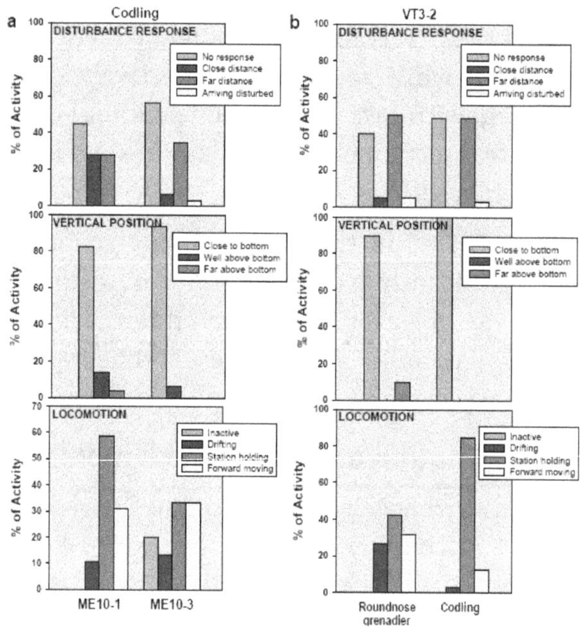

Fig. 5. Disturbance responses, vertical positioning above bottom and locomotion behavior in (a) codling during two ROV transects on the northern Mid-Atlantic Ridge and (b) roundnose grenadier and codling during a ROV transect in Belle Isle Canyon, Bay of Biscay

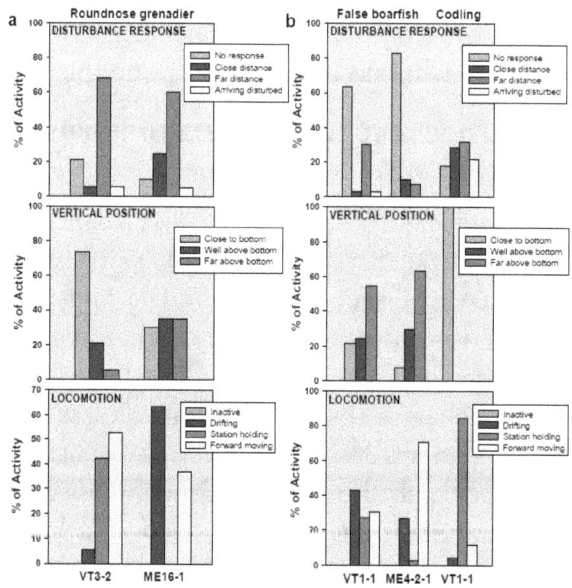

Fig. 6. Disturbance responses, vertical positioning above bottom and locomotion behavior in (a) roundnose grenadier during two ROV transects in Belle Isle Canyon, Bay of Biscay (left), and on the northern Mid-Atlantic Ridge (right) and (b) false boarfish (left and middle) and codling (right) during two transects, one on Mériadzek Terrace, Bay of Biscay (VT1-1) and the other one on the Mid-Atlantic Ridge (ME4-2-1)

4. DISCUSSION

Deep-sea fish disturbance responses

The underwater vehicles examined in this study provoked disturbance responses in deep-sea fishes seen during bottom transects, which can be most accurately described as avoidance or flight behavior. No evident indications of attraction to the UVs, as shown in other studies (e.g., Stoner et al. 2007; Moore et al. 2008), were noted. Vehicle stops and targeted long-term observations, such as tracking individual fish, were not done during the dive transects. Behavioral recordings in the research reported were conducted solely during the fish's appearance on the forward-facing camera screen during transects. It is well possible that additional disturbance responses occurred at larger distances before appearance or after the fish disappeared on the screen, but those were not recorded. Apart from these obvious restrictions, the registration and subsequent quantitative comparison of disturbance responses recorded during UV video transects is a solid method to investigate the influences of various factors such as different vehicles, habitats, or species on the frequency and intensity of evoked reactions (see also, Lorance et al. 2002, Uiblein et al. 2002, 2003).

While the manned submersible did not evoke any response in codling (first case study), they responded considerably disturbed when encountered in the same area with an ROV. A large portion of the disturbance responses happened at far distance or even before encounter indicating early detection, before the main illumination focus reached the fish. Sound may therefore be seen as a main source of disturbance. No exact comparative measurements are however available of the light and sound intensity produced by the two vehicles during those dives. Also, the possibilities cannot be ruled out that the signals acted in combination and that other disturbance sources such as, e.g., pressure waves produced by the moving vehicle body were involved, too. The present findings provide however no evidence that the much larger- bodied manned submersible elicited a comparatively higher disturbance response than any of the four ROV's used, whereas an opposite effect was demonstrated in the first case study.

In orange roughy, light may play an important role in addition to sound in eliciting disturbance responses, because a considerable portion of the reactions occurred at short distances only. Interestingly, the responsiveness to the ROV Bathysaurus decreased between the two adjacent habitats on the ridge. No differences in natural behavior (vertical positioning and locomotion) were observed. One additional difference, however, was a much higher density of orange roughy during the first transect, indicating aggregation formation. Does orange roughy remain particularly vigilant when residing in dense conspecific aggregations? During transects with the manned submersible Nautile an aggregation of orange roughy in the central St. Nazaire canyon did not differ in disturbance responses from conspecifics encountered in the peripheral area (Lorance et al. 2002, Uiblein et al. 2003). Aggregation formation in this species may be related to rather different activities such as resting, spawning or feeding (Lorance et al. 2002). More detailed studies of this ROV dive in the area of the northern Mid-Atlantic Ridge are planned that shall also include comparisons with roundnose grenadier and associated habitat conditions encountered during these transects.

Depth may be an important factor influencing disturbance responses, as can be concluded from the behavior of codling during ROV transects in the Bay of Biscay. These results corroborate with behavioral observations of the northern cutthroat eel Synaphobranchus kaupii which also showed more frequent disturbance responses at a deeper located dive in the Bay of Biscay (Uiblein et al. 2002, 2003). The latter species shows a deeper-bigger pattern, hence larger fish living at greater depth have a larger sensory surface that should facilitate signal perception. Also, as food becomes scarcer with larger depths, fish need to pay more attention to environmental stimuli. Both these argumentations may also apply to codling, however, more field and biological data would be necessary to test these assumptions.

Species differences in disturbance responses during single dive transects provide the best evidence for the importance of intrinsic, organism-dependent factors that need to be considered when studying anthropogenic disturbance. Codling showed no response during the manned submersible dive in the Bay of Biscay (OB22), while roundnose grenadier responded considerably and hence may be more sensitive to the signals emitted from this vehicle. It reacted mainly at far distance or immediately before encounter what points towards the perception of rather far-ranging signals (e.g., rather noise than light). On the other hand, codling showed considerable disturbance responses when confronted with an ROV. In the same situation, false boarfish responded to a lesser extent. These three taxa differ fundamentally from each other in their biology: the codling typically holds station close to soft bottoms, the false boarfish prefers to swim or drift closely to shelter provided by hard bottom structures and corals, while the roundnose grenadier is more flexible showing different locomotion behavior and vertical positioning depending on habitat context. Among these three species/species groups, false boarfish appears least prone to predation risk, also given their rather high body (see also Moore et al. 2008). Probably the response to UV's reveals also something about a species' vigilance and assessment of predation risk.

Responses of deep-sea fish to disturbances and their natural behaviors: a comprehensive overview

When disturbance responses are properly identified, recorded and analyzed, natural behaviour can be studied separately thus allowing to gain insights into the ecology of deep- sea fishes even in the presence of anthropogenic influences. To illustrate this, four case studies were conducted, three elaborating different aspects of natural behavior (locomotion, vertical positioning) with disturbance effects remaining constant and one with all three behaviors varying. In the first two instances only locomotion varied for codling between two separated transects during an ROV dive on the Mid-Atlantic Ridge and for roundnose grenadier and codling during a single ROV transect in the Bay of Biscay. These data indicate that while species clearly differ among each other ("species-specific" behavior), it is also of high importance to understand their behavioral flexibility in adaptation to different habitats. Behavioral flexibility or plasticity allows a choice among different locomotion modes and to select those that fit best to the prevailing conditions in the respective habitat. For instance, less station holding and increased inactivity ("sit and wait") as exemplified by codling in one of two ridge habitats (Fig. 5a) should allow efficient, energy-saving foraging when currents are weak or absent and food abundance is relatively high.

As deep-sea fishes are behaviorally flexible, one can expect to find considerable differences among contrasting habitats, as demonstrated for the roundnose grenadier by ROV dives in the Bay of Biscay and the Mid-Atlantic Ridge. While disturbance responses remained rather similar in both areas, the fish displayed more drifting and no station holding and were positioned significantly higher in the water column on the ridge. This reflects obviously behavioral adjustment to typical ridge conditions (see also, Zaferman 1992) with food particles arriving at the bottom mainly through the water column, while food input deriving from the productive shelf areas is lacking.

A rather complex picture of deep-sea fish behavioral ecology is obtained when all behaviors differ and different habitats are contrasted with different species or species groups, like in the last case study. False boarfish from habitats in the Bay of Biscay and the Mid-Atlantic Ridge were compared showing less disturbance responses, a slightly higher vertical position, less station holding, and more forward movement on the ridge site. The boarfish's behavior in the Bay of Biscay clearly contrasts with codling during the same transect, the latter showing a higher disturbance response, a position on or very close to the bottom, and more station holding. Interpretations are however complicated through one (or several) additional factor(s) that need to be considered in this as well as in the anterior case study featuring roundnose grenadier, because two different UV's were used.

Towards optimizing *in situ* behavioral ecology of deep-sea fishes and related research

A promising approach towards reaching best possible interpretations of what deep-sea fishes do, why they do it, and how they respond to human-induced environmental changes is to consider all influential external and internal factors in the data analysis and in the interpretation of the results. The central method to approach this goal is to analyze video- recordings made during UV transects based on detailed description, categorization and registration of the entire behavior observed with special emphasis to separate human-induced responses from natural behavior, followed by statistical comparisons. Additional data on the biology and ecology of the target species, the physical and biological environment, and the effects and possible impacts of anthropogenic disturbance need to be integrated, too.

To reduce complexity, the number of influential variables should be minimized whenever possible. Optimally, the same design models of UV's should be used during all dives that need to be compared. Dive transects, video recordings, and data analysis should follow standardized protocols. During each transect representative size measurements combined with estimates of absolute swimming speed should be obtained from each studied fish species. Visually well identifiable species should be preferably selected for study so to minimize possible informational noise introduced by species differences within composed groups. Groups of closely related species should be used only exceptionally, when in situ species identification is impossible and the species have a very similar body structure, hence similar behavior can be expected. Short video or photographic close-ups of each individual fish from problematic species groups should be taken (preferably by a second camera) to visualize diagnostic details helpful for species identification. Advice and assistance from taxonomists specialized in problematic fish groups should be gathered.

Use of different UV's in comparative studies cannot be recommended, because it may turn out to be difficult, if not impossible, to adjust for disturbance effects. Most certainly more than a single signal source of disturbance needs to be considered. Experimental manipulation of light, sound, and vehicle velocity, but possibly also the magnetic field, singly or in combination, might certainly assist to better understand the relative importance of these potential sources of disturbance (see also Stoner et al. 2007). However, for full control of disturbance effects from UV's, one would also need to investigate the receiver bias and in particular the sensory equipment (Popper & Hastings 2009) and reaction norms (Tuomainen & Candolin 2010) which may differ considerably among fish species, populations, size classes, and ontogenetic stages.

The longer the encounter with an UV the more increases the chance of interactions and evocation of disturbance responses. During longer UV stops, odor plumes deriving from collected organisms or bait brought along may be formed and scavengers may be attracted (Trenkel & Lorance 2011). If point observation are made during longer stops of an UV, these data should be treated separately from transect data. Also, when stationary, the vehicle itself may be perceived in quite different ways than when transitionally encountered during transects and disturbance responses may change and in some cases shift from avoidance to attraction or even to aggression (see for instance, Moore et al. 2008). Observations of deep-sea fishes deriving from longer-term interactions with UV's are certainly interesting per se, but may not always contribute to properly understand natural behavior. To minimize interactions, it may be beneficial to securely set the vehicle on the ground and deactivate the engines for behavioral observations near the bottom. In point observations of open water and at the bottom, it is advisable to disable illumination and utilize infrared light in conjunction with infrared-sensitive cameras (Widder et al. 2005). The examination of ultraviolet effects on deep-sea fish behavior holds significant implications for various other studies concerning deep-sea fishes, including in situ evaluations of abundance, population dynamics, habitat associations, community structure, and biological diversity patterns (Stoner et al. 2008). Consequently, the proposals and recommendations for optimizing in situ behavioral ecology may potentially be beneficial for wider applications in deep-sea fish study and management.

5. CONCISE SUMMARY

A crucial condition for in situ ecological studies of deep-sea fishes utilizing underwater vehicles (UVs) is the ability to differentiate between disturbance reactions triggered by the vehicles and undisturbed natural behavior. Nine case studies based on ten video transects of the North Atlantic's deep bottoms (Bay of Biscay, Mid-Atlantic Ridge) utilizing a manned submersible and three remotely operated vehicles (ROVs) are presented to illustrate variations in behavioral disturbance among vehicles, habitats, and species. Three species were examined: roundnose grenadier (Coryphaenoides rupestris), orange roughy (Hoplostethus atlanticus), and false boarfish (Neocyttus helgae), together with codling, a group of closely related species, with North Atlantic codling (Lepidion eques) being the most prevalent. Throughout each UV transect, recordings were made of disturbance responses and two activity patterns exhibited by undisturbed fish: vertical placement in the water column and mode of movement. Each behavior was categorized and quantitatively examined with sample sizes exceeding 18 individuals per species or species group and transect. Codling exhibited minimal disturbance responses to a manned submersible, yet reacted vigorously to a remotely operated vehicle during two transects conducted in the same region. Utilizing the identical ultraviolet (UV) conditions revealed distinct variations in disturbance responses between the adjacent diving transects and species/species groups, showing habitat- and species-specific reactions to signals released by the vehicle, including sound and light, as well as perhaps other stimuli. In three supplementary case studies, disturbance responses remained rather consistent across transects or species, however natural behavior varied. The final study offers a comprehensive overview, revealing distinctions across the three behaviors; nevertheless, the interpretations are confounded by the employment of different vehicles across various ecosystems. The results highlight the importance of in situ quantitative behavioral studies utilizing UV-based video transects in deep-sea fish ecology and other study domains. Comprehensive suggestions and recommendations for the optimization of vehicle disturbance control and observation methodologies are presented.

REFERENCES

Beale, C.M. 2007. The Behavioral Ecology of Disturbance Responses. International Journal of Comparative Psychology 20: 111-120

Beebe, W. 1933. Preliminary Account of Deep Sea Dives in the Bathysphere with Especial Reference to One of 2200 Feet. Proceedings of the National Academy of Sciences of the USA. 1933 January; 19(1): 178–188.

Heezen, B.C. & C.D. Hollister. 1971. The Face of the Deep. Oxford UP, New York.

Lorance, P., Trenkel, V.M. & F. Uiblein, 2005. Characterizing natural and reaction behaviour of large mid-slope species: consequences for catchability. In: Shotton, R. (Ed). 2005. Deep Sea 2003: Conference on the Governance and Management of Deep-sea Fisheries. Part1: Conference Reports. FAO Fisheries Proceedings 3/1: 162-164

Lorance, P. & V. Trenkel, 2006. Variability in natural behaviour, and observed reactions to an ROV, by mid-slope fish species. Journal of Experimental Marine Biology and Ecology 332 : 106-119.

Lorance, P., Uiblein, F., & D. Latrouite, 2002: Habitat, behaviour and colour patterns of orange roughy *Hoplostethus atlanticus* (Pisces: Trachichthyidae) in the Bay of Biscay. Journal of the Marine Biological Association of the UK 82: 321-331

Moore, J.A., Auster, P.J., Calini, D., Heinonen, K., Barber, K. & B. Hecker, 2008. False boarfish *Neocyttus helgae* in the Western North Atlantic. Bulletin of the Peabody Museum of Natural History 49: 31-41.

Popper, A.N. & M.C. Hastings 2009; The effects of human-generated sound on fish. Integrative Zoology 4: 43-52

Sokal, R.R. & F.J. Rohlf, 1981. Biometry. WH Freeman, New York.

Stoner, A.W., Ryer, C.H., Parker, S.J., Auster, P.J. & W.W. Wakefield, 2008. Evaluating the role of fish behavior in surveys conducted with underwater vehicles. Canadian Journal of Fisheries and Aquatic Sciences 65: 1230-1241

Trenkel V.M., Francis, R.I.C.C., Lorance, P. Mahévas, S., Rochet M-J. & D.M. Tracey. 2004a. Availability of deep-water fish to trawling and visual observation from a remotely operated vehicle (ROV). Marine Ecology Progress Series 284:293–303.

Trenkel V.M., Lorance P. & S. Mahévas 2004b. Do visual transects provide true population density estimates for deep-water fish? ICES Journal of Marine Science 62:1050–1056.

Trenkel V.M., Lorance P., 2011. Estimating *Synaphobranchus kaupii* densities; Contribution of fish behavior to differences between bait experiments and visual strip transects. Deep-Sea Research I 58:63-71.

Tuomainen, U. & U. Candolin, 2011. Behavioural responses to human-induced environmental change. Biological Reviews, 86:640-657

Uiblein, F., Lorance, P. & D. Latrouite, 2002: Variation in locomotion behaviour in northern cutthroat eel (*Synaphobranchus kaupi*) on the Bay of Biscay continental slope. Deep- Sea Research I 49: 1689-1703

Uiblein, F., Lorance, P. & D. Latrouite 2003: Behaviour and habitat utilization of seven demersal fish species on the Bay of Biscay continental slope, NE Atlantic. Marine Ecology Progress Series 257: 223–232

Uiblein, F., Bordes, F., Lorance, P., Nielsen, J.G., Shale, D., Youngbluth, M. & R. Wienerroither, 2010: Behavior and habitat selection of deep-sea fishes: a methodological perspective. In: S. Uchida (ed.): Proceedings of an International Symposium Into the Unknown, researching mysterious deep-sea animals 2007, Okinawa Churaumi Aquarium, Okinawa, Japan, 5-21.

Widder, E.A., Robison, B.H., Reisenbichler, K.R. & S.H.D. Haddock, 2005. Using red light for in situ observations of deep-sea fishes. Deep-Sea Research I 52: 2077–2085.

Yoklavich, M.M., Love, M.S. & K.A. Forney, 2007. A fishery-independent assessment of an overfished rockfish stock, cowcod (*Sebastes levis*), using direct observations from an occupied submersible. Canadian Journal of Fisheries and Aquatic Sciences 64: 1795- 1804.

11
Chapter

MAPPING AND DILUTION ASSESSMENT OF WASTEWATER DISCHARGES UTILIZING GEOSTATISTICS VIA AN AUTONOMOUS UNDERWATER VEHICLE

1. INTRODUCTION: AN OVERVIEW

Wastewaters are frequently released into coastal seas via outfall diffusers that effectively dilute effluent and typically confine any environmental damage to a limited area. Predicting this impact is challenging due to the intricacy of hydrodynamic processes that amalgamate the wastewater and the fluctuation in oceanic circumstances. Notwithstanding significant advancements in the comprehension of these mixing processes, as models now exist that can yield plausible predictions under steady-state settings (Hunt et al., 2010), other facets remain enigmatic and unpredictable. Consequently, considerable effort has recently been dedicated to enhancing methods for monitoring and analyzing sewage plumes across diverse oceanographic circumstances.

1.1 MARES Autonomous Underwater Vehicles (AUVs)

Autonomous Underwater Vehicles (AUVs) have been used efficiently in a wide range of applications. They were first developed with military applications in mind, for example for mine hunting missions. Later on, scientists realized their true potential and started to use them as mobile sensors, taking measurements in difficult scenarios and at a reasonable cost (Bellingham, 1997; Bellingham et al., 1992; Fernandes et al., 2000; Nadis, 1997; Robinson et al., 1999; Yu et al., 2002). MARES (Modular Autonomous Robot for Environment Sampling) AUV has been successfully used to monitor sea outfalls discharges (Abreu et al., 2010; Abreu & Ramos, 2010; Ramos & Abreu, 2010; 2011a;b;c) (see Fig. 1). MARES is 1.5 m long, has a diameter of 8-inch and weighs about 40 kg in air. It features a plastic hull with a dry mid body (for electronics and batteries) and additional rings to accommodate sensors and actuators. Its modular structure simplifies the system's development (the case of adding sensors, for example). It is propelled by two horizontal thrusters located at the rear and two vertical thrusters, one at the front and the other at the rear. This configuration allows for small operational speeds and high maneuverability, including pure vertical motions. It is equipped with an omnidirectional acoustic transducer and an electronic system that allows for long baseline navigation. The vehicle can be programmed to follow predefined trajectories while collecting relevant data using the onboard sensors. A Sea-Bird Electronics 49 FastCAT CTD had already been installed onboard the MARES AUV to measure conductivity, temperature and depth. MARES' missions for environmental monitoring of wastewater discharges are conducted using a GUI software that fully automates the operational procedures of the campaign (Abreu et al., 2010). By providing visual and audio information, this software guides the user through a series of steps which include: (1) real time data acquisition from CTD and ADCP sensors, (2) effluent plume parameter modeling using the CTD and ADCP data collected, (3) automatic path creation using the plume model parameters, (4) acoustic buoys and vehicle deployment, (5) automatic acoustic network setup and (6) real time tracking of the AUV mission.

Fig. 1. Autonomous Underwater Vehicle MARES.

1.2 Data processing

Data processing constitutes the final phase of a sewage outfall discharge monitoring effort. This procedure entails the capacity to extrapolate from monitored samples to unmonitored areas. Despite being highly chaotic due to turbulent diffusion, the dispersion process of the effluent converges to a state of natural variability once the plume ceases to ascend and the severity of turbulent fluctuations diminishes to zero (Hunt et al., 2010). It is probable that beyond this juncture, the pollutant chemicals exhibit spatial correlation. In this instance, geostatistics seems to be a suitable method for modeling the spatial distribution of the effluent. Geostatistics has been effectively employed to analyze and characterize the spatial variability of soil properties, assess water and wind resources, devise sampling strategies for monitoring estuarine sediments, investigate the thickness of effluent-affected sediment near wastewater discharges, and ascertain the spatial distribution of sewage pollution in coastal sediments, among other applications. In addition to providing estimated values, geostatistics offers a measure of the estimate's accuracy through the kriging variance. Geostatistics offers a distinct edge over conventional methods for evaluating pollution. This study use conventional block kriging to model and map the spatial distribution of temperature and salinity values collected by an AUV during a monitoring campaign of a Portuguese sea outfall. The objective is to differentiate the effluent plume from the receiving waters, evaluate its spatial variability near the discharge, and assess dilution.

2. GEOSTATISTICAL ANALYSIS

2.1 Stationary random function models

The predominant geostatistical estimate methods employ stationary random function models. A random function is a collection of random variables associated with specific spatial locations, with their interdependence defined by a probabilistic mechanism. A random function is considered stationary if all random variables possess identical probability distributions, and if the joint probability distribution of any pair of random variables relies solely on the distance between the two points, independent of their specific positions. If the stochastic function is stationary, the expected value and variance can effectively summarize the univariate characteristics of the set of random variables. The parameters typically employed to encapsulate the bivariate behavior of a stationary random function are its covariance function, correlogram, and variogram. The comprehensive definition of the probabilistic generating mechanism of a random function is typically challenging, even in one dimension. Fortunately, for several common issues we face, it is unnecessary to understand the probability generating method. Typically, we utilize a stable random function as our model, defining solely its covariance or variogram (Isaaks & Srivastava, 1989; Kitanidis, 1997; Wackernagel, 2003).

2.2 Ordinary kriging

Ordinary kriging method is often referred with the acronym BLUE which stands for "Best Linear Unbiased Estimator". "Linear" because its estimates are weighted linear combinations of the available data; "Unbiased" since it tries to have the mean error equal to 0; and "Best"because it aims at minimizing the variance of the errors. Let us then see how the concept of a random function model can be used to decide how to weight the nearby samples so that our estimates are unbiased. For any point at which we want to estimate the unknown value, our model is a stationary random function that consists of n random variables, one for the value at each of the n sample locations, $Z(\mathbf{x}_1), Z(\mathbf{x}_2), \ldots, Z(\mathbf{x}_n)$, and one for the unknown value at the point we are trying to estimate $Z(\mathbf{x}_0)$. Each of these random variables has the same probability law; at all locations, the expected value of the random variable is m and the variance is σ^2. Every value in this model is seen as an outcome (or realization) of the random variable. Our estimate is also a random variable since it is a weighted linear combination of the random variables at the n sampled locations (Cressie, 1993; Goovaerts, 1997; Isaaks & Srivastava, 1989; Kitanidis, 1997; Stein, 1999; Wackernagel, 2003; Webster & Oliver, 2007):

$$\hat{Z}(\mathbf{x}_0) = \sum_{i=1}^{n} w_i \cdot Z(\mathbf{x}_i). \tag{1}$$

The estimation error is defined as the difference between the random variable modeling the true value and the estimate:

$$\varepsilon(\mathbf{x}_0) = Z(\mathbf{x}_0) - \hat{Z}(\mathbf{x}_0). \tag{2}$$

The estimation error is also a random variable. Its expected value, often referred to as the bias, is

$$E\left[\varepsilon(\mathbf{x}_0)\right] = m\left(1 - \sum_{i=1}^{n} w_i\right). \tag{3}$$

Setting this expected value to 0, to ensure an unbiasedness estimate results in:

$$\sum_{i=1}^{n} w_i = 1. \tag{4}$$

This is known as the condition of unbiasedness (Isaaks & Srivastava, 1989; Kitanidis, 1997; Wackernagel, 2003). The expression of the variance of the modeled error is

$$\sigma^2_{\varepsilon(\mathbf{x}_0)} = var\left[Z(\mathbf{x}_0)\right] - 2\,cov\left[\hat{Z}(\mathbf{x}_0), Z(\mathbf{x}_0)\right] + var\left[\hat{Z}(\mathbf{x}_0)\right]. \tag{5}$$

Since we have already assumed that all of the random variables have the same variance σ^2, then $var\left[Z(\mathbf{x}_0)\right] = \sigma^2$. The second term in Equation 5 can be written as

$$\begin{aligned} -2\,cov\left[\hat{Z}(\mathbf{x}_0), Z(\mathbf{x}_0)\right] &= -2\sum_{i=1}^{n} w_i \cdot cov\left[Z(\mathbf{x}_i), Z(\mathbf{x}_0)\right] \\ &= -2\sum_{i=1}^{n} w_i C_{i0}. \end{aligned} \tag{6}$$

The third term of Equation 5 can be expressed as

$$var\left[\hat{Z}(\mathbf{x}_0)\right] = \sum_{i=1}^{n}\sum_{j=1}^{n} w_i \cdot w_j \cdot cov\left[Z(\mathbf{x}_i), Z(\mathbf{x}_j)\right]$$

$$= \sum_{i=1}^{n}\sum_{j=1}^{n} w_i w_j C_{ij}. \tag{7}$$

Then, the expression of the error variance comes in the following way:

$$\sigma^2_{\varepsilon(\mathbf{x}_0)} = \sigma^2 - 2\sum_{i=1}^{n} w_i C_{i0} + \sum_{i=1}^{n}\sum_{j=1}^{n} w_i w_j C_{ij}. \tag{8}$$

Equation 8 expresses the error variance as function of the n weights, once chosen the random model function parameters, namely the variance σ^2 and all the covariances C_{ij}. The minimization of $\sigma^2_{\varepsilon(x0)}$ is constrained by the unbiasedness condition imposed earlier, which can be solved using the method of Lagrange multipliers. We start by introducing a new parameter μ, called the Lagrange multiplior, in Equation 8 in the following way:

$$\sigma^2_{\varepsilon(\mathbf{x}_0)} = \sigma^2 - 2\sum_{i=1}^{n} w_i C_{i0} + \sum_{i=1}^{n}\sum_{j=1}^{n} w_i w_j C_{ij} + \underbrace{2\mu\left(\sum_{i=1}^{n} w_i - 1\right)}_{=0}. \tag{9}$$

Then we minimize $\sigma^2_{\varepsilon(x0)}$ by calculating the $n + 1$ partial first derivatives of Equation 9 with respect to the n weights and the Lagrange multiplier, and setting each one to 0, which produces the following system of equations:

$$\frac{\partial(\sigma^2_{\varepsilon(\mathbf{x}_0)})}{\partial(w_1)} = -2C_{10} + 2\sum_{j=1}^{n} w_j C_{1j} + 2\mu = 0 \quad \Rightarrow \quad \sum_{j=1}^{n} w_j C_{1j} + \mu = C_{10}$$

$$\vdots \qquad\qquad\qquad\qquad \vdots$$

$$\frac{\partial(\sigma^2_{\varepsilon(\mathbf{x}_0)})}{\partial(w_i)} = -2C_{i0} + 2\sum_{j=1}^{n} w_j C_{ij} + 2\mu = 0 \quad \Rightarrow \quad \sum_{j=1}^{n} w_j C_{ij} + \mu = C_{i0}$$

$$\vdots \qquad\qquad\qquad\qquad \vdots$$

$$\frac{\partial(\sigma^2_{\varepsilon(\mathbf{x}_0)})}{\partial(w_n)} = -2C_{n0} + 2\sum_{j=1}^{n} w_j C_{nj} + 2\mu = 0 \quad \Rightarrow \quad \sum_{j=1}^{n} w_j C_{nj} + \mu = C_{n0}$$

$$\frac{\partial(\sigma^2_{\varepsilon(\mathbf{x}_0)})}{\partial(\mu)} = 2\left(\sum_{i=1}^{n} w_i - 1\right) = 0 \quad \Rightarrow \quad \sum_{i=1}^{n} w_i = 1$$

which can also be written in a compact way as

$$\sum_{j=1}^{n} w_j C_{ij} + \mu = C_{i0}, \quad \forall i = 1, 2, \ldots, n; \qquad \sum_{i=1}^{n} w_i = 1. \tag{10}$$

This system of equations, often referred to as the ordinary kriging system, can be written in matrix notation as

$$\underbrace{\begin{bmatrix} C_{11} & \ldots & C_{1n} & 1 \\ \vdots & \vdots & \ddots & \vdots \\ C_{n1} & \ldots & C_{nn} & 1 \\ 1 & \ldots & 1 & 0 \end{bmatrix}}_{(n+1)\times(n+1)}^{\mathbf{C}} \cdot \underbrace{\begin{bmatrix} w_1 \\ \vdots \\ w_n \\ \mu \end{bmatrix}}_{(n+1)\times 1}^{\mathbf{W}} = \underbrace{\begin{bmatrix} C_{10} \\ \vdots \\ C_{n0} \\ 1 \end{bmatrix}}_{(n+1)\times 1}^{\mathbf{D}} \tag{11}$$

The set of weights and the Lagrange multiplier that will produce an unbiased estimate of $Z(\mathbf{x}_0)$ with the minimum error variance are then given by

$$\mathbf{W} = \mathbf{C}^{-1} \cdot \mathbf{D}. \tag{12}$$

The value of $\sigma^2_{\varepsilon\,(\mathbf{x}0)}$ can be obtained in a quicker way using an alternative expression to Eq. 8. Multiplying each of the n equations given in Eq. 10 by wi and summing these *n* equations leads to the following:

$$\sum_{i=1}^{n}\sum_{j=1}^{n} w_i w_j C_{ij} = \sum_{i=1}^{n} w_i C_{i0} - \mu.$$

Substituting this into Equation 8 the minimized error variance comes as follows:

$$\sigma^2_{\varepsilon(\mathbf{x}_0)} = \sigma^2 - \left(\sum_{i=1}^{n} w_i C_{i0} + \mu\right), \tag{13}$$

or, in terms of matrices as

$$\sigma^2_{\varepsilon(\mathbf{x}_0)} = \sigma^2 - \mathbf{W}^T \cdot \mathbf{D}. \tag{14}$$

The minimized error variance is usually called the *ordinary kriging variance.*

2.3 Block kriging

In numerous environmental applications, it has been noted that standard kriging typically demonstrates substantial prediction errors (Bivand et al., 2008). This results from the greater variability in the observations. When forecasting averages across extensive regions, such as blocks, a significant portion of the variability diminishes, resulting in reduced prediction errors for block mean values. If the blocks are sufficiently small, the spatial patterns remain intact. The block kriging system resembles the point kriging system as delineated in Equation 11. The matrix C remains unchanged as it is unaffected by the position of the block estimate. The covariances for vector D are point-to-block covariances. Supposing that the mean value over a block V is approximated by the arithmetic average of the N point variables contained within that block (Goovaerts, 1997; Isaaks & Srivastava, 1989), i.e.

$$Z_V \approx \frac{1}{N}\sum_{j=1}^{N} Z(\mathbf{x}_j), \tag{15}$$

the point-to-block covariances required for vector **D** are

$$\overline{C}_{iV} = cov\left[Z(\mathbf{x}_i), Z_V\right] = \frac{1}{N}\sum_{j=1}^{N} C_{ij}, \quad \forall i = 1, 2, \ldots, n. \tag{16}$$

The block kriging variance is

$$\sigma_V^2 = \overline{C}_{VV} - \left(\sum_{i=1}^{n} w_i \overline{C}_{iV} + \mu\right), \tag{17}$$

where $\overline{C}_{VV}$ is the average covariance between pairs of points within V:

$$\overline{C}_{VV} = \frac{1}{N^2}\sum_{i=1}^{N}\sum_{j=1}^{N} C_{ij}. \tag{18}$$

An equivalent procedure, that can be computationally more expensive than block kriging, is to obtain the block estimate by averaging the N kriged point estimates within the block (Goovaerts, 1997; Isaaks & Srivastava, 1989).

2.4 Spatial continuity

Spatial continuity is seen in the majority of Earth science datasets. Contour maps and related representations exhibit a non-random distribution of data, with low values clustering near one another and high values similarly grouped together. Measurements in proximity are more likely to exhibit similar values than those that are distantly located (Isaaks & Srivastava, 1989). To calculate the weights and the Lagrange multiplier that yield each estimate and the corresponding reduced error variance, it is essential to ascertain the covariances of the C and D matrices. As previously stated, due to the stationarity of our random function, all pairs of random variables separated by a distance and direction h (referred to as lag) possess identical joint probability distributions. The covariance function, C(**h**) is the covariance between random variables separated by a lag **h** (Isaaks & Srivastava, 1989; Kitanidis, 1997; Wackernagel, 2003). For a stationary random function, the covariance function C(**h**) is:

$$C(\mathbf{h}) = E\left[Z(\mathbf{x})\,Z(\mathbf{x}+\mathbf{h})\right] - \{E\left[Z(\mathbf{x})\right]\}^2. \tag{19}$$

The covariance between random variables at identical locations is the variance of the random function:

$$C(0) = E\left[\{Z(\mathbf{x})\}^2\right] - \{E\left[Z(\mathbf{x})\right]\}^2 = var\left[Z(\mathbf{x})\right] = \sigma^2. \tag{20}$$

The semivariogram, or simply variogram, is half the expected squared difference between random variables separated by a lag **h**:

$$\gamma(\mathbf{h}) = \frac{1}{2}E\left[\{Z(\mathbf{x}) - Z(\mathbf{x}+\mathbf{h})\}^2\right] = \frac{1}{2}var\left[Z(\mathbf{x}) - Z(\mathbf{x}+\mathbf{h})\right]. \tag{21}$$

The quantity γ (**h**) is known as the semivariance at lag h. The "semi"refers to the fact that it is half of a variance. The variogram between random variables at identical locations is zero, i.e. γ(**0**)= 0. Using Equations 19, 20 and 21, we can relate the variogram with the covariance function as:

$$\gamma(\mathbf{h}) = C(0) - C(\mathbf{h}) = \sigma^2 - C(\mathbf{h}). \tag{22}$$

In practice, the pattern of spatial continuity chosen for the random function is usually taken from the spatial continuity evident in the sample data set. Geostatisticians usually define the spatial continuity of the sample data set through the variogram and solve the ordinary kriging system using covariance (Isaaks & Srivastava, 1989). The maximum value reached by the variogram is called the sill. The distance at which the sill is reached is called the range. The vertical jump from zero at the origin to the value of semivariance at extremely small separation distances is called the nugget effect. The estimator of the variogram usually used, known as Matheron's method-of-moments estimator (MME) is (Matheron, 1965; Webster & Oliver, 2007)

$$\gamma(\mathbf{h}) = \frac{1}{2N(\mathbf{h})}\sum_{i=1}^{N(\mathbf{h})}\left[Z(\mathbf{x}_i) - Z(\mathbf{x}_i+\mathbf{h})\right]^2, \tag{23}$$

where $z(\mathbf{x}_i)$ is the value of the variable of interest at location $\mathbf{x}_i$ and $N(\mathbf{h})$ is the number of pairs of points separated by the particular lag vector **h**. Cressie and Hawkins (Cressie & Hawkins, 1980) developed an estimator of the variogram that should be robust to the presence of outliers and enhance the variogram spatial continuity, having also the advantage of not spreading the effect of outliers in computing the maps. This estimator (CRE) is defined as follows (Cressie & Hawkins, 1980):

$$\gamma(\mathbf{h}) = \frac{1}{2} \times \frac{\left\{ \frac{1}{N(\mathbf{h})} \sum_{i=1}^{N(\mathbf{h})} |Z(\mathbf{x}_i) - Z(\mathbf{x}_i + \mathbf{h})|^{1/2} \right\}^4}{0.457 + \frac{0.494}{N(\mathbf{h})} + \frac{0.045}{[N(\mathbf{h})]^2}}. \qquad (24)$$

After calculating the sample variogram, a function known as the variogram model must be fitted to it. Initially, the matrices C and D may require semivariance values for lags that are absent from the sample data. Furthermore, the application of the variogram does not ensure the existence and uniqueness of a solution to the standard kriging system. The predominant variogram models are the spherical model, the exponential model, the Gaussian model, and the Matern model (Isaaks & Srivastava, 1989).

2.5 Cross-validation

Cross-validation is a procedure used to compare the performance of several competing models (Webster & Oliver, 2007). It starts by splitting the data set into two sets: a modelling set and a validation set. Then the modelling set is used for variogram modelling and kriging on the locations of the validation set. Finally the measurements of the validation set are compared to their predictions (Bivand et al., 2008). If the average of the cross-validation errors (or Mean Error, ME) is close to 0,

$$\mathrm{ME} = \frac{1}{m} \sum_{i=1}^{m} \left[Z(\mathbf{x}_i) - \hat{Z}(\mathbf{x}_i) \right]. \qquad (25)$$

we may say that apparently the estimates are unbiased ($Z(\mathbf{x}_i)$ and $\hat{Z}(\mathbf{x}_i)$ are, respectively, the measurement and estimate at point xi and m is the number of measurements of the validation set). A significant negative (positive) mean error can represent systematic overestimation (underestimation). The magnitude of the Root Mean Squared Error (RMSE) is particularly interesting for comparing different models (Wackernagel, 2003; Webster & Oliver, 2007):

$$\mathrm{RMSE} = \sqrt{\frac{1}{m} \sum_{i=1}^{m} \left[Z(\mathbf{x}_i) - \hat{Z}(\mathbf{x}_i) \right]^2}. \qquad (26)$$

The RMSE value should be as small as possible indicating that estimates are close to measurements. The kriging standard deviation represents the error predicted by the estimation method. Dividing the cross-validation error by the corresponding kriging standard deviation allows to compare the magnitudes of both actual and predicted error (Wackernagel, 2003; Webster & Oliver, 2007). Therefore, the average of the standardized squared cross-validation errors (or Mean Standardized Squared Error, MSSE)

$$\text{MSSE} = \frac{1}{m}\sum_{i=1}^{m}\frac{\left[Z(\mathbf{x}_i) - \hat{Z}(\mathbf{x}_i)\right]^2}{\sigma^2_{R(\mathbf{x}_i)}}, \tag{27}$$

should be about one, indicating that the model is accurate. A scatterplot of true versus predicted values provides additional evidence on how well an estimation method has performed. The coefficient of determination R^2 is a good index for summarizing how close the points on the scatterplot come to falling on the 45-degree line passing through the origin (Isaaks & Srivastava, 1989). R2 should be close to one.

3. RESULTS

3.1 Study site

A map of the study site is shown in Fig. 2(a). Foz do Arelho outfall is located off the Portuguese west coast near Óbidos lagoon. In operation since June 2005, is presently discharging about 0.11 m3/s of mainly domestic wastewater from the WWTPs of Óbidos, Carregal, Caldas da Rainha, Gaeiras, Charneca and Foz do Arelho, but it can discharge up to 0.35 m^3/s. The total length of the outfall, including the diffuser, is 2150 m. The outfall pipe, made of HDPE, has a diameter of 710 mm. The diffuser, which consists of 10 ports spaced 8 or 12 meters apart, is 93.5 m long. The ports, nominally 0.175 m in diameter, are discharging upwards at an angle of 90° to the pipe horizontal axis; the port height is about 1 m. The outfall direction is southeast-northwest (315.5° true bearing) and is discharging at a depth of about 31 m. In that area the coastline itself runs at about a 225° angle with respect to true north and the isobaths are oriented parallel to the coastline. A seawater quality monitoring program for the outfall has already started in May 2006. Its main purposes are to evaluate the background seawater quality both in offshore and nearshore locations around the vicinity of the sea outfall and to follow the impacts of wastewater discharge in the area. During the campaign the discharge remained fairly constant with an average flowrate of approximately 0.11 m^3/s. The operation area specification was based on the outputs of a plume prediction model (Hunt et al., 2010) which include mixing zone length, spreading width, maximum rise height and thickness. The model inputs are, besides the diffuser physical characteristics, the water column stratification, the current velocity and direction, and the discharge flowrate. Information on density stratification was obtained from a vertical profile of temperature and salinity acquired in the vicinity of the diffuser two weeks before the campaign (see Fig. 3).

The water column was weakly stratified due to both low-temperature and salinity variations. The total difference in density over the water column was about 0.13 σ-unit. The current direction of 110° was estimated based on predictions of wind speed and direction of the day of the campaign. A current velocity of 0.12 m/s was estimated based on historic data. The effluent flowrate consider for the plume behavior simulation was 0.11 m^3/s. According to the predictions of the model, the plume was spreading 1 m from the surface, detached from the bottom and forming a two-layer flow. The end of the mixing zone length was predicted to be 141 m downstream from the diffuser. Fig. 2(b) shows the diffuser and a plan view of the AUV operation area (specified according to the model predictions), mainly in the northeast direction from the diffuser, covering about 20000 m^2.

The vehicle collected CTD data at 1.5 m and 3 m depth, in accordance to the plume minimum dilution height prediction. During the mission transited at a fairly constant velocity of 1 m/s (2 knots) recording data at a rate of 16 Hz. Maximum vertical oscillations of the AUV in performing the horizontal trajectories were less than 0.5 m (up and down).

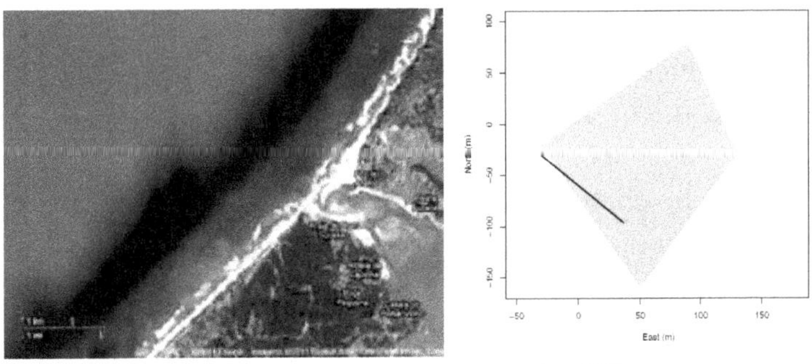

(a) Map of the study site (©2011 Google - Images).

(b) AUV operation area.

Fig. 2. Vicinity of Foz do Arelho sea outfall.

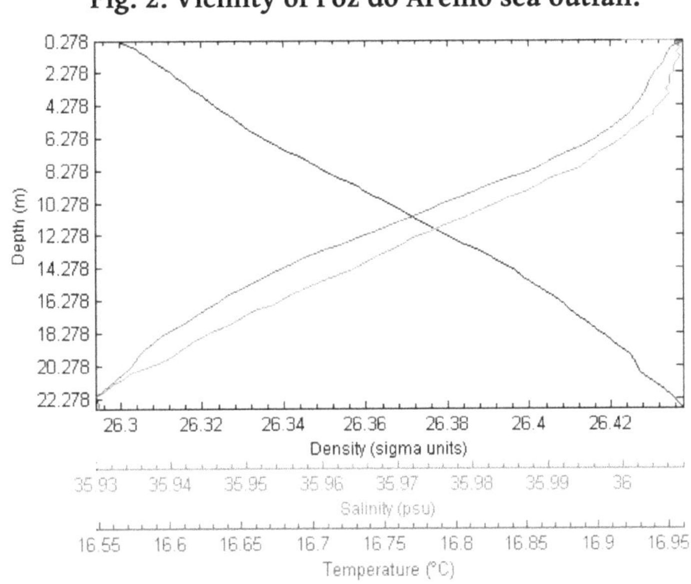

Fig. 3. Vertical STD profile used in the plume behavior simulation.

3.2 Exploratory analysis

In order to obtain elementary knowledge about the temperature and salinity data sets, conventional statistical analysis was conducted (see the results in Table 1 and Table 2). At the depth of 1.5 m the temperature ranged from 15.359°C to 15.562°C and at the depth of 3 m the temperature ranged from 15.393°C to 15.536°C. The mean value of the data sets was 15.463°C and 15.469°C, respectively at the depths of 1.5 m and 3 m, which was very close to the median value that was respectively 15.466°C and 15.472°C. The coefficient of skewness is relatively low (-0.309) for the 1.5 m data set and not very high (-0.696) for the 3 m data set, indicating that in the first case the histogram is approximately symmetric and in the second case that distribution is only slightly asymmetric. The very low values of the coefficient of variation (0.002 and 0.001) reflect the fact that the histograms do not have a tail of high values. At the depth of 1.5 m the salinity ranged from 35.957 psu to 36.003 psu and at the depth of 3 m the salinity ranged from 35.973 psu to 36.008 psu. The mean value of the data sets was 35.991 psu and 35.996 psu, respectively at the depths of 1.5 m and 3 m, which was very close to the median value that was respectively 35.990 and 35.998 psu. The coefficient of skewness is not to much high in both data sets (-0.63 and -1.1) indicating that distributions are only slightly asymmetric. The very low values of the coefficient of variation (0.0002 and 0.0001) reflect the fact that the histograms do not have a tail of high values. The ordinary kriging method works better if the distribution of the data values is close to a normal distribution. Therefore, it is interesting to see how close the distribution of the data values comes to being normal. Fig. 4 shows the plots of the normal distribution adjusted to the histograms of the temperature measured at depths of 1.5 m and 3 m, and Fig. 5 shows the plots of the normal distribution adjusted to the histograms of the salinity measured at depths of 1.5 m and 3 m. The density value in the histogram is the ratio between the number of samples in a bin and the total number of samples divided by the width of the bin (constant). Apart from some erratic high values it can be seen that the histograms are reasonably close to the normal distribution.

Table 1. Summary statistics of temperature measurements.

	Temperature@1.5 m	Temperature@3.0 m
Samples	20,026	10,506
Mean	15.463°C	15.469ºC
Median	15.466°C	15.472°C
Minimum	15.359°C	15.393ºC
Maximum	15.562°C	15.536ºC
Coefficient of skewness	-0.31	-0.70
Coefficient of variation	0.002	0.001

Table 2. Summary statistics of salinity measurements.

	Salinity@1.5 m	Salinity@3.0 m
Samples	20,026	10,506
Mean	35.991 psu	35.996 psu
Median	35.990 psu	35.998 psu
Minimum	35.957 psu	35.973 psu
Maximum	36.003 psu	36.008 psu
Coefficient of skewness	-0.63	-1.1
Coefficient of variation	0.0002	0.0001

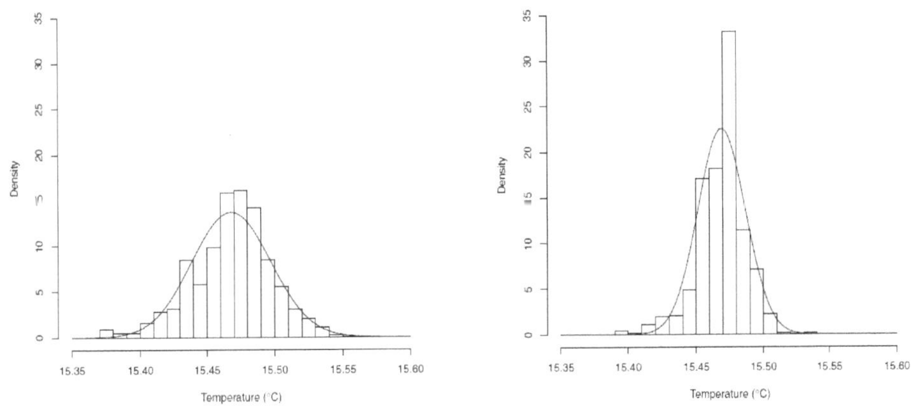

Fig. 4. Histograms of temperature measurements at depths of 1.5 m (left) and 3 m (right).

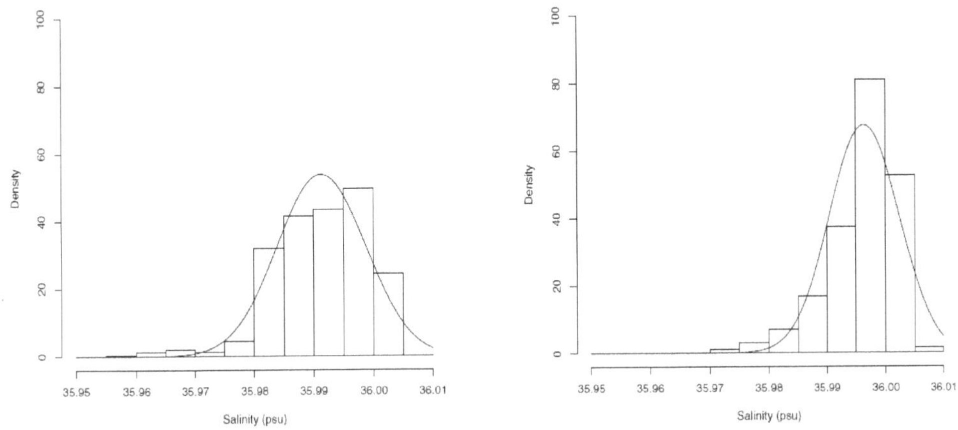

Fig. 5. Histograms of salinity measurements at depths of 1.5 m (left) and 3 m (right).

3.3 Variogram modeling

For the purpose of this analysis, the temperature and the salinity measurements were divided into a modeling set (comprising 90% of the samples) and a validation set (comprising 10% of the samples). Modeling and validation sets were then compared, using Student's-t test, to check that they provided unbiased sub-sets of the original data. Furthermore, sample variograms for the modeling sets were constructed using the MME estimator and the CRE estimator. This robust estimator was chosen to deal with outliers and enhance the variogram's spatial continuity. An estimation of semivariance was carried out using a lag distance of 2 m. Table 3 and Table 4 show the parameters of the fitted models to the omnidirectional sample variograms constructed using MME and CRE estimators. All the variograms were fitted to Matern models (for several shape parameters ν) with the exception to the salinity data measured at the depth of 3 m. The range value (in meters) is an indicator of extension where autocorrelation exists. The variograms of salinity show significant differences in range. The autocorrelation distances are always larger for the CRE estimator which may demonstrate the enhancement of the variogram's spatial continuity. All variograms have low nugget values which indicates that local variations could be captured due to the high sampling rate and to the fact that the variables under study have strong spatial dependence. Anisotropy was investigated by calculating directional variograms. However, no anisotropy effect could be shown.

3.4 Cross-validation

The block kriging method was preferred since it produced smaller prediction errors and smoother maps than the point kriging. Using the 90% modeling sets of the two depths, a two-dimensional ordinary block kriging, with blocks of 10 × 10 m^2, was applied to estimate temperature at the locations of the 10% validation sets. The validation results for both parameters measured at depths of 1.5 m and 3 m depths are shown in Table 5 and Table 6. At both depths temperature was best estimated by the variogram constructed using CRE. Salinity at the depth of 1.5 m was best estimated by the variogram constructed using CRE and at the depth of 3 m was best estimated using the Gaussian model with the MME. The difference in performance between the two estimators: block kriging using the MME estimator (MBK) or block kriging using the CRE estimator (CBK) is not substantial. Fig. 6 shows the omnidirectional sample variograms for temperature at the depth of 1.5 m and 3 m fitted by the preferred models. Fig. 7 shows the omnidirectional sample variograms for salinity at the depth of 1.5 m and 3 m fitted by the preferred models.

Table 3. Parameters of the fitted variogram models for temperature measured at depths of 1.5 and 3.0 m.

Depth	Variogram Estimator	Model	Nugget	Sill	Range
1.5	MME	Matern (ν = 0.4)	0.000	0.001	75.0
	CRE	Matern (ν = 0.5)	0.000	0.002	80.1
3.0	MME	Matern (ν = 0.3)	0.000	0.0002	101.3
	CRE	Matern (ν = 0.7)	0.000	0.002	107.5

Table 4. Parameters of the fitted variogram models for salinity measured at depths of 1.5 and 3 m.

Depth	Variogram Estimator	Model	Nugget	Still	Range
1.5	MME	Matern (v = 0.6)	0.436	11.945	134.6
	CRE	Matern (v = 0.6)	0.153	10786.109	51677.1
3.0	MME	Matern (v = 0.8)	0.338	11.724	181.6
	CRE	Gaussian	0.096	120.578	390.1

Table 5. Cross-validation results for the temperature maps at depths of 1.5 and 3 m.

Depth	Mehod	R^2	ME	MSE	RMSE
1.5	MBK	0.9184	2.0174e-4	8.0530e-5	8.9739e-3
	CBK[a]	0.9211	1.6758e-4	7.7880e-5	8.8248e-3
3.0	MBK	0.8748	1.0338e-4	3.6295e-5	6.0244e-3
	CBK[a]	0.8827	0.6538e-4	3.4008e-5	5.8316e-3
[a] The preferred model.					

Fig. 8 and Fig. 9 show the scatterplots of true versus estimated values for the most satisfactory models. The dark line is the 45° line passing through the origin and the discontinuous line is the OLS (Ordinary Least Squares) regression line. These plots show that observed and predicted values are highly positively correlated. The R^2 value for the temperature at the depth of 1.5 m was 0.9211 and the RMSE was 0.0088248 °C, and at the depth of 3 m was 0.8827 and the RMSE was 0.0058316 °C (Table 5). The R^2 value for the salinity at the depth of 1.5 m was 0.9513 and the RMSE was 0.0016435 psu, and at the depth of 3 m was 0.8982 and the RMSE was 0.0019793 psu (Table 6).

Table 6. Cross-validation results for the salinity maps at depths of 1.5 and 3 m.

Depth	Method	R^2	ME	MSE	RMSE
1.5	MBK	0.9471	3.1113e-5	2.8721e-6	1.6947e-3
	CBK[a]	0.9513	-3.1579e-5	2.7010e-6	1.6435e-3
3.0	MBK[a]	0.8982	-7.1735e-5	3.9175e-6	1.9793e-3
	CBK	0.7853	-8.1264e-5	8.2589e-6	2.8738e-3

[a] The preferred model.

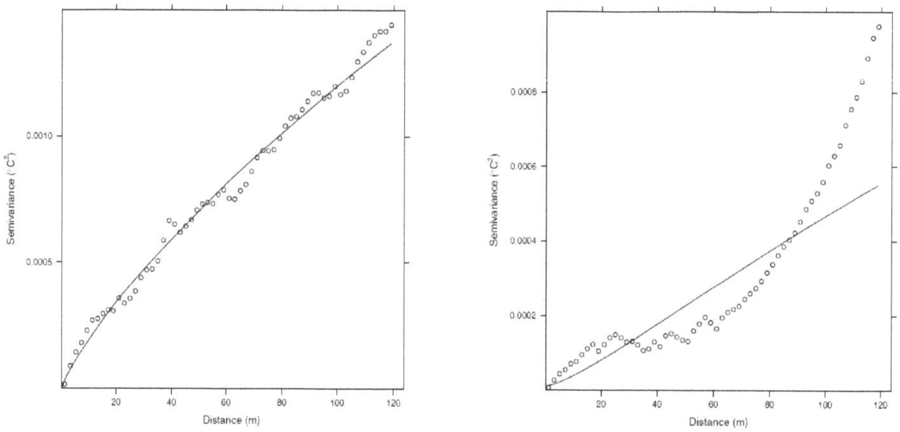

Fig. 6. Variograms for temperature at depths of 1.5 m (left) and 3 m (right).

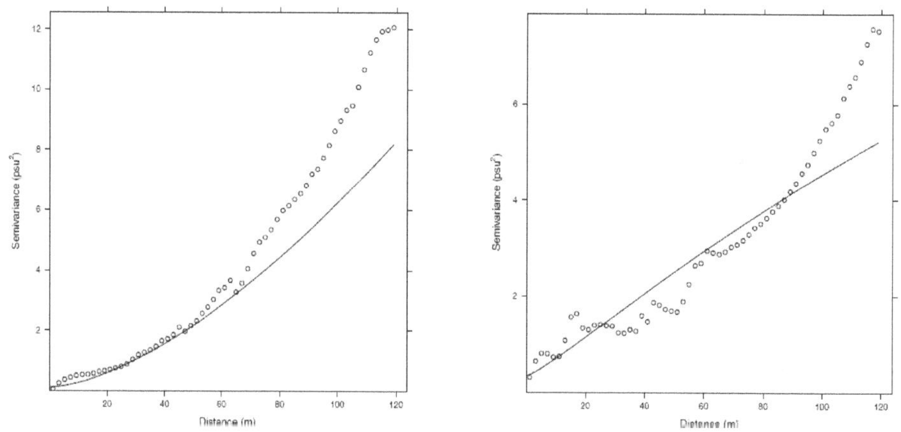

Fig. 7. Variograms for salinity at depths of 1.5 m (left) and 3 m (right).

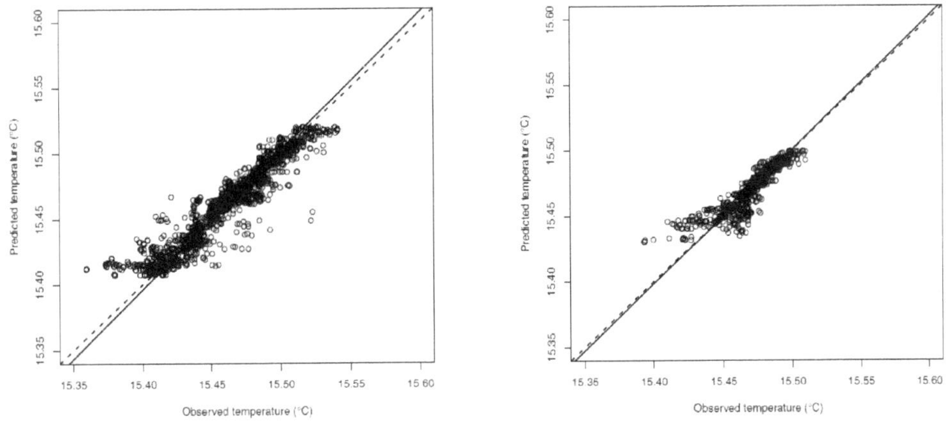

Fig. 8. Predicted versus observed temperature at the depths of 1.5 m (left) and 3 m (right) using the preferred models.

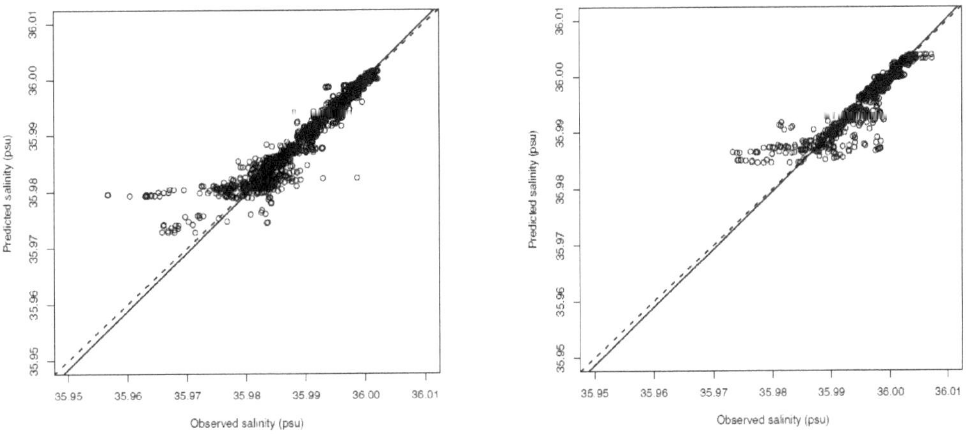

Fig. 9. Predicted versus observed salinity at the depths of 1.5 m (left) and 3 m (right) using the preferred models.

3.5 Mapping

Fig. 10 shows the block kriged maps of temperature on a 2 × 2 m^2 grid using the preferred models. Fig. 13 shows the block kriged maps of salinity on a 2× 2 m^2 grid using the preferred models. In the 1.5 m kriged map the temperature ranges between 15.407°C and 15.523°C and the average value is 15.469°C (the measured range is 15.359°C–15.562°C and the average value is 15.463°C). In the 3 m kriged map the temperature ranges between 15.429°C and 15.502°C and the average value is 15.467°C (the measured range is 15.393°C–15.536°C and the average value is 15.469°C). We may say that estimated values are in accordance with the measurements since their distributions are similar (identical average values, medians, and quartiles). The difference in the ranges width is due to only 5.0% of the samples in the 1.5 m depth map (2.5% on each side of the distribution) and only 5.3% of the samples in the 3.0 m depth map (3.1% on the left side and 2.2% on the rigth side of the distribution). These samples should then be identified as outliers not representing the behaviour of the plume in the established area. In the 1.5 m kriged map the salinity ranges between 35.960 psu and 36.004 psu and the average value is 35.992 psu, which is in accordance with the measurements (the measured range is 35.957psu – 36.003psu and the average value is 35.991 psu). In the 3 m kriged map the salinity ranges between 35.977 psu and 36.004 psu and the average value is 35.995 psu, which is in accordance with the measurements (the measured range is 35.973psu – 36.008psu and the average value is 35.996 psu). As predicted by the plume prediction model, the effluent was found dispersing close to the surface. From the temperature and salinity kriged maps it is possible to distinguish the effluent plume from the background waters. It appears as a region of lower temperature and lower salinity when compared to the surrounding ocean waters at the same depth. At the depth of 1.5 m the major difference in temperature compared to the surrounding waters is about -0.116°C while at the depth of 3 m this difference is about -0.073°C. At the depth of 1.5 m the major difference in salinity compared to the surrounding waters is about -0.044 psu while at the depth of 3 m this difference is about -0.027 psu. It is important to note that these very small differences in temperature and salinity were detected due to the high resolution of the CTD sensor. (Washburn et al., 1992) observed temperature and salinity anomalies in the plume in the order, respectively of -0.3ºC and -0.1 psu, when compared with the surrounding waters within the same depth range. The small plume-related anomalies observed in the maps are evidence of the rapid mixing process. Due to the large differences in density between the rising effluent plume and ambient ocean waters, entrainment and mixing processes are vigorous and the properties within the plume change rapidly (Petrenko et al., 1998; Washburn et al., 1992). The effluent plume was found northeast from the diffuser beginning, spreading downstream in the direction of current. Using the navigation data, we could later estimate current velocity and direction and the values found were, respectively, 0.4 m/s and 70ºC, which is in accordance with the location of the plume.

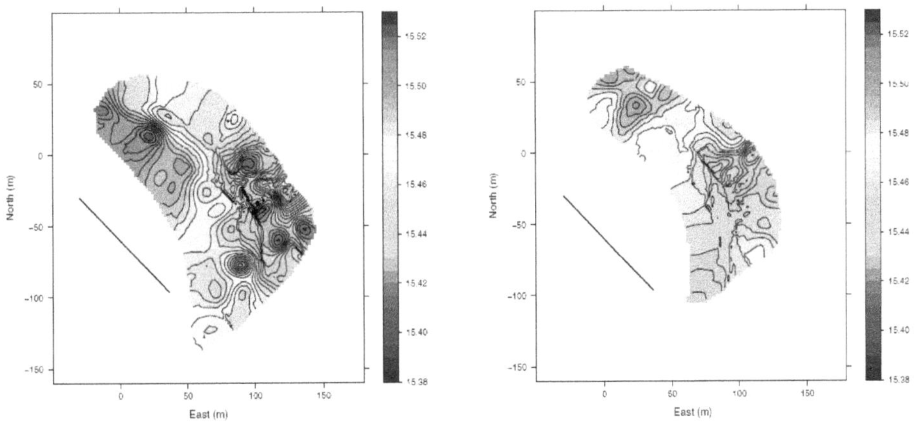

Fig. 10. Prediction map of temperature distribution (ºC) at depths of 1.5 m (left) and 3 m (right).

Fig. 12 shows the variance of the estimation error (kriging variance) for the maps of temperature distribution at depths of 1.5 m and 3 m. The standard deviation of the estimation error is less than 0.0195°C at the depth of 1.5 m and less than 0.0111°C at the depth of 3 m. Results of the same order were obtained for salinity. It's interesting to observe that, as expected, the variance of the estimation error is less the closer is the prediction from the trajectory of the vehicle. The dark blue regions correspond to the trajectory of MARES AUV.

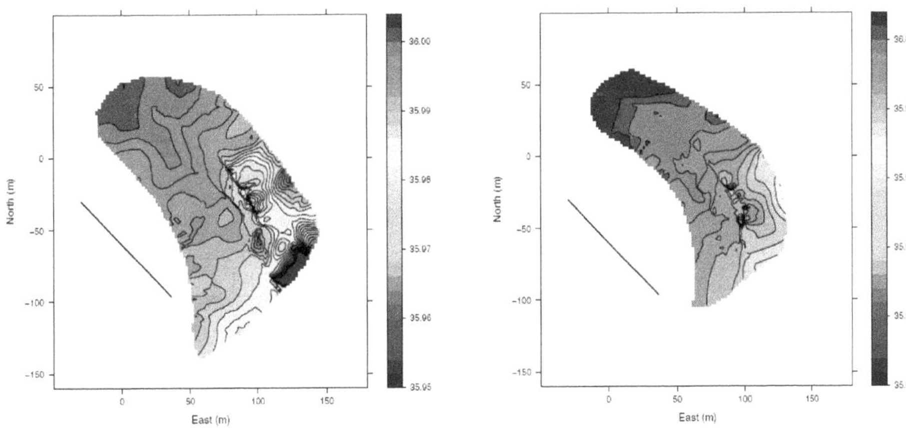

Fig. 11. Prediction map of salinity distribution (psu) at depths of 1.5 m (left) and 3 m (right).

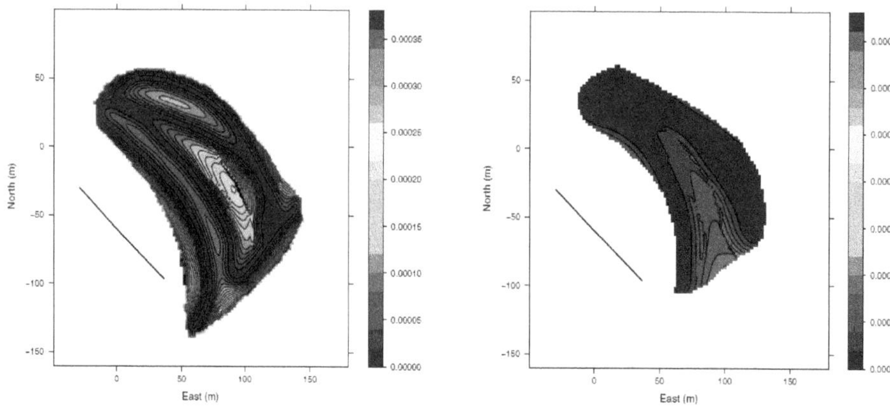

Fig. 12. Variance of the estimation error for the maps of temperature distribution at depths of 1.5 m (left) and 3 m (right)

3.6 Dilution estimation

Environmental effects are all related to concentration C of a particular contaminant X. Defining C_a as the background concentration of substance X in ambient water and C_0 as the concentration of X in the effluent discharge, the local dilution comes as follows (Fischer et al., 1979):

$$S = \frac{C_0 - C_a}{C - C_a}, \qquad (28)$$

which can be rearranged to give $C = C_a\left(\frac{S-1}{S}\right) + \left(\frac{1}{S}\right)C_0$. In the case of variability of the background concentration of substance X in ambient water the local dilution is given by

$$S = \frac{C_0 - C_{a0}}{C - C_a}, \qquad (29)$$

where C_{a0} is the background concentration of substance X in ambient water at the discharge depth. This expression in 29 can be arranged to give $C = C_a + \left(\frac{1}{S}\right)(C_0 - C_{a0})$, which in simple terms means that the increment of concentration above background is reduced by the dilution factor S from the point of discharge to the point of measurement of C. Using salinity distribution at depths of 1.5 m and 3 m we estimated dilution using Equation 29 (see the contour maps in Fig. 13). We assumed C_0 = 2.3 psu, C_{a0} = 35.93 psu, C_a = 36.008 psu at 1.5 m depth and Ca – 36.006 psu at 3 m depth. The minimum dilution estimated at the depth of 1.5 m was 705 and at the depth of 3.0 m was 1164 which is in accordance with Portuguese legislation that suggests that outfalls should be designed to assure a minimum dilution of 50 when the plume reaches surface (INAG, 1998). (Since dilution increases with the plume rising we should expect that the minimum values would be greater if the plume reached surface (Hunt et al., 2010)).

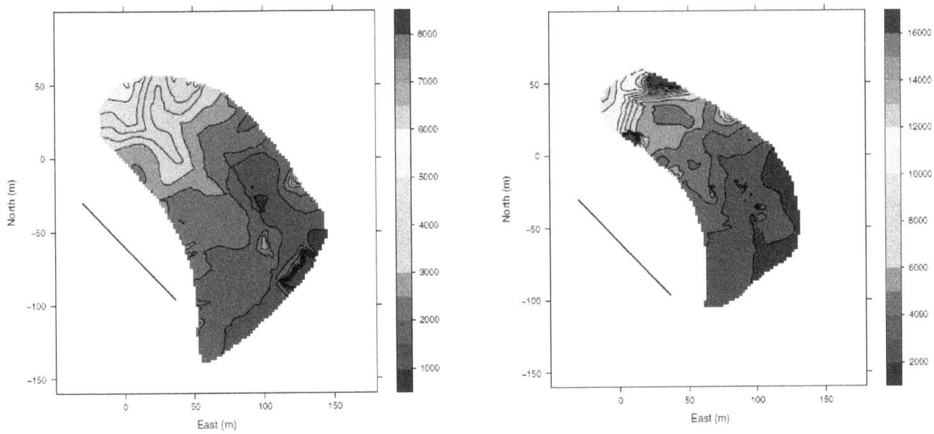

Fig. 13. Dilution maps at depths of 1.5 m (left) and 3 m (right).

4. CONCISE SUMMARY

Geostatistical analysis of temperature and salinity data collected by an AUV at depths of 1.5 m and 3 m during an ocean outfall monitoring operation enabled the creation of kriged maps depicting sewage dispersion in the area. The spatial variability of the sampling data has been examined, revealing an approaching normal distribution of the temperature and salinity values, which is favorable. The conventional estimator by Matheron and the robust estimator by Cressie and Hawkins were employed to calculate the omnidirectional variograms, which were subsequently fitted to Matern models with various shape parameters and a Gaussian model. The performance of each competing model was compared using a split-sample approach. In the case of temperature, the validation results, using a two-dimensional ordinary block kriging, suggested the Matern model ($\nu = 0.5 - 1.5$ m and $\nu = 0.7 - 3.0$ m) with semivariance estimated by CRE. In the case of salinity, the validation results, using a two-dimensional ordinary block kriging, suggested the Matern model ($\nu = 0.6 - 1.5$ m and $\nu = 0.8 - 3.0$ m) with semivariance estimated by CRE, for the depth of 1.5 m, and with semivariance estimated by MME, for the depth of 3 m. The difference in performance between the two estimators was not substantial. Block kriged maps of temperature and salinity at depths of 1.5 m and 3 m illustrate the spatial variation of these parameters in the studied area, allowing for the identification of the effluent plume, which manifests as a region of reduced temperature and salinity relative to the surrounding waters, located northeast of the diffuser's origin and extending downstream in the current's direction. We evaluated dilution at depths of 1.5 m and 3 m using salinity distribution. The values identified comply with Portuguese legislation. The results indicate that geostatistical approach can yield accurate estimates of effluent dispersion, which are essential for evaluating environmental effect and managing marine outfalls.

REFERENCES

Abreu, N., Matos, A., Ramos, P. & Cruz, N. (2010). Automatic interface for AUV mission planning and supervision, MTS/IEEE *International Conference Oceans 2010,* Seattle, USA.

Abreu, N. & Ramos, P. (2010). An integrated application for geostatistical analysis of sea outfall discharges based on R software, *MTS/IEEE International Conference Oceans 2010,* Seattle, USA.

Bellingham, J. (1997). New Oceanographic Uses of Autonomous Underwater Vehicles, *Marine Technology Society Journal* 31(3): 34–47.

Bellingham, J., Goudey, C., Consi, T. R. & Chryssostomidis, C. (1992). A Small Long Range Vehicle for Deep Ocean Exploration, *Proceedings of the International Offshore and Polar Engineering Conference, San Francisco,* pp. 151–159.

Bivand, R. S., Pebesma, E. J. & Gómez-Rubio, V. (2008). *Applied spatial data analysis with R.* ISBN: 97 -0-387-78170-9.

Cressie, N. (1993). *Statistics for spatial data,* New York.

Cressie, N. & Hawkins, D. M. (1980). Robust estimation of the variogram, I, *Jour. Int. Assoc. Math. Geol.* 12(2): 115–125.

Fernandes, P. G., Brierley, A. S., Simmonds, E. J., Millard, N. W., McPhail, S. D., Armstrong, F., Stevenson, P. & Squires, M. (2000). Fish do not Avoid Survey Vessels, *Nature* 404: 35–36.

Fischer, H. B., List, J. E., Koh, R. C. Y., Imberger, J. & Brooks, N. H. (1979). *Mixing in Inland and Coastal Waters,* Academic Press.

Goovaerts, P. (1997). *Geostatistics for natural resources evaluation, Applied Geostatistics Series.* ISBN13: 9780195115383, ISBN10: 0195115384.

Hunt, C. D., Mansfield, A. D., Mickelson, M. J., Albro, C. S., Geyer, W. R. & Roberts, P. J. W. (2010). Plume tracking and dilution of effluent from the Boston sewage outfall, *Marine Environmental Research* 70: 150–161.

INAG (1998). *Linhas de Orientação Metodológia para a Elaboração de Estudos Técnicos Necessários para Cumprir o Artï¿½ 7ï¿½ do D.L. 152/97. Descargas em Zonas Menos Sensíveis,* Instituto da Água. Ministério do Ambiente, Lisboa.

Isaaks, E. H. & Srivastava, R. M. (1989). *Applied geostatistics,* New York Oxford. ISBN 0-19-505012-6-ISBN 0-19-505013-4 (pbk.).

Kitanidis, P. (1997). *Introduction to geostatistics: applications in hydrogeology, New York (USA). Matheron, G. (1965). Les variables régionalisées et leur estimation: une application de la théorie des fonctions aléatoires aux sciences de la nature,* Paris.

Nadis, S. (1997). Real Time Oceanography Adapts to Sea Changes, *Science* 275: 1881–1882.

Petrenko, A. A., Jones, B. H. & Dickey, T. D. (1998). Shape and Initial Dilution of Sand Island, Hawaii Sewage Plume, *Journal of Hydraulic Engineering, ASCE* 124(6): 565–571.

Ramos, P. & Abreu, N. (2010). Spatial analysis of sea outfall discharges using block kriging, *6th International Conference on Marine Waste Water Discharges and Coastal Environment, MWWD2010,* Langkawi, Malaysia.

Ramos, P. & Abreu, N. (2011a). Environmental impact assessment of Foz do Arelho sewage plume using MARES AUV, *IEEE/OES International Conference Oceans 2011,* Santander, Spain.

Ramos, P. & Abreu, N. (2011b). Spatial analysis and dilution estimation of Foz do Arelho Outfall using observations gathered by an AUV, *International Symposium on Outfall Systems,* Mar del Plata, Argentina.

Ramos, P. & Abreu, N. (2011c). Using an AUV for Assessing Wastewater Discharges Impact: An Approach Based on Geostatistics, *Marine Technology Society Journal* 45(2).

Robinson, A., Bellingham, J., Chryssostomidis, C., Dickey, T., Levine, E., Petrikalakis, N., Porter, D., Rothschild, B., Schmidt, H., Sherman, K., Holliday, D. & Atwood, D. (1999). Real-time forecasting of the multidisciplinary coastal ocean with the littoral ocean observing and predicting system (LOOPS), *Proceedings of the Third Conference on Coastal Atmospheric and Oceanic Prediction Processes,* American Meteorological Society, New Orleans, LA.

Stein, M. L. (1999). *Interpolation of spatial data: some theory for kriging,* New York.

Wackernagel, H. (2003). *Multivariate geostatistics: an introduction with applications,* Berlin.

Washburn, L., Jones, B. H., Bratkovich, A., Dickey, T. D. & Chen, M.-S. (1992). Mixing, Dispersion, and Resuspension in Vicinity of Ocean Wastewater Plume, *Journal of Hydraulic Engineering, ASCE* 118(1): 38–58.

Webster, R. & Oliver, M. (2007). *Geostatistics for environmental scientists, 2nd Edition.* ISBN-13: 978-0-470-02858-2(HB).

Yu, X., Dickey, T., Bellingham, J., Manov, D. & Streitlien, K. (2002). The Application of Autonomous Underwater Vehicles for Interdisciplinary Measurements in Massachusetts and Cape Cod Bayes, *Continental Shelf Research* 22(15): 2225–2245.

INDEX

www.ingramcontent.com/pod-product-compliance
Lightning Source LLC
LaVergne TN
LVHW080056220926
848484LV00007BA/1697
* 9 7 8 1 7 8 7 1 5 4 0 9 4 *